MW00814663

PHYSICS RESEARCH AND TECHNOLOGY

LAYERED STRUCTURE EFFECTS AS REALISATION OF ANIZOTROPY IN MAGNETIC, GALVANOMAGNETIC AND THERMOELECTRIC PHENOMENA

Physics Research and Technology

Additional books in this series can be found on Nova's website
under the Series tab.

Additional e-books in this series can be found on Nova's website
under the e-book tab.

PHYSICS RESEARCH AND TECHNOLOGY

LAYERED STRUCTURE EFFECTS AS REALISATION OF ANIZOTROPY IN MAGNETIC, GALVANOMAGNETIC AND THERMOELECTRIC PHENOMENA

PETER V. GORSKYI

New York

Copyright © 2014 by Nova Science Publishers, Inc.

All rights reserved. No part of this book may be reproduced, stored in a retrieval system or transmitted in any form or by any means: electronic, electrostatic, magnetic, tape, mechanical photocopying, recording or otherwise without the written permission of the Publisher.

For permission to use material from this book please contact us:
Telephone 631-231-7269; Fax 631-231-8175
Web Site: http://www.novapublishers.com

NOTICE TO THE READER

The Publisher has taken reasonable care in the preparation of this book, but makes no expressed or implied warranty of any kind and assumes no responsibility for any errors or omissions. No liability is assumed for incidental or consequential damages in connection with or arising out of information contained in this book. The Publisher shall not be liable for any special, consequential, or exemplary damages resulting, in whole or in part, from the readers' use of, or reliance upon, this material. Any parts of this book based on government reports are so indicated and copyright is claimed for those parts to the extent applicable to compilations of such works.

Independent verification should be sought for any data, advice or recommendations contained in this book. In addition, no responsibility is assumed by the publisher for any injury and/or damage to persons or property arising from any methods, products, instructions, ideas or otherwise contained in this publication.

This publication is designed to provide accurate and authoritative information with regard to the subject matter covered herein. It is sold with the clear understanding that the Publisher is not engaged in rendering legal or any other professional services. If legal or any other expert assistance is required, the services of a competent person should be sought. FROM A DECLARATION OF PARTICIPANTS JOINTLY ADOPTED BY A COMMITTEE OF THE AMERICAN BAR ASSOCIATION AND A COMMITTEE OF PUBLISHERS.

Additional color graphics may be available in the e-book version of this book.

Library of Congress Cataloging-in-Publication Data

ISBN: 978-1-62808-875-5

Published by Nova Science Publishers, Inc. † New York

CONTENTS

PREFACE

A model of electron spectrum of layered crystal was proposed by R. Fivaz [1] in 1967. In this model, electron motion in-plane of layers is described by effective mass method, and across them, or as it is generally called, along the C-axis, by strong coupling method. In so doing, most theoretical works consider as layered crystals only those which satisfy the condition $\zeta/\Delta > 2$ and, moreover, $\zeta/\Delta >> 1$ (ζ is the Fermi energy referred to the bottom of a narrow conduction miniband, Δ is half-width of this miniband). This case corresponds to open Fermi surface. Therefore, when as far back as 1976 the author of this book proposed to his scientific supervisor Dr. V.M. Nitsovich to consider the case of closed or transient Fermi surface, i.e. such when $\zeta/\Delta \le 2$, the latter was utterly amazed, but, on second thoughts, he agreed. Everything seemed to go well, and the author maintained his diploma paper and then his thesis for a degree corresponding to Doctor of Philosophy in the USA and Europe. The thesis mainly dealt with a research on electron gas diamagnetism in layered crystals with closed Fermi surfaces for the case of a quantizing magnetic field normal to layers, as well as in thin films on their basis. Then, during the approbation, a chapter was added on the conductivity of such crystals in the case when the electric and quantizing magnetic fields are parallel to each other and normal to layers.

Later on, the author's life circumstances were such that over 13 years, namely from 1986 till 2000, he found himself beyond the process of research in condensed matter physics. And when after a long interval the author returned to his home Chernivtsi National University and resumed these studies he was much surprised to find that theoreticians still considered as layered only crystals with $\zeta/\Delta > 2$. Moreover, when in the period from 2000 till 2006 he again started writing papers this time dealing with longitudinal galvanomagnetic effects in layered crystals with $\zeta/\Delta \le 2$, reviewers of a number of scientific journals strongly objected to publication of these papers, alleging as their reasons that such crystals do not belong to layered at all.

From 2007 till 2010 the author again was divorced from scientific research process, but in late 2010 he published a paper [2] where, from his point of view, he clearly demonstrated by the example of longitudinal conductivity in a quantizing magnetic field that layered structure effects, i.e. real anisotropy of electron spectrum of layered crystals can also manifest themselves with closed Fermi surfaces. Thanks in no small part to the efforts of the editorial board of Ukrainian Journal of Physics and the members of N.N. Bogolyubov Institute for Theoretical Physics, National Academy of Sciences, Ukraine to whom the author extends his appreciation, this paper got into arXiv. This marked the beginning of the author's cooperation with "Nova Publishers", on behalf of which he was invited to take part in writing the book "Anisotropy Research: New Developments" for which he prepared a chapter. In this chapter, by the example of a longitudinal thermoEMF in a quantizing magnetic field he again demonstrated how layered structure effects manifest themselves with closed Fermi surfaces. The author feels it his pleasant duty to note that in the preparation of this chapter he was rendered an invaluable technical assistance by the editor of the book, Dr. Hirpa Lemu from the University of Stavanger (Norway) who proposed a more rational chapter organization, and Mrs. T. Venkel from Chernivtsi National University who edited the English text. Then the author was asked by the publishers whether he had some ideas of writing or editing a new book. On mailing an appropriate form to the publishers, a contract was concluded with the author, in execution of which this book grew out.

It should be noted that in this book the author does not set oneself a goal to give an exhaustive description of magnetic, galvanomagnetic and thermoelectric properties of layered crystals and a detailed analysis of the related works of his colleagues, but focuses attention on the manifestation of layered structure effects, i.e. anisotropy in the dependences of diamagnetic susceptibility, longitudinal conduction, longitudinal thermoEMF and longitudinal power factor on a quantizing magnetic field at helium temperatures. In so doing, unlike the theoretical works of previous authors, the book is primarily concerned with the case of closed and transient Fermi surfaces. Exactly these circumstances account for a relatively small list of references at the end of the book and a considerable number of the author's references to his own works. At the same time, the case of interlayer charge ordering is considered. Such ordering brings about a topological phase transition of the 2nd kind from a closed or transient to a highly open Fermi surface. Subsequently, in a quantizing magnetic field due to Fermi surface compression in the direction of a magnetic field coinciding with the C-axis there occurs an inverse topological transition from a highly open connected to a disconnected Fermi surface consisting of several closed sheets. This fact is immediately reflected on the dependences of the above mentioned physical characteristics of layered crystals on a quantizing magnetic field.

In parallel with this, a longitudinal conductivity of a layered crystal with a two-sheeted Fermi surface consisting of closed and strongly open sheets is considered in quasi-classical approximation.

In the book, the nonequilibrium characteristics of layered crystals, such as longitudinal conductivity, longitudinal thermoEMF and longitudinal power factor are considered in the approximation of relaxation time for different dependences of this time on the longitudinal quasi-momentum. It should be noted that this approach not always yields the results coinciding with those of the Kubo formalism. However, at the present time the approach based on the Kubo formalism is developed only for layered crystals with the highly open Fermi surfaces for which the relaxation time and the Dingle temperature to a high degree of accuracy can be considered constant values. Modification of this approach for the case of closed Fermi surfaces is involved with considerable difficulties of mathematical nature whose solution now is not the aim of this book. But, sooner or later, such modification is likely to be done.

There is also a brief outline of the conductivity of a layered crystal in a quantizing magnetic field for the nondegenerate case and two kinds of longitudinal Kapitsa effect: low-temperature due to current carrier condensation on the bottom of the lowest Landau subband

and high-temperature due to the field dependence of relaxation time with induced current carrier scattering by acoustic phonons.

It should be noted that many physicists (of those who are not engaged in galvanomagnetic effects in cryogenic area) still believe that longitudinal galvanomagnetic effects should exist under no circumstances, so one of the paragraphs in this book is specially dedicated to general conditions of such effects origination.

The author does not consider deliberately the spin effect of physical characteristics of layered crystals in a quantizing magnetic field, because for a simple, i.e. not renormalized by spin-orbit interaction conductivity band the Lande factor is constant and equal to two, just like for a free electron. Therefore, account of spin does not present any essential difficulty for those wishing to do it. Nevertheless, it is worth mentioning that a number of works on this issue have been published by Askerov, Figarova and Makhmudov [3].

Naturally, the list of what did not enter the book, but, probably would be appropriate in it, could be continued endlessly, but the author hopes that the readers will evaluate the book for what it does contain, rather than it does not.

The author understands that popularity of the book will be strongly restrained by lack of ample amount of experimental facts proving its theoretical propositions and calculations. However, he hopes that this book will give an impetus to organization of new experimental works. Meanwhile, the author is aware of only two experimental works [4, 5] related to bromine-intercalated graphite which considered a topological transition from an open to closed Fermi surface. Nevertheless, the author will be glad if there are readers who care about the issues raised in the book as he himself.

For the time of direct writing the book the author came to work at Institute of Thermoelectricity of the National Academy of Sciences and Ministry of Education, Science, Youth and Sports of Ukraine. The members of the institute immediately supported his initiative related to writing the book. In this connection, he expresses his appreciation to academician L.I. Anatychuk and chief research scientist L.N. Vikhor for the understanding, support, and statement of some problems as well as to Mrs. T. Zibachynska for preparation of the English text. Thanks are also due to the author's wife Natalia and his junior son Gennady for the stoic attitude to the load he laid on his family when writing the book. My special congratulations to my father Vladimir Fedorovich Gorskyi for moral and financial support of book writing process.

Chapter 1

INTRODUCTION

Layered, or, as they are frequently referred to, quasi-two-dimensional systems, are being studied intensively today both theoretically and experimentally. The most commonly encountered model of the energy band spectrum of these crystals was proposed in [1] and can be represented as:

$$\varepsilon(\vec{k})=\frac{\hbar^2}{2m^*}\left(k_x^2+k_y^2\right)+\Delta(1-\cos ak_z) \qquad (1.1)$$

In this formula, k_x, k_y are components of quasi-momentum in layer plane, k_z is component of quasi-momentum normal to layers, m^* is effective mass of electron in layer plane, Δ is half-width of a mini-band describing translational motion of electrons in a plane normal to layers, a is the distance between translation-equivalent layers. Since the advent of this model, a large number of theoretical and experimental works have been published concerned with a research on different physical characteristics of such crystals. For instance, quantum theory of light absorption by free carriers in a semiconductor with a superlattice described by spectrum (1) was constructed in [6], quantum dimensional effect in such crystals was considered in [7], high-frequency properties of semiconductors with a superlattice, in particular, their conduction in an alternating electric high-frequency field were dealt with in monograph [8].

Magnetic and galvanomagnetic effects in these crystals were also studied theoretically and experimentally, including a quantizing magnetic field. Thus, D.Grecu and V.Protopescu seem to be the first to calculate the oscillating part of the thermodynamic potential of such crystals in a quantizing magnetic field normal to layers. These calculations were performed for the case when condition $\zeta/\Delta >> 1$ is met, where ζ is chemical potential or the Fermi energy referred to the bottom of conduction band. In fact, it can be easily perceived in this case that according to k_z the Fermi surface occupies the entire first Brillouin zone $-\pi/a \le k_z \le \pi/a$ and with a periodic continuation it becomes connected, i.e. represents a corrugated cylinder. It can be also readily appreciated that within the first Brillouin zone this cylinder has three extreme plane sections normal to its longitudinal axis, namely one maximum plane section $k_z = 0$ and two equal in area minimum plane sections $k_z = \pm\pi/a$. If

$\zeta/\Delta >> 1$, then these two sections practically coincide, so de Haas- van Alphen oscillations (DHvA) and Shubnikov-de Haas (SdH) oscillations in a quantizing magnetic field normal to layers look as high-frequency oscillations superimposed with low-frequency beats. Experiments showed [9] that, for instance, in synthetic metals based on organic compounds of the type β-$(ET_2)IBr_2$ and some others, high frequencies and beat frequencies can differ even 100 or more times. It is exactly on the basis of essential difference in high frequencies and beat frequencies that the majority of theoreticians and experimenters concerned with magneto-oscillation effects started to qualify crystals as layered or common, three-dimensional. Thus, it is considered that layered structure effects do not manifest themselves if condition $\zeta/\Delta >> 1$ or at least $\zeta/\Delta > 2$ is not met, despite the fact that crystal band spectrum is described by relation (1). Therefore, the overwhelming majority of modern theoretical works studying galvanomagnetic effects in such crystals in a quantizing magnetic field was performed for the case of $\zeta/\Delta >> 1$. Some of these works will be mentioned and briefly described further as the text goes. In opposition to this, the author of proposed book thinks that neglect of layered structure effects is justified only in the case of $\zeta/\Delta << 1$.

From relation (1) it is not difficult to perceive that at $\zeta/\Delta < 2$ the Fermi surface occupies according to k_z only part of the first Brillouin zone and with a periodic continuation proves to be disconnected, i.e. consisting of a series of lens shaped bodies. Such surface possesses only one nonzero extreme section, namely maximum plane section $k_z = 0$. Therefore, to this day it is considered that from the standpoint of magnetic oscillation effects such surface is indistinguishable from ellipsoid, hence analysis of these effects in layered crystals with such, i.e. closed Fermi surfaces is of no particular interest nowadays. A transient surface is the Fermi surface corresponding to the case of $\zeta/\Delta = 2$. In this situation it occupies the entire first Brillouin zone, with a periodic continuation it is connected, but within $-\pi/a \le k_z \le \pi/a$ it has only one extreme, namely maximum plane section $k_z = 0$.

The aforesaid, however, does not necessarily mean that layered crystals with closed Fermi surfaces were not studied at all. One of case studies of a layered crystal with closed Fermi surface is a crystal of conventional thermoelectric material Bi_2Te_3 the band structure of which is described by a six-ellipsoid Drabble-Wolfe model [10]. This crystal can be referred to layered ones if only because of its pronounced cleavage planes along which it easily cleaves. However, the anisotropy of its physical characteristics, such as thermal conductivity and electric conductivity is relatively low. Thus, its thermal conductivity anisotropy does not exceed 3, and electric conductivity anisotropy – 2.7 for *p*-type material and 4÷6 for *n*-type material. This, of course, is not much as compared to, for instance, dichalcogenides of transient metals or synthetic metals based on intercalated graphite for which the effective mass anisotropy can reach 10^5, and electric conductivity anisotropy – 10^7. Magnetic oscillation studies of the band structure of this material using dHvA and SdH effects were performed in [11]. This yielded components of effective mass tensor and the number of ellipsoids in the band. However, it turned out that in heavily doped samples the shape of the Fermi surface can differ considerably from ellipsoid. This fact prompted a number of researchers, such as D.M. Bercha and V.T. Maslyuk, to replace in their works [12, 13] an ellipsoid by a more complex constant-energy surface described by relation (1). For such surface they, in particular, calculated the Landau diamagnetic susceptibility and electron gas dielectric permeability. However, a rather complicated procedure for calculation of diamagnetic susceptibility based on the calculation of integrals from Bessel functions in a complex plane prevented them from making these calculations completely and with a reasonable degree of accuracy. Thus, for the simplest case of a quantizing magnetic field normal to layers these researchers for the case of closed Fermi surface obtained just a small correction to a well-known Landau formula for a parabolic band spectrum. Besides, V.T. Maslyuk in his work [12] used the effective Hamiltonian method to calculate the energy spectrum of a layered crystal in a quantizing magnetic field parallel to layers. Thereafter, V.G. Peschansky, when constructing a theory of galvanomagnetic effects in layered crystals with a spectrum (1.1), calculated the band spectrum and electric conductivity of such crystals with a random orientation of a quantizing magnetic field with respect to layer plane, using, similar to other researchers, the effective Hamiltonian method [14]. The author of this book is rather sceptical about this method, since until now nobody has demonstrated that its results always coincide with those that would be due to a valid inclusion of a magnetic field to the initial Hamiltonian of crystal electron subsystem with a periodic potential or lattice pseudo-potential. So, in his studies he deliberately restricts himself to the case of a quantizing magnetic field normal to layers, since in this case a quasi-classical solution of the problem of electron spectrum in a quantizing magnetic field coincides with a precise (or conventionally precise) solution, if, according to (1.1), electron mass in layer plane should be understood to mean its effective mass.

Chapter 2

ENERGY SPECTRUM OF CHARGE CARRIERS IN A LAYERED CRYSTAL AND THIN FILM ON ITS BASIS IN A QUANTIZING MAGNETIC FIELD

In the most general form the energy spectrum of current carriers in a layered crystal can be written as:

$$\varepsilon(\vec{k})=\frac{\hbar^2}{2m^*}\left(k_x^2+k_y^2\right)+W(k_z). \tag{2.1}$$

This formula employs the same notation as in (1), and $W(k_z)$ is a certain periodic, hence expandable in a Fourier series, function of its argument describing electron motion in the direction normal to layers. Then formula (1) corresponds to a limiting case of a strong coupling, when wave functions of adjacent layers are weakly overlapped, and lattice periodic potential as a coordinate function varies rather slowly in layer plane, and, on the contrary, rather quickly in the perpendicular direction. Therefore, the Hamiltonian corresponding to spectrum (2) in coordinate representation is given by:

$$\hat{H}=\frac{\hbar^2}{2m^*}\left(\nabla_x^2+\nabla_y^2\right)+\frac{\hbar^2}{2m_0}\nabla_z^2+V(z). \tag{2.2}$$

In (2.2) $V(z)$ is a periodic lattice potential depending only on coordinate in the direction normal to layers, m_0 is free electron mass. Slow variation of a periodic potential in layer plane, i.e. smallness of forces acting on electrons in this plane is exactly taken into consideration in the substitution of free electron mass by the effective mass in this plane, though many authors, in particular, the authors of monograph [8], make no distinction between masses m^* and m_0.

From the Hamiltonian representation in the form of (2.2) it is clear that on introducing into this Hamiltonian of a magnetic field normal to layers, the Schrödinger equation for it falls into two equations: the equation for a harmonic oscillator in the coordinates x and y in layer plane and the one-dimensional problem of conventional band theory in the coordinate z in the direction normal to layers. Thus it is evident that the energy spectrum of electrons in a layered crystal in a quantizing magnetic field normal to layers will be as follows:

$$\varepsilon(n,k_z)=\mu^* B(2n+1)+W(k_z). \tag{2.3}$$

In this formula, n is the Landau level number, $\mu^*=\mu_B m_0/m^*$, μ_B is the Bohr magneton, the rest of notation were explained above. There is no difficulty in understanding that the same spectrum results from the rule of quantization of electron orbit areas in k-space, proposed by Lifshits and Kosevich when constructing their theory of dHvA and SdH oscillations for crystals with an arbitrary law of current carrier dispersion in quasi-classical approximation [15-17]. It is in this respect that quasi-classical solution of the problem of layered crystal electron spectrum in a quantizing magnetic field normal to layers coincides with the exact solution. The fact of such coincidence for other magnetic field orientations is not proved. Next, the majority of researchers for function $W(k_z)$ are restricted to a strong coupling approximation. With the exception of some cases, the author of proposed book does in a similar fashion.

Now we pass on to finding the energy spectrum of conduction electrons in a thin film based on a layered crystal. In so doing, we will consider a layer plane to coincide with a film plane, and crystal itself to possess the inversion symmetry. Then the respective Hamiltonian of the problem in coordinate representation will acquire the form as follows:

$$\hat{H}_{tf}=\frac{\hbar^2}{2m^*}\left(\nabla_x^2+\nabla_y^2\right)+\frac{\hbar^2}{2m_0}\nabla_z^2+V(z)+U(z). \tag{2.4}$$

Potential $U(z)$ takes into account the impossibility of spontaneous liberation of electrons beyond the film and is of the form:

$$U(z)=\begin{cases}0(-d/2\le z\le d/2)\\ \infty(|z|>d/2)\end{cases}. \tag{2.5}$$

In this formula, d is film thickness. From the Hamiltonian representation in the form of (2.4) it is again clear that the Schrödinger equation for it, when introducing a magnetic field normal to layers, hence to film plane, falls into the Schrödinger equation for a free electron with effective mass m^* and the band theory equation with a modified potential $V(z)+U(z)$, i.e.:

$$\left(\frac{\hbar^2}{2m_0}\nabla_z^2+V(z)+U(z)\right)\Psi(z)=E\Psi(z). \tag{2.6}$$

A similar model was considered in [18], but for $V(z)=0$. Such an assumption, though it can be used for the interpretation of physical properties of thin films, comprises an essential contradiction. On the one hand, it is supposed that no forces inside the film act on the electron. On the other hand, solution of the problem of dielectric permittivity of electron gas in a film leads to a strong Coulomb repulsion of electrons, disregarded in the beginning [19]. Therefore, to describe correctly the properties of conduction electrons in a film, one should take into account lattice periodic potential $V(z)$ that reflects a real character of forces affecting the electron inside the film. Assuming that crystal possesses the inversion symmetry, we will represent $V(z)$ as follows:

$$V(z)=\sum_{m=-\infty}^{\infty} V_m \cos(2m\pi z/a). \tag{2.7}$$

Wave functions satisfying equation (2.6), periodicity condition in region $|z| \le d/2$ and possessing inversion symmetry can be represented as follows:

$$\Psi(z) \equiv \Psi_{k_s}^{\pm}(z)=\sum_{m=-\infty}^{\infty} \begin{pmatrix} A_m \\ B_m \end{pmatrix} \begin{matrix} \cos \\ \sin \end{matrix} \left[\left(k_s^{\pm} + \frac{2m\pi}{a} \right) z \right]. \tag{2.8}$$

In this formula $k_s^{\pm}$ are size-quantized quasi-momentums, s are the numbers of film subbands.

To meet the boundary conditions

$$\Psi_{k_s}^{\pm}\left(\pm \frac{d}{2} \right) = 0, \tag{2.9}$$

we assume:

$$\begin{cases} k_s^{+} + \dfrac{2m\pi}{a} = \dfrac{\pi}{d}(2s+1) \\ k_s^{-} + \dfrac{2m\pi}{a} = \dfrac{2\pi s}{d} \end{cases}. \tag{2.10}$$

Substitution of functions (2.8) into equation (2.6) with regard to (2.7) and equating coefficients of like harmonics yields the following systems of equations for coefficients A_m and B_m:

$$\left[\frac{\hbar^2}{2m_0} \left(k_s^{+} + \frac{2m\pi}{a} \right)^2 - E \right] A_m = \sum_{m'} V_{m'} A_{m+m'}, \tag{2.11}$$

$$\left[\frac{\hbar^2}{2m_0}\left(k_s^- + \frac{2m\pi}{a}\right)^2 - E\right]B_m = \sum_{m'} V_{m'} B_{m+m'} \,. \tag{2.12}$$

From (2.11) and (2.12) it can be seen that the equations for coefficients A_m and B_m in the thin and bulk samples are of the same form. Assuming that in going from a bulk sample to a thin film the form of a periodic potential (coefficients V_m) is not changed, we come to the conclusion that the type of the dependence of E on $k_s^{\pm}$ is the same as in the bulk sample. Therefore, we finally obtain that magneto-size-quantized levels of electron in a thin film are defined as:

$$\varepsilon_{ns} = \mu^* B(2n+1) + W(\pi s/d). \tag{2.13}$$

With a large film thickness, the neighboring values of $k_s^{\pm}$ are close to each other, so that introducing a conventional quasi-momentum, we come back to relation (2.3).

Chapter 3

ELECTRON DENSITY OF STATES AND STATISTIC PROPERTIES OF ELECTRON GAS IN LAYERED CRYSTALS AND THIN FILMS ON THEIR BASIS IN THE ABSENCE OF A MAGNETIC FIELD

To calculate the density of states in a layered crystal with electron spectrum (2), we first find the volume restricted by constant-energy surface $\varepsilon(\vec{k})=\varepsilon$ with the arbitrary value of ε with regard to spin degeneracy. Taking into account that section of this surface by arbitrary plane $k_z = const$ is a circle, for the desired volume with regard to spin degeneracy and $W(k_z)$ function parity we get the following expression:

$$V(\varepsilon)=\frac{2\cdot 2\cdot 2m^{*}\pi}{(2\pi)^3\hbar^2}\int_0^{k_z(\varepsilon)}[\varepsilon - W(k_z)]dk_z \,. \tag{3.1}$$

Electron density of states on a per-unit volume basis in this case is equal to a derivative of expression (3.1) with respect to energy ε :

$$g(\varepsilon)=\frac{4m^{*}}{h^2}k_z(\varepsilon). \tag{3.2}$$

In formulae (3.1) and (3.2), $k_z(\varepsilon)$ is a positive root of equation $W(k_z)=\varepsilon$.

In a particular case, when the form of $W(k_z)$ corresponds to strong-coupling approximation, as in (1.1), we have:

$$g(\varepsilon)=\frac{4m^{*}}{ah^2}\arccos\left(1-\frac{\varepsilon}{\Delta}\right). \tag{3.3}$$

If in (3.3) we restrict ourselves to effective mass approximation, we get the following formula:

$$g(\varepsilon)=\frac{4m^*}{ah^2}\sqrt{\frac{2\varepsilon}{\Delta}}\ . \tag{3.4}$$

For the open or transient constant-energy surface, i.e. when $\varepsilon \geq 2\Delta$, the density of states is maximum and equal to $g_{\max} = 4m^*\pi/(ah^2)$. Quite similarly one can find the electron density of states for the thin film. This density of states is:

$$g_{tf}(\varepsilon)=\frac{4m^*\pi}{h^2 d}s_\varepsilon\ . \tag{3.5}$$

In this formula, s_ε – is the largest integer part meeting the inequality $W(\pi s/d)\leq\varepsilon$. Therefore, for the case of $W(k_z)$ corresponding to strong-coupling approximation we get:

$$g_{tf}(\varepsilon)=\frac{4m^*\pi}{h^2 d}E\left(\frac{d\arccos\left(1-\frac{\varepsilon}{\Delta}\right)}{\pi a}\right). \tag{3.6}$$

In effective mass approximation we get, respectively:

$$g_{tf}(\varepsilon)=\frac{4m^*\pi}{h^2 d}E\left(\frac{d\sqrt{2\varepsilon/\Delta}}{\pi a}\right). \tag{3.7}$$

In formulae (3.6) and (3.7), $E(x)$ is an integer part of x number. With large values of ratio d/a the difference between a number and its integer part can be ignored, hence we get formulae (3.3) and (3.4) typical of the bulk samples. Normalized for a maximum, the density of states for the bulk samples in conformity with formulae (3.3) and (3.4) is depicted in Figure 1.

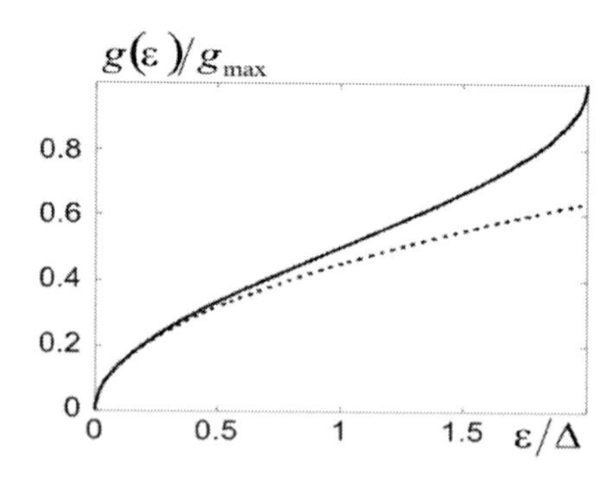

Figure 1. Electron density of states in the bulk samples for a real layered crystal (solid curve) and in effective mass approximation (dashed curve).

The figure demonstrates that the energy dependence of electron density of states in a real layered crystal almost coincides with that in effective mass approximation if $\varepsilon/\Delta \leq 0.25$. Thereafter said densities of states start to differ and the difference for closed and transient constant-energy surfaces reaches a maximum at $\varepsilon/\Delta = 2$. It also shows that as a whole at all the energies in the region of $\varepsilon/\Delta \leq 2$ the density of states for a real layered crystal is always higher than in effective mass approximation. And it means that at the same current carrier concentration and temperature electron gas chemical potential in a real layered crystal will be less than in effective mass approximation. This fact will in some way or other be reflected on all investigated physical characteristics of crystal electron subsystem. Moreover, the figure also demonstrates that electron density of states in a real layered crystal has two singularities: at $\varepsilon = 0$, corresponding to "empty" crystal lattice with no conduction electrons, and at $\varepsilon/\Delta = 2$, corresponding to transient constant-energy surface. In effective mass approximation the second singularity is absent. This fact is closely related to a difference between geometries of constant-energy surfaces in a real layered crystal and in effective mass approximation. In effective mass approximation at $0 < \varepsilon/\Delta \leq 2$ the constant-energy surface has two tangential planes normal to k_z axis. Whereas in a layered crystal at $\varepsilon/\Delta = 2$ such planes are absent. Therefore, it is clear that the greatest difference between the physical characteristics of electron subsystem of a layered crystal with closed constant-energy surface and the same characteristics calculated in effective mass approximation will occur in the case of $\varepsilon/\Delta = 2$.

We now pass on to investigation of the temperature dependence of electron gas chemical potential in layered crystals and thin films on their basis. If ζ_0 is electron gas chemical potential in a bulk layered crystal at absolute zero temperature and electron concentration in all cases under study is constant, then the equation governing electron gas chemical potential ζ in a real layered crystal at arbitrary temperature T will be given by:

$$(\gamma_0 - 1)\arccos(1-\gamma_0) + \sqrt{2\gamma_0 - \gamma_0^2} = t\int_0^{\pi} \ln\left[1 + \exp\left(\frac{\gamma - 1 + \cos x}{t}\right)\right] dx. \qquad (3.8)$$

In effective mass approximation this equation will be as follows:

$$(\gamma_0 - 1)\arccos(1-\gamma_0) + \sqrt{2\gamma_0 - \gamma_0^2} = t\int_0^{\infty} \ln\left[1 + \exp\left(\frac{\gamma - 0.5x^2}{t}\right)\right] dx. \qquad (3.9)$$

In these formulae, $\gamma_0 = \zeta_0/\Delta, \gamma = \zeta/\Delta, t = kT/\Delta$. The results of solving these equations are depicted in Figure 2.

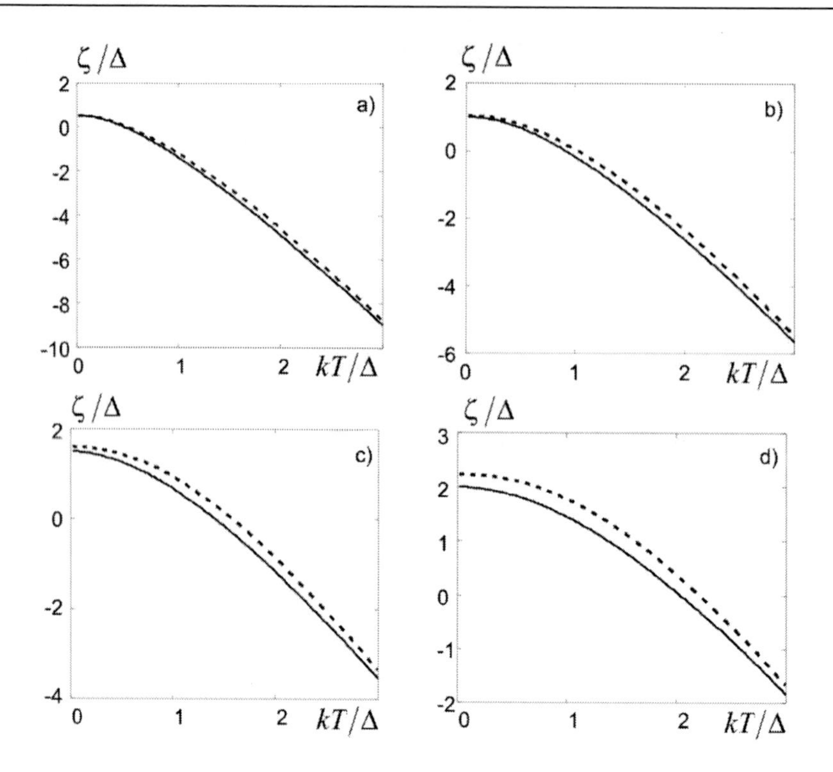

Figure 2. Temperature dependence of electron gas chemical potential in a bulk crystal at $0 \le kT/\Delta \le 3$ and: a) $\zeta_0/\Delta = 0.5$; b) $\zeta_0/\Delta = 1$; c) $\zeta_0/\Delta = 1.5$; d) $\zeta_0/\Delta = 2$.

Solid curves are for a real layered crystal, dashed curves – for effective mass approximation.[1] The figure demonstrates that in all the cases electron gas chemical potential, as could be expected, drops with a rise in temperature, and also the quicker, the less is ζ_0, i.e. current carrier concentration. At all temperatures and current carrier concentrations electron gas chemical potential in a real layered crystal is less than in effective mass approximation. The difference between the solid and dashed curves for reasons specified above increases with increasing the ratio ζ_0/Δ and reaches a maximum at $\zeta_0/\Delta = 2$.

For the thin film case equations (3.8) and (3.9) take on the form, respectively:

$$(\gamma_0 - 1)\arccos(1-\gamma_0) + \sqrt{2\gamma_0 - \gamma_0^2} = \frac{\pi t}{N}\sum_{s=1}^{N}\ln\left[1+\exp\left(\frac{\gamma - 1 + \cos\frac{\pi s}{N}}{t}\right)\right]. \tag{3.10}$$

$$(\gamma_0 - 1)\arccos(1-\gamma_0) + \sqrt{2\gamma_0 - \gamma_0^2} = \frac{\pi t}{N}\sum_{s=1}^{N}\ln\left[1+\exp\left(\frac{\gamma - 0.5\frac{\pi^2 s^2}{N^2}}{t}\right)\right]. \tag{3.11}$$

[1] The same meaning of figure parts notation and tracing of curves is preserved on the remainder of figures, unless otherwise specified.

Solutions of these equations at different γ_0 and $N = d/a$ are shown in Figures 3 to 5.

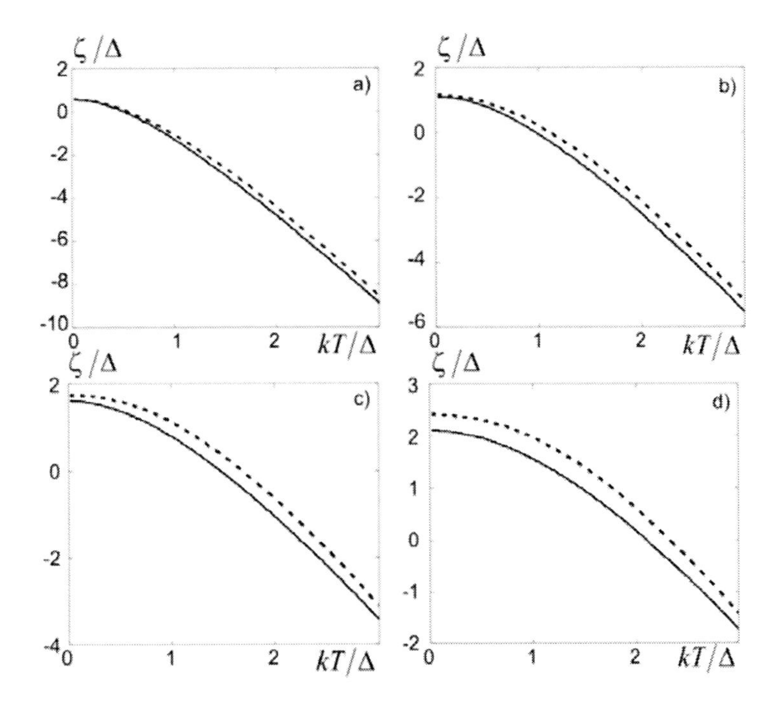

Figure 3. Temperature dependence of electron gas chemical potential in a thin film at $0 \le kT/\Delta \le 3$ and $d/a = 10$.

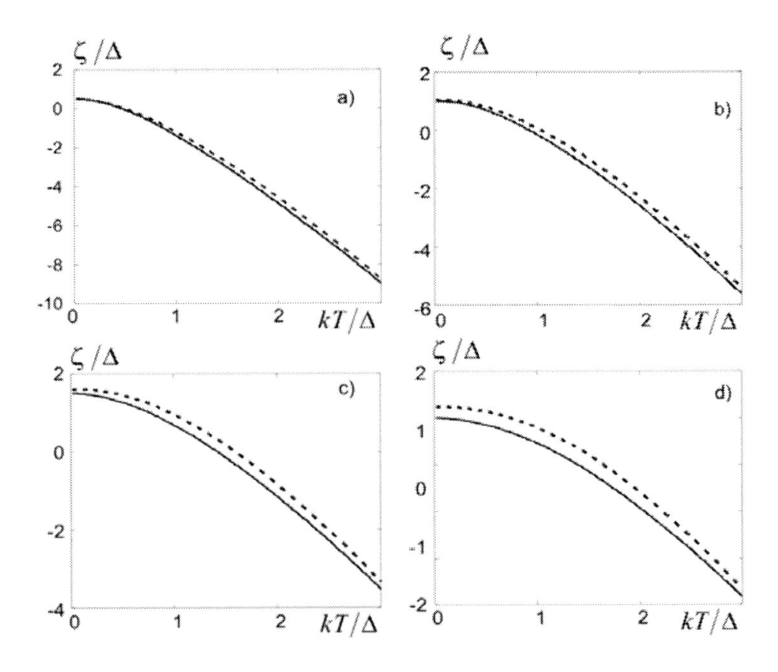

Figure 4. Temperature dependence of electron gas chemical potential in a thin film at $0 \le kT/\Delta \le 3$ and $d/a = 100$.

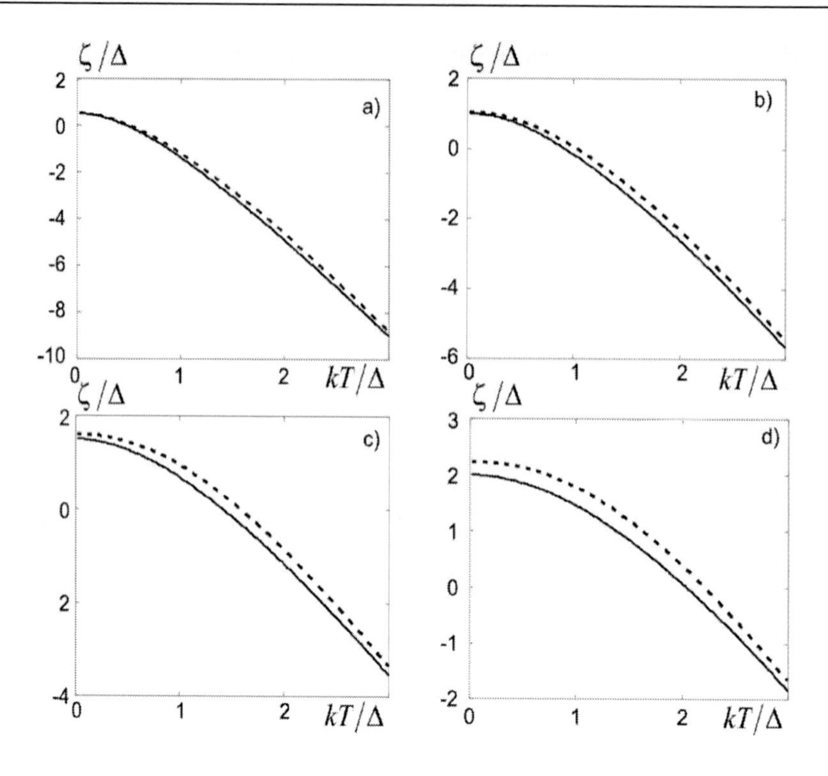

Figure 5. Temperature dependence of electron gas chemical potential in a thin film at $0 \le kT/\Delta \le 3$ and $d/a = 1000$.

Figures 3 – 5 demonstrate that the temperature dependence of chemical potential in the absence of a magnetic field and the difference between the case of a real layered crystal and effective mass approximation in the thin films and bulk crystals are fundamentally equal. Moreover, at $d/a \ge 100$ there are no essential distinctions between the thin films and bulk crystals. Appreciable differences are observed only at $d/a \le 10$, i.e. in rather thin films, these differences becoming greater with increasing the ratio ζ_0/Δ.

Chapter 4

DENSITY OF STATES AND STATISTICAL PROPERTIES OF ELECTRON GAS IN LAYERED CRYSTALS IN THE PRESENCE OF INTERLAYER CHARGE ORDERING

We now consider interlayer charge ordering in layered crystals with the energy band spectrum (1.1). It is noteworthy that until now the majority of researchers believe that such ordering is impossible or rather weak. Quite the opposite, relying on the results of X-ray structural analysis and neutron scattering showing the absence of lattice distortions in the direction normal to layers, they consider only intralayer charge ordering [20,21]. Meanwhile, in 1978 Pashitsky and Spiegel in their paper [22] theoretically proved possible coexistence of superconductivity and ferromagnetism in such layered crystals. In so doing, they thought electron gas chemical potentials in consecutive layers to be alternating. And this is identical to interlayer charge ordering that can be considered as a simple alternation of layers with different electron density. By virtue of this, the surface density n_i of charge carriers in i-th layer can be represented as:

$$n_i = n_0 a\left[1 + (-1)^i \delta\right]. \tag{4.1}$$

In this formula, n_0 is bulk concentration of charge carriers in a crystal, δ is order parameter describing the nonuniformity of filling the layers with electrons, the rest of notation is described above. The physical reason for the emergence of such charge ordering is interaction of charge carriers with phonon mode, the wave vector of which is equal to π/a, i.e. a period of charge-density wave (CDW). The physical reasons opposing such charge ordering are the Coulomb repulsion of like charge carriers, band motion of electrons in layer plane increasing the internal energy of charge carrier system and band motion of electrons between the layers tending to smooth over the irregularity of their layered distribution. The presence of such charge ordering is matched by the following band spectrum $\varepsilon_{CO}(\vec{k})$:

$$\varepsilon_{CO}(\vec{k}) = \frac{\hbar^2}{2m^*}\left(k_x^2 + k_y^2\right) \pm \sqrt{W_0^2\delta^2 + \Delta^2\cos^2 ak_z}\,. \tag{4.2}$$

In this formula, W_0 is effective attracting interaction of carriers due to a competition between electron-phonon interaction and the Coulomb repulsion, the rest of notation is generally accepted or explained above. In this case the energy is referred to the middle of pseudo-gap between minibands.

From relation (4.2) it follows that in the origination of said charge ordering the character of charge carriers motion in layer plane is not changed, and the narrow conduction miniband is split into two subbands of smaller width, divided by a pseudo-gap of width $2W_0\delta$ depending on order parameter δ. Narrowing of conduction miniband with charge ordering follows from the inequality $\sqrt{W_0^2\delta^2+\Delta^2}<W_0\delta+\Delta$ valid at $\delta>0$. Besides, from the form itself of spectrum (4.2) it is evident that in ordering the translation period is doubled.

This spectrum is matched by the following density of states:

$$g_{CO}(\varepsilon)=\frac{4m}{ah^2}\begin{cases}\arccos\left(\sqrt{\varepsilon^2-W_0^2\delta^2}/\Delta\right), -\sqrt{\Delta^2+W_0^2\delta^2}\le\varepsilon\le -W_0\delta\\ \pi/2, -W_0\delta\le\varepsilon\le W_0\delta,\\ \pi-\arccos\left(\sqrt{\varepsilon^2-W_0^2\delta^2}/\Delta\right), W_0\delta\le\varepsilon\le\sqrt{\Delta^2+W_0^2\delta^2},\\ \pi, \varepsilon\ge\sqrt{\Delta^2+W_0^2\delta^2}.\end{cases} \qquad (4.3)$$

This density of states is plotted in Figure 6.

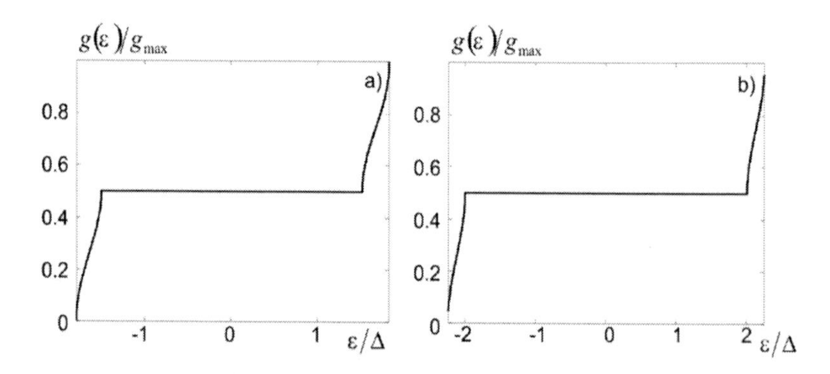

Figure 6. Density of states in the presence of charge ordering: a) at $W_0\delta/\Delta=1.5$, b) at $W_0\delta/\Delta=2$.

The figure demonstrates that in each of narrowed minibands a change in the density of states as a function of energy is similar to a change in the density of states in a single miniband in the absence of charge ordering. But the energy range where a change in the density of states is considerable is the narrower, the larger is the ratio $W_0\delta/\Delta$. Maximum value of the density of states in a charge–ordered crystal is the same as in the absence of charge ordering. Moreover, it is seen that in a charge-ordered layered crystal the density of states has a plateau in the field of a pseudo-gap between the minibands, and it is the wider, the larger is $W_0\delta$. In the field of a pseudo-gap the density of states is equal to half its maximum value. From such a behaviour of the density of states it follows that interlayer charge ordering must be energetically advantageous for any layered crystal whose parameters meet the condition of charge ordering existence at finite temperatures.

A system of self-consistent equations defining electron gas chemical potential ζ and order parameter δ in a charge-ordered crystal are of the form:

$$\begin{cases} \dfrac{t}{2\pi\gamma_{02D}}\displaystyle\int_0^{\pi}\left\{\begin{aligned}&\ln\left[1+\exp\left(t^{-1}\left(\gamma+\sqrt{w^2\delta^2+\cos^2 x}\right)\right)\right]+\\ &+\ln\left[1+\exp\left(t^{-1}\left(\gamma-\sqrt{w^2\delta^2+\cos^2 x}\right)\right)\right]\end{aligned}\right\}dx=1 \\ \dfrac{tw\delta}{2\pi\gamma_{02D}}\displaystyle\int_0^{\pi}\frac{\left\{\begin{aligned}&\ln\left[1+\exp\left(t^{-1}\left(\gamma+\sqrt{w^2\delta^2+\cos^2 x}\right)\right)\right]-\\ &-\ln\left[1+\exp\left(t^{-1}\left(\gamma-\sqrt{w^2\delta^2+\cos^2 x}\right)\right)\right]\end{aligned}\right\}}{\sqrt{w^2\delta^2+\cos^2 x}}dx=\delta \end{cases} . \qquad (4.4)$$

In (4.4), in addition to the above, the following notation is introduced: $w = W_0/\Delta$, $\gamma_{02D} = \zeta_{02D}/\Delta$, $\zeta_{02D} = n_0 a h^2/4\pi m^*$ is chemical potential of an ideal two-dimensional Fermi-gas at the absolute zero temperature. The second equation of system (4.4) has always a trivial solution $\delta \equiv 0$ corresponding to the disordered state. Besides, it is clear that a nontrivial solution of this system $\delta > 0$ corresponding to charge ordering is possible only at $w > 0$, i.e. when electron-phonon interaction dominates over the Coulomb repulsion, as it occurs, for instance, in superconductors. Moreover, analysis of the system of equations (4.4) shows that condition $w/\gamma_{02D} > 1$ is necessary, but, broadly speaking, insufficient for charge ordering existence. It means that electron-phonon interaction must dominate not merely over the Coulomb interaction, but over the sum of this interaction and the Fermi energy of ideal two-dimensional gas at the absolute zero temperature. The physical reason for this is increased kinetic energy of electrons motion in layers at their irregular distribution that must be compensated together with the Coulomb repulsion for the interlayer charge ordering to be energetically advantageous. And the insufficiency of this condition is attributable to the fact that the band motion of electrons between the layers at $\Delta > 0$ tends to smooth over the irregularity of their layered distribution.

From the system of equations (4.4) it also follows that at $w/\gamma_{02D} = 2$ and $\gamma = 0$ the second equation of the system turns to identity and order parameter must be found from the first equation.

The temperature dependence of electron gas chemical potential in a charge-ordered layered crystal found as a result of numerical solution of system (4.4) at $\gamma_{02D} = 1$ and different values of w is represented in Figure 7.

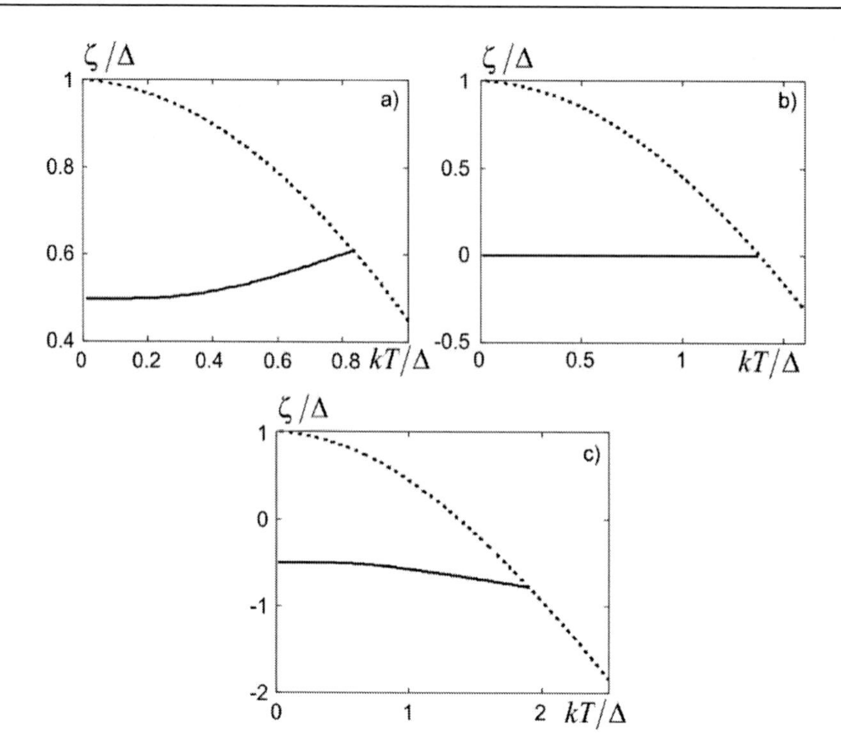

Figure 7. Temperature dependence of electron gas chemical potential in a layered charge-ordered crystal at: a) $w/\gamma_{02D} = 1.5$; b) $w/\gamma_{02D} = 2$; c) $w/\gamma_{02D} = 2.5$. Dashed curves correspond to the disordered state.

The figure demonstrates that in the charge-ordered state the system chemical potential is lower than in the disordered state at the same temperature. This testifies to the energy advantage of interlayer charge ordering when crystal parameters meet the condition for the existence of such ordering. In the figure it is also seen that in the temperature region of charge ordering existence at $1 < w/\gamma_{02D} < 2$ the level of electron gas chemical potential in a layered crystal is above the middle of a pseudo-gap between the minibands and chemical potential increases with a rise in temperature; at $w/\gamma_{02D} = 2$ the level of electron gas chemical potential lies in the middle of a gap between the minibands and chemical potential is temperature-independent; at $w/\gamma_{02D} > 2$ the level of electron gas chemical potential in a layered crystal lies below the middle of a pseudo-gap between the minibands and chemical potential drops with a rise in temperature. The intersection point of solid and dashed curves in each figure corresponds to a critical temperature whereby charge ordering destruction takes place. At the intersection point, electron gas chemical potential is broken, which is typical of second-order phase transitions. In the disordered state, as it must be, electron gas chemical potential drops with a rise in temperature. From such a dependence of chemical potential it follows that, for instance, electron gas heat capacity during phase transition must undergo a finite jump, which is also typical of second-order phase transitions.

The temperature dependence of order parameter δ with the same values of crystal parameters is shown in Figure 8.

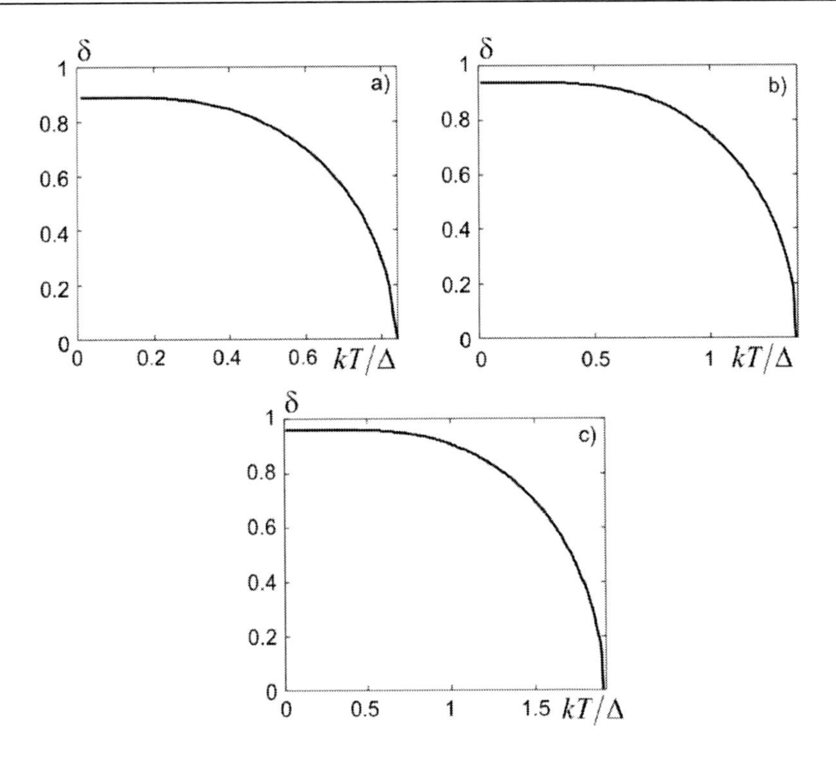

Figure 8. Temperature dependence of order parameter in a layered charge-ordered crystal at: a) $w/\gamma_{02D} = 1.5$; b) $w/\gamma_{02D} = 2$; c) $w/\gamma_{02D} = 2.5$.

The figure demonstrates that with each value of effective interaction the temperature dependence of order parameter is also typical of second-order phase transitions. However, as long as half-width of miniband $\Delta > 0$, then, as was expected, even at $T = 0$ charge ordering is incomplete, i.e. $\delta < 1$. Comparison of Figure 7 to Figure 8 shows that with increase in effective attracting interaction W_0, the critical temperature of phase transition and the value of order parameter, as was expected, increase. Besides, it is seen that in the charge-ordered state at $T = 0$ the level of chemical potential always lies in a pseudo-gap between minibands. Therefore, phase transition to the charge-ordered state can be considered as topological, i.e. such whereby the Fermi surface changes its connectivity being converted from a closed or transient to an open one (the case $\zeta_{02D}/\Delta = 1$ corresponds at $\delta = 0$ to a transient Fermi surface). Thus, from the energy dependence of density of states in a charge-ordered crystal depicted in Figure 6 it follows that at low temperatures order parameter does not have a direct influence on the electron heat capacity and the Landau diamagnetic susceptibility of electron gas in a crystal, which is at least in qualitative agreement with the experimental data [23].

However, it should be noted that charge ordering of said type is by no means in all layered crystals so strongly pronounced as it is implied by Figure 8. Moreover, analysis of the system of equations (4.4) shows that for each set of W_0, ζ_{02D} values from the range of $W_0/\zeta_{02D} > 1$ there is such a critical width of miniband Δ_c whereby at $\Delta \geq \Delta_c$ the interlayer charge ordering is impossible even at $T = 0$. Apparently, this fact makes many researchers to think possible only intralayer charge ordering. However, if charge ordering can be only intralayer, then the electric conductivity anisotropy of layered crystals in going to the charge-ordered state must always decrease. Nevertheless, in many cases, for instance, in graphite compounds intercalated with alkali metals, it is increased [24, 25].

Chapter 5

Statistical Properties of Electron Gas in Layered Crystals and Thin Films on their Basis in the Presence of a Quantizing Magnetic Field under Strong Degeneracy Conditions

For the calculation of physical characteristics of layered crystals and thin films on their basis in a quantizing magnetic field of fundamental importance is dependence of electron gas chemical potential on magnetic field induction. To study this dependence, we first derive an equation defining electron gas chemical potential with a band spectrum (2) in a quantizing magnetic field normal to layers. For this purpose we will proceed from the following relation:

$$n_0 = \sum_{\alpha} g_{\alpha} f^0(\varepsilon_{\alpha}). \tag{5.1}$$

In this formula, n_0 is the bulk concentration of electrons in a crystal, $\alpha \equiv (n, k_z)$ is a set of quantum numbers characterizing, according to (4), the state and energy of electron in a quantizing magnetic field normal to layers, g_{α} – statistical weight of the energy level corresponding to this set, calculated per unit volume, ε_{α} – electron energy, $f^0(\varepsilon_{\alpha})$ – the Fermi-Dirac equilibrium distribution function depending on the desired electron gas chemical potential ζ in a quantizing magnetic field.

To derive the required equation, we will use the formula of infinite geometric progression with a denominator less than unity, and represent $f^0(\varepsilon_{\alpha})$ as below:

$$f^0(\varepsilon_{\alpha}) = \begin{cases} \sum_{l=0}^{\infty} (-1)^l \exp\left(l \frac{\varepsilon_{\alpha} - \zeta}{kT} \right), \varepsilon_{\alpha} \le \zeta \\ \sum_{l=1}^{\infty} (-1)^{l-1} \exp\left(l \frac{\zeta - \varepsilon_{\alpha}}{kT} \right), \varepsilon_{\alpha} \ge \zeta \end{cases}. \tag{5.2}$$

Representation (5.2) allows, due to a linear dependence of electron energy ε_α on the Landau level number n, using the Poisson summation formula and making summation over the Landau levels to the end with any degree of electron gas degeneracy. As a result, we get the following equation defining the desired electron gas chemical potential ζ [26]:

$$n_0 = \frac{4m^*}{ah^2} \int_{W(x)\le\zeta} [\zeta - W(x)]dx + \frac{8m^*\pi kT}{ah^2}\sum_{l=1}^{\infty}\frac{(-1)^l}{\sinh(\pi^2 lkT/\mu^* B)}\int_{W(x)\le\zeta} \sin\left(\pi l\frac{\zeta - W(x)}{\mu^* B}\right)dx + \\ + \frac{eB}{\pi ah}\sum_{l=1}^{\infty}\frac{(-1)^{l-1}}{\sinh(\mu^* Bl/kT)}\left[\int_{W(x)\le\zeta} \exp\left(l\frac{W(x)-\zeta}{kT}\right) + \int_{W(x)\ge\zeta} \exp\left(l\frac{\zeta - W(x)}{kT}\right)dx\right] . \quad (5.3)$$

This formula introduces the designation of integration variable $ak_z \equiv x$, the rest of notation is generally accepted or explained above.

This equation undergoes formal limiting passage to the case of a weak magnetic field, as well as to the case of a non-degenerate electron gas, when summation over the Landau level numbers can be done explicitly.

In the derivation of this equation we used the Poisson summation formula of the form:

$$\sum_{n=0}^{\infty} F(2n+1) = 0.5\int_0^{\infty} F(y)dy + \sum_{s=1}^{\infty}(-1)^s\int_0^{\infty} F(y)\cos(\pi sy)dy . \quad (5.4)$$

On substitution to equation (5.1) of spectrum (2.3) with regard to statistical weight of the Landau level equal to $2eB/\pi ah$ and representation (5.2), proper arrangement of integration limits with respect to x and y, integration with respect to y is made in an elementary way. Afterwards, integration with respect to indexes that do not occur under the integral sign over x is done using the relation:

$$\sum_{m=1}^{\infty}(-1)^{m-1}\frac{m^2}{m^2+\beta^2} = \frac{\pi\beta}{2\sinh(\pi\beta)}, \quad (5.5)$$

which is valid for all β.

From this equation it follows that under strong degeneracy chemical potential is an oscillating function of a magnetic field, however, when the degeneracy is removed, i.e. when the Fermi surface disappears, oscillations disappear as well, i.e. the oscillating term in Eq.(32) goes to null equation. Exactly such form of this equation allows logical consideration of not only open Fermi surfaces discussed in the majority of works dealing with the physical characteristics of layered crystals, but also closed Fermi surfaces.

In the case of the thin film Eq.(5.3) goes over to the following:

$$n_0 = \frac{4m^*}{dh^2}\sum_{s=1}^{s_\zeta}[\zeta - W(s)] + \frac{8m^*\pi^2 kT}{dh^2}\sum_{l=1}^{\infty}\frac{(-1)^l}{\sinh(\pi^2 lkT/\mu^* B)}\sum_{s=1}^{s_\zeta}\sin\left(\pi l\frac{\zeta - W(s)}{\mu^* B}\right) + \frac{eB}{dh}\sum_{l=1}^{\infty}\frac{(-1)^{l-1}}{\sinh(\mu^* Bl/kT)}\left[\sum_{s=1}^{s_\zeta}\exp\left(l\frac{W(s)-\zeta}{kT}\right) + \sum_{s=s_\zeta+1}^{N}\exp\left(l\frac{\zeta - W(s)}{kT}\right)\right]. \quad (5.6)$$

Under conditions of strong degeneracy in a quantizing magnetic field in Eq. (32) and Eq. (35) we take into account the first and the second terms, since the third term is exponentially small.

In the case when $W(x) = \Delta(1-\cos x)$, under strong degeneracy in a quantizing magnetic field Eq.(5.3) takes on the form [2,26]:

$$(\gamma_0 - 1)\arccos(1-\gamma_0) + \sqrt{2\gamma_0 - \gamma_0^2} = (\gamma - 1)\arccos(1-\gamma) + \sqrt{2\gamma - \gamma^2} + 2\pi t\sum_{l=1}^{\infty}\frac{(-1)^l}{\sinh(\pi^2 ltb^{-1})}\times \left\{\sin\left(\pi l\frac{\gamma-1}{b}\right)\left[C_0 J_0(\pi lb^{-1}) + 2\sum_{r=1}^{\infty}(-1)^r C_{2r}J_{2r}(\pi lb^{-1})\right] + 2\cos\left(\pi l\frac{\gamma-1}{b}\right)\sum_{r=0}^{\infty}C_{2r+1}J_{2r+1}(\pi lb^{-1})\right\}. \quad (5.7)$$

In this equation, $J_n(x)$ – the Bessel functions of the first kind with an actual argument, $b = \mu^* B/\Delta$, $C_0 = \arccos(1-\gamma)$, $C_m = \sin(mC_0)/m$.

In the case when in the expansion of function $W(x)$ in powers of x only a quadratic term is retained which corresponds to effective mass approximation, Eq. (5.3) acquires the form [26]:

$$(\gamma_0 - 1)\arccos(1-\gamma_0) + \sqrt{2\gamma_0 - \gamma_0^2} = \frac{2}{3}\gamma\sqrt{2\gamma} + 2\pi t\sqrt{b}\sum_{l=1}^{\infty}\frac{(-1)^l}{l^{1/2}\sinh(\pi^2 ltb^{-1})}\times \left[\sin(\pi lb^{-1}\gamma)C\left(\sqrt{2l\gamma b^{-1}}\right) + \cos(\pi lb^{-1}\gamma)S\left(\sqrt{2l\gamma b^{-1}}\right)\right]. \quad (5.8)$$

In this equation, $C(x)$ and $S(x)$ – the Fresnel sine integral and the Fresnel cosine integral, respectively, the rest of notation is the same. The left-hand sides of Eq. (5.7) and Eq.(5.8) are the same, because we consider the influence of layered structure effects on the dependence of electron gas chemical potential on a quantizing magnetic field on condition of constant bulk concentration of electrons in a crystal.

Eq.(5.3) and Eq.(5.6) can be also solved numerically. The results of this solution are shown in Figures 9, 10.

The figures demonstrate that in all cases considered chemical potential is an oscillating function of a magnetic field. However, in the thin film the form of oscillations is essentially different from their form in the thick film and the bulk crystal. It occurs because considerable change in the free energy of crystal electron subsystem takes place not only when the Landau level crosses the Fermi level, but also at intersection of the Fermi level with any boundary of populated or free film subband.

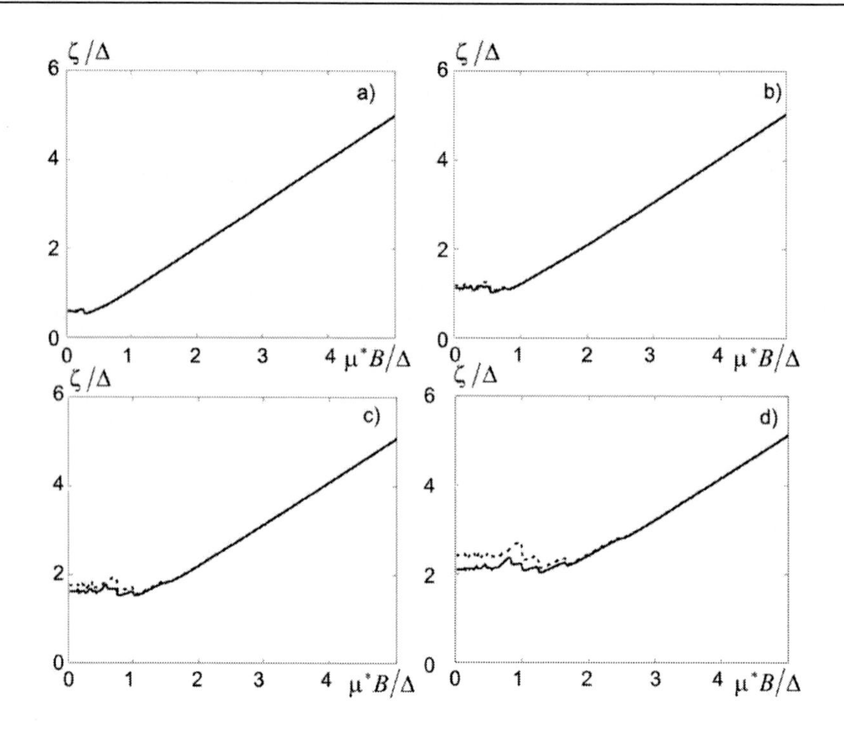

Figure 9. Field dependence of chemical potential of strongly degenerate electron gas in the thin film at $d/a = 10, kT/\Delta = 0.03$, and $0 \le \mu^* B/\Delta \le 5$.

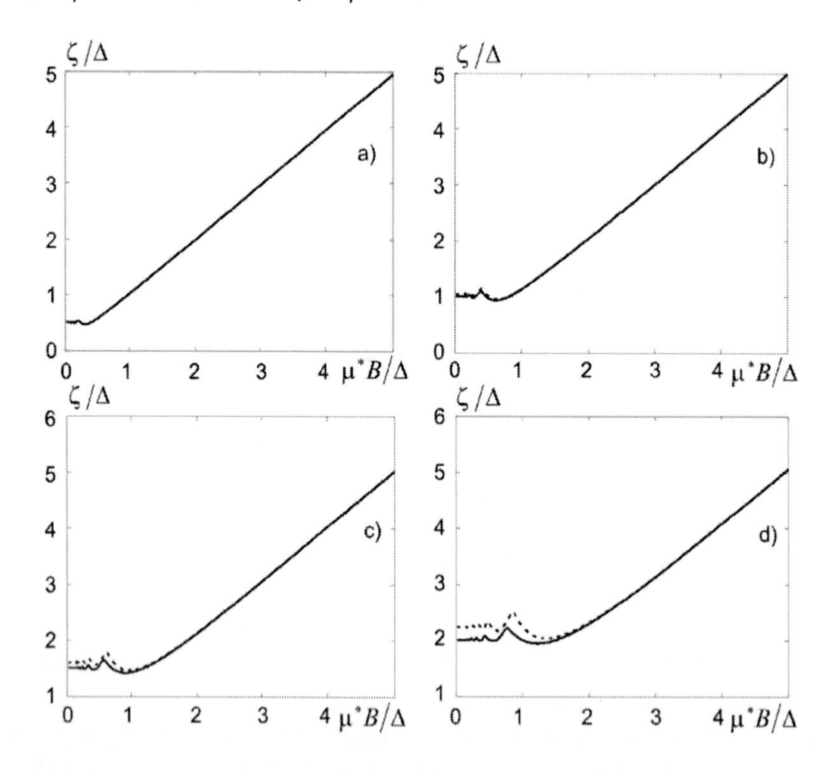

Figure 10. Field dependence of chemical potential of strongly degenerate electron gas in the thin film at $d/a = 100$, $kT/\Delta = 0.03$ and $0 \le \mu^* B/\Delta \le 5$.

Moreover, it is seen that the difference between the dashed and solid curves with increase in electron concentration, as was to be expected, increases, due to increase in the degree of filling a narrow miniband describing an interlayer motion of electrons. As long as chemical potential oscillations take place, the dashed curves in all figures lie above the solid ones. It occurs because at all film thicknesses the density of electron states in a real layered crystal is larger than the same density in effective mass approximation. Moreover, analysis shows that at $d/a \geq 100$ the field dependence of chemical potential in the thin film almost coincides with that in the bulk crystal, this coincidence taking place both in a real layered crystal and in effective mass approximation.

However, after the last oscillation maximum at all film thicknesses and all electron concentrations the solid and dashed curves approach each other rather quickly, and the magnetic field dependence of chemical potential becomes almost linear. It occurs because, for instance, Eq.(5.7) in a strong quantizing magnetic field, i.e. in the ultraquantum limit, has the following asymptotic solution:

$$\zeta = \mu^* B + \Delta\left[1 - \cos\left(\frac{f(\gamma_0)\Delta}{2\mu^* B}\right)\right], \tag{5.9}$$

where

$$f(\gamma_0) = (\gamma_0 - 1)\arccos(1 - \gamma_0) + \sqrt{2\gamma_0 - \gamma_0^2}\,. \tag{5.10}$$

In effective mass approximation formula (5.10) takes on the form:

$$\zeta = \mu^* B + \frac{\Delta^3 f^2(\gamma_0)}{8(\mu^* B)^2}\,. \tag{5.11}$$

From formulae (5.9) and (5.11) it follows that in a quantizing magnetic field the Fermi surface of a crystal is shaped as a single "magnetic tube" which, as the magnetic field increases, is squeezed along its axis which coincides with the magnetic field direction. Such deformation of the Fermi surface results from condensation of electrons at the bottom of single filled Landau subband with the number $n = 0$. It can be explained as follows. A change in a quantizing magnetic field cannot change the number of electrons in a crystal. However, with magnetic field increase due to increased statistic weight of the Landau level, the "magnetic tube" is expanded in directions normal to the magnetic field. But as long as the "magnetic tube's" "volume" is unvaried due to constant number of electrons in a crystal, it must be squeezed in the direction of a magnetic field. This fact has to be taken into account in the calculation of physical characteristics of crystal electron subsystem.

Chapter 6

STATISTICAL PROPERTIES OF ELECTRON GAS IN LAYERED CRYSTALS AND THIN FILMS ON THEIR BASIS IN THE PRESENCE OF A QUANTIZING MAGNETIC FIELD UNDER WEAK AND INTERMEDIATE DEGENERACY CONDITIONS

We now assume that the system temperature is sufficiently high for the oscillations of chemical potential in a quantizing magnetic field to be smoothed and the electron condensation in the lowest Landau subband with the number $n=0$ not to manifest itself. In this case equations defining electron gas chemical potential in a layered crystal and thin film on its basis are, respectively, as follows:

$$n_0 = \frac{4m^*}{ah^2} \int_{W(x)\le\zeta} [\zeta - W(x)]dx + \\ + \frac{eB}{\pi ah} \sum_{l=1}^{\infty} \frac{(-1)^{l-1}}{\sinh(\mu^* Bl/kT)} \left[\int_{W(x)\le\zeta} \exp\left(l\frac{W(x)-\zeta}{kT} \right) + \int_{W(x)\ge\zeta} \exp\left(l\frac{\zeta - W(x)}{kT} \right) dx \right], \tag{6.1}$$

$$n_0 = \frac{4m^*}{dh^2} \sum_{s=1}^{s_\zeta} [\zeta - W(s)] + \\ + \frac{eB}{dh} \sum_{l=1}^{\infty} \frac{(-1)^{l-1}}{\sinh(\mu^* Bl/kT)} \left[\sum_{s=1}^{s_\zeta} \exp\left(l\frac{W(s)-\zeta}{kT} \right) + \sum_{s=s_\zeta+1}^{N} \exp\left(l\frac{\zeta - W(s)}{kT} \right) \right]. \tag{6.2}$$

The results of numerical solving these equations, both for a real layered crystal, and in effective mass approximation, are shown in Figures 11-13.

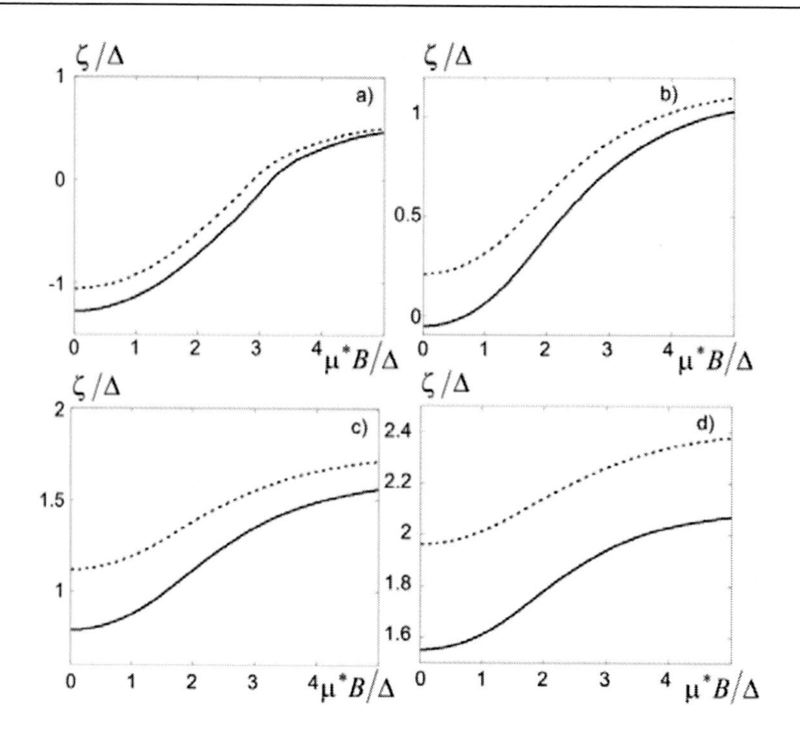

Figure 11. Field dependence of electron gas chemical potential in the thin film at $d/a = 10$, $kT/\Delta = 1$ and $0 \le \mu^* B/\Delta \le 5$.

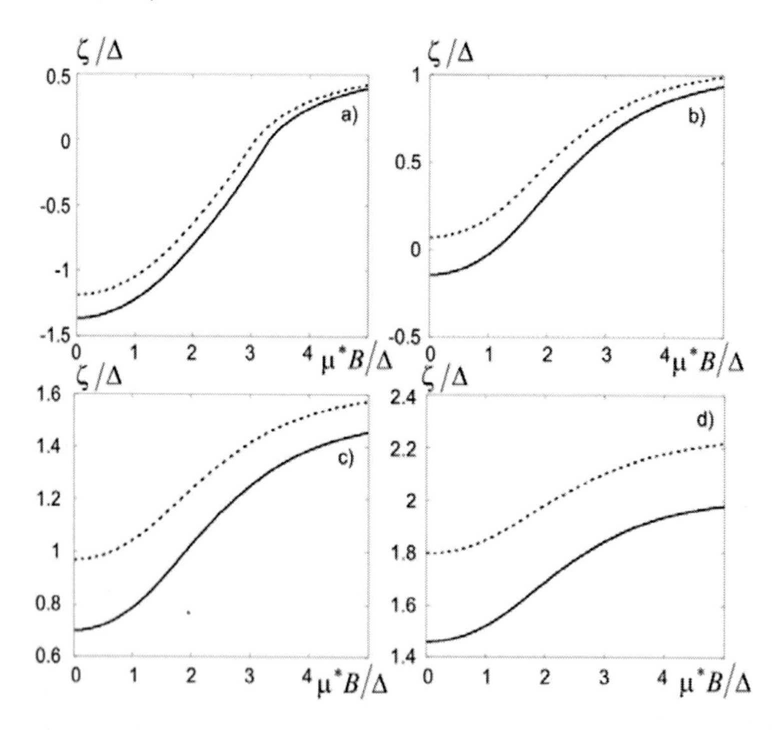

Figure 12. Field dependence of electron gas chemical potential in the thin film at $d/a = 100$, $kT/\Delta = 1$ and $0 \le \mu^* B/\Delta \le 5$.

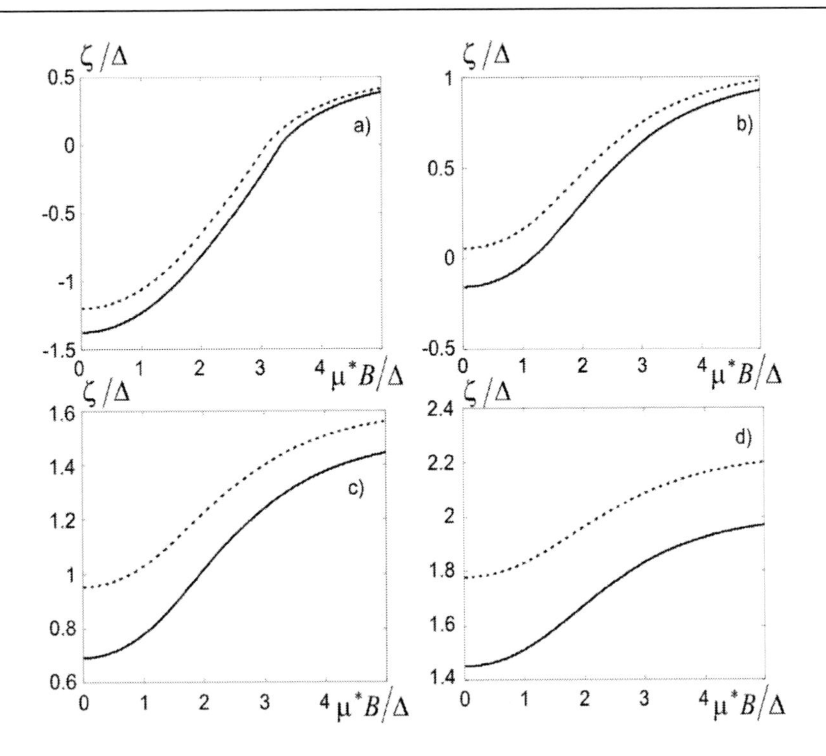

Figure 13. Field dependence of electron gas chemical potential in the thin film at $d/a = 1000$, $kT/\Delta = 1$ and $0 \leq \mu^* B/\Delta \leq 5$.

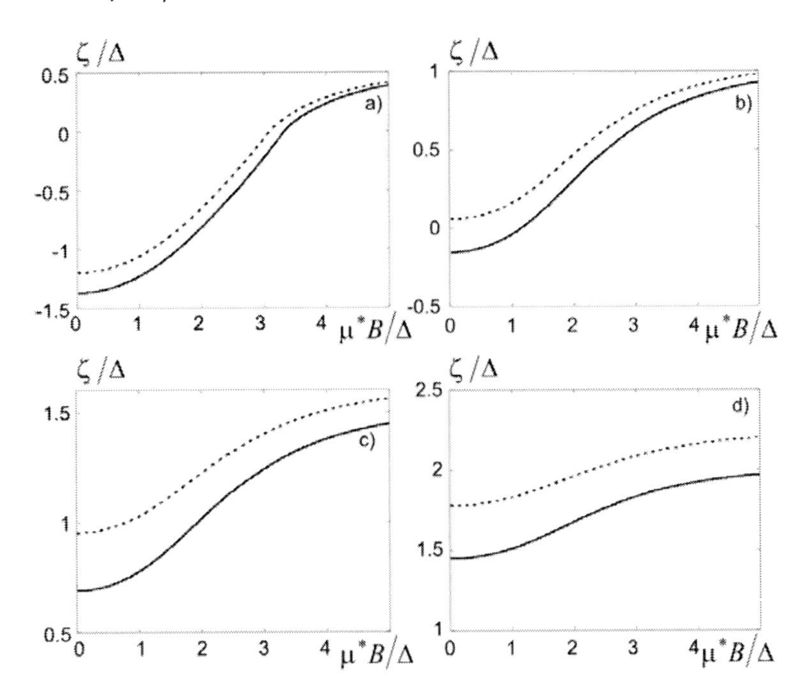

Figure 14. Field dependence of electron gas chemical potential in the bulk crystal at $kT/\Delta = 1$ and $0 \leq \mu^* B/\Delta \leq 5$.

The figures demonstrate that in all cases the dashed curves lie above the solid ones, i.e. electron gas chemical potential in a real layered crystal is lower than in effective mass approximation, which, as before, is due to a higher density of electron states in a layered crystal as compared to effective mass approximation. The difference between the dashed and solid curves, as was to be expected, increases with electron concentration, hence, the degree of miniband filling. As chemical potential oscillations are smoothed, it is a monotonically increasing function of magnetic field. This is because the system free energy in the presence of the Landau levels increases with magnetic field growth. It is also evident that as the film thickness increases, the system chemical potential decreases by virtue of the fact that the distance between neighbouring size-quantized levels is reduced and lower energy states are populated. With high d/a ratios the field dependence of electron gas chemical potential in the thin film approaches the dependence typical of the bulk crystal.

Chapter 7

STATISTICAL PROPERTIES OF ELECTRON GAS IN LAYERED CHARGE-ORDERED CRYSTALS IN A QUANTIZING MAGNETIC FIELD UNDER STRONG DEGENERACY CONDITIONS

To consider the statistical properties of electron gas in layered charge-ordered crystals in a quantizing magnetic field under strong degeneracy conditions, let us write a system of equations defining the dependence of chemical potential ζ and order parameter δ on the induction of a quantizing magnetic field normal to layers [27]. This system is given by:

$$\begin{cases} \dfrac{1}{2\pi\gamma_{02D}}\left[\displaystyle\int\limits_{\gamma+R\geq 0, 0\leq x\leq\pi}(\gamma+R)dx+\int\limits_{\gamma-R\geq 0, 0\leq x\leq\pi}(\gamma-R)dx\right]+ \\ +\dfrac{t}{\gamma_{02D}}\displaystyle\sum_{l=1}^{\infty}\frac{(-1)^l}{\sinh(\pi^2 ltb^{-1})}\left[\int\limits_{\gamma+R\geq 0, 0\leq x\leq\pi}\sin\left[\pi lb^{-1}(\gamma+R)\right]dx+\int\limits_{\gamma-R\geq 0, 0\leq x\leq\pi}\sin\left[\pi lb^{-1}(\gamma-R)\right]dx\right]=1 \\ \dfrac{w\delta}{2\pi\gamma_{02D}}\left[\displaystyle\int\limits_{\gamma+R\geq 0, 0\leq x\leq\pi}(\gamma+R)R^{-1}dx-\int\limits_{\gamma-R\geq 0, 0\leq x\leq\pi}(\gamma-R)R^{-1}dx\right]+ \\ +\dfrac{tw\delta}{\gamma_{02D}}\displaystyle\sum_{l=1}^{\infty}\frac{(-1)^l}{\sinh(\pi^2 ltb^{-1})}\left[\int\limits_{\gamma+R\geq 0, 0\leq x\leq\pi}R^{-1}\sin\left[\pi lb^{-1}(\gamma+R)\right]dx-\int\limits_{\gamma-R\geq 0, 0\leq x\leq\pi}R^{-1}\sin\left[\pi lb^{-1}(\gamma-R)\right]dx\right]=\delta \end{cases}. \quad (7.1)$$

In (43), an additional designation $R=\sqrt{w^2\delta^2+\cos^2 x}$ is introduced, the rest of notation is explained above. It is apparent from the system that its second equation has always a trivial solution $\delta\equiv 0$ corresponding to the disordered state. However, if $w>0$, i.e. effective electron-electron interaction is attractive, then the nontrivial solution proves to be energetically more advantageous than the trivial one. The results of numerical solution of system (43) are shown in Figures 15 and 16.

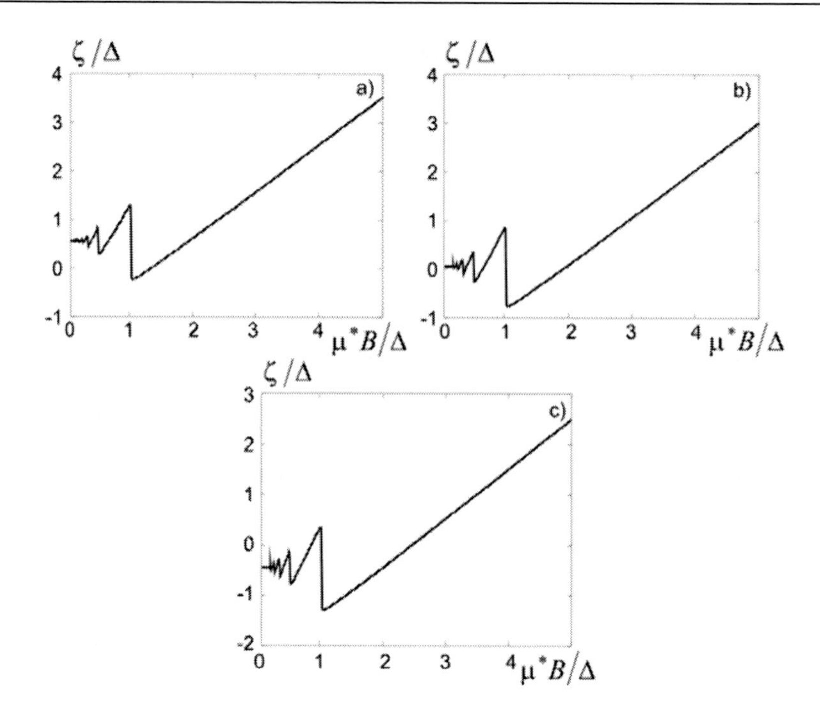

Figure 15. Field dependence of chemical potential at $0 \le \mu^* B/\Delta \le 5, \zeta_{02D}/\Delta = 1$, $kT/\Delta = 0.03$ and: a) $W_0/\zeta_{02D} = 1.5$, b) $W_0/\zeta_{02D} = 2$, c) $W_0/\zeta_{02D} = 2.5$.

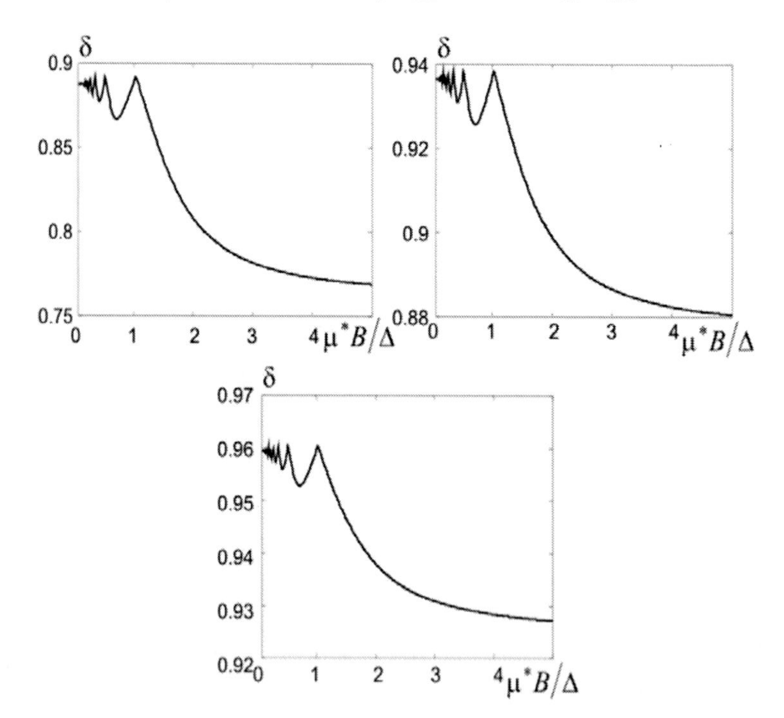

Figure 16. Field dependence of order parameter at $0 \le \mu^* B/\Delta \le 5, \zeta_{02D}/\Delta = 1$, $kT/\Delta = 0.03$ and: a) $W_0/\zeta_{02D} = 1.5$, b) $W_0/\zeta_{02D} = 2$, c) $W_0/\zeta_{02D} = 2.5$.

Figure 15 demonstrates that chemical potential is an oscillating function of magnetic field, and in a strong magnetic field it varies almost in a jump-like fashion. Note that in reality these jumps are somewhat smeared due to the Dingle factor [28] whose role will not be discussed in this book. Despite its importance as a tool for studying the mechanisms of charge carrier scattering in solids, the Dingle factor will be almost everywhere considered to be equal to unity. Conditions under which it can be done will be considered in more detail in that part of the book which covers manifestation of layered structure effects in the galvanomagnetic phenomena. From Figure 15 it is also seen that with increasing the value of effective attractive interaction, the frequency of chemical potential oscillations grows. This testifies to the fact that the degree of Fermi surface openness grows with increasing the value of effective attractive interaction, which is attributable to a contraction of conduction minibands in the charge ordering process. The last jump of chemical potential with each value of effective interaction corresponds to a topological transition from an open Fermi surface to a closed one in a quantizing magnetic field. Comparison of Figure 15 to Figure 10d with a respective displacement of energy reference point shows that charge ordering is also energetically advantageous in a quantizing magnetic field, since it reduces electron gas chemical potential. After the point of topological transition the dependence of chemical potential on magnetic field induction becomes almost linear. This may be due to the fact that in the ultraquantum limit the dependence of chemical potential on magnetic field induction is as follows:

$$\zeta = \mu^* B - \sqrt{W_0^{\,2}\delta^2 + \Delta^2 \cos^2 \frac{\pi\zeta_{02D}}{4\mu^* B}}\,. \tag{7.2}$$

The reason for this form of solution is that the Fermi surface of charge-ordered crystal in a strong quantizing magnetic field consists of three magnetic tubes, or the Landau tubes. One of them has symmetry plane $k_z = 0$, its section by this plane is maximum, and this tube gets narrow either side of this plane. Two other tubes that are halves of the first one, have small plane sections $k_z = \pm\pi/a$ and get narrow inside the one-dimensional Brillouin zone. As the magnetic field grows, these tubes are squeezed along their common axis parallel to the magnetic field.

Figure 16 demonstrates that order parameter is an oscillating function of magnetic field, the oscillation frequency increasing with the value of effective attractive interaction. Like in the absence of a magnetic field, the order parameter grows with the value of effective attractive interaction. The order parameter reaches one of its maximums at point of topological transition from an open Fermi surface to a closed one, following which charge ordering is destroyed, though rather slowly [27,29].

Chapter 8

STATISTICAL PROPERTIES OF ELECTRON GAS IN LAYERED CHARGE-ORDERED CRYSTALS IN A QUANTIZING MAGNETIC FIELD UNDER WEAK AND INTERMEDIATE DEGENERACY CONDITIONS

As before, we now assume that the system temperature is sufficiently high to ignore the oscillations of chemical potential and order parameter and electron condensation in the lowest Landau subband. In this case the system of equations defining chemical potential and order parameter is given by:

$$\begin{cases} \dfrac{1}{2\pi\gamma_{02D}}\left[\displaystyle\int_{\gamma+R\ge 0,0\le x\le\pi}(\gamma+R)dx+\int_{\gamma-R\ge 0,0\le x\le\pi}(\gamma-R)dx\right]+ \\ +\dfrac{b}{2\pi\gamma_{02D}}\displaystyle\sum_{l=1}^{\infty}\frac{(-1)^l}{\sinh(lbt^{-1})}\left[\int_0^{\pi}\exp\left[-lt^{-1}|\gamma+R|\right]dx+\exp\left[-lt^{-1}|\gamma-R|\right]dx\right]=1 \\ \dfrac{w\delta}{2\pi\gamma_{02D}}\left[\displaystyle\int_{\gamma+R\ge 0,0\le x\le\pi}(\gamma+R)R^{-1}dx-\int_{\gamma-R\ge 0,0\le x\le\pi}(\gamma-R)R^{-1}dx\right]+ \\ +\dfrac{bw\delta}{2\pi\gamma_{02D}}\displaystyle\sum_{l=1}^{\infty}\frac{(-1)^l}{\sinh(lbt^{-1})}\left[\int_0^{\pi}R^{-1}\exp\left[-lt^{-1}|\gamma+R|\right]dx-\int_0^{\pi}R^{-1}\exp\left[-lt^{-1}|\gamma-R|\right]dx\right]=\delta \end{cases} . \quad (8.1)$$

Like in the absence of a magnetic field, the second equation of this system has always a trivial solution $\delta\equiv 0$ corresponding to the disordered state. However, at $w>1$ even in a quantizing magnetic field at lower than critical temperature the nontrivial solution proves to be energetically morc advantageous than the trivial one. At $w=2$ $\gamma=0$, i.e. $\zeta=0$ and $W_0/\zeta_0=2$, the second equation of system (45) changes into identity and order parameter must be found from the first equation. In this case, in the region of existence of interlayer charge ordering, electron gas chemical potential does not depend on magnetic field induction.

The results of numerical solution of system (8.1) are depicted in Figures 17 and 18.

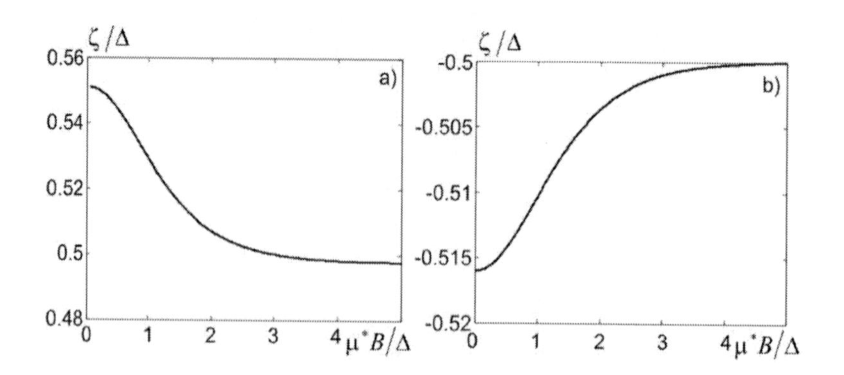

Figure 17. Field dependence of chemical potential in a layered charge-ordered crystal at $kT/\Delta = 0.6$, $0 \le \mu^* B/\Delta \le 5$, and: a) $W_0/\zeta_0 = 1.5$, b) $W_0/\zeta_0 = 2.5$.

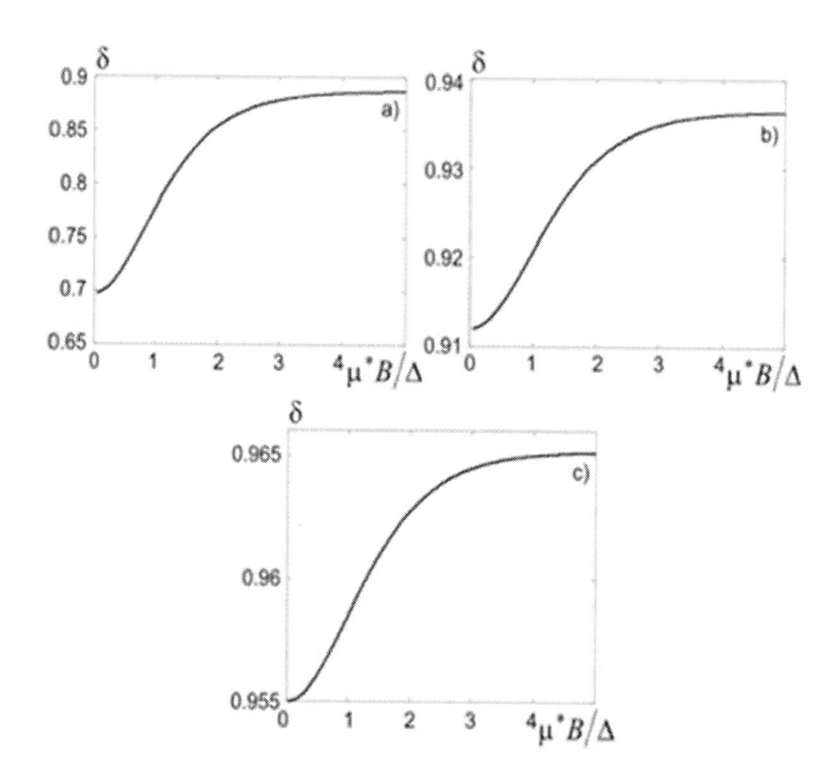

Figure 18. Field dependence of order parameter in a layered charge-ordered crystal at $kT/\Delta = 0.6$, $0 \le \mu^* B/\Delta \le 5$, and: a) $W_0/\zeta_0 = 1.5$, b) $W_0/\zeta_0 = 2$, c) $W_0/\zeta_0 = 2.5$.

The figures show that at $W_0/\zeta_0 < 2$ the system chemical potential decreases with a growth of quantizing magnetic field induction, and at $W_0/\zeta_0 > 2$ – increases with a growth of quantizing magnetic field induction. It is easy to see, since it follows from system (8.1) that growth of quantizing magnetic field induction due to increase in the distance between the Landau subbands in a certain sense is equivalent to crystal cooling. The same accounts for clearly visible in Figure 18 increase in order parameter with a growth of quantizing magnetic field induction. The figures also show that the rate of change in chemical potential and order parameter with a growth of quantizing magnetic field induction decreases with increasing the value of effective attractive interaction between electrons. Moreover, it is seen that in the considered range of change in magnetic field induction at given temperature the energy advantage of charge ordering persists, as long as chemical potential of electron gas in the ordered state even in a quantizing magnetic field proves to be lower than in the disordered state in the absence of a magnetic field.

Chapter 9

General Formulae for the Diamagnetic Susceptibility of Electron Gas in Layered Crystals and Thin Films on their Basis

To calculate the diamagnetic susceptibility of electron gas, we will use the grand partition function method and evaluate the thermodynamic potential of electron gas unit volume with a band spectrum (2) in a quantizing magnetic field normal to layers. For this purpose the following formula will be used:

$$-\Omega = kT\sum_{\alpha} g_{\alpha} \ln\left[1+\exp\left(\frac{\zeta-\varepsilon_{\alpha}}{kT}\right)\right]. \tag{9.1}$$

This formula employs the same notation as in formula (5.1).

Then we expand the logarithmic function as a power series:

$$-\Omega = \sum_{\alpha(\varepsilon_{\alpha}\le\zeta)} g_{\alpha}(\zeta-\varepsilon_{\alpha}) + kT\sum_{l=1}^{\infty}(-1)^{l-1}l^{-1}\left[\sum_{\alpha(\varepsilon_{\alpha}\le\zeta)}\exp\left(l\frac{\varepsilon_{\alpha}-\zeta}{kT}\right)+\sum_{\alpha(\varepsilon_{\alpha}\ge\zeta)}\exp\left(l\frac{\zeta-\varepsilon_{\alpha}}{kT}\right)\right]. \tag{9.2}$$

Now, using (9.2), we will do the same as in the derivation of Eq.(5.3). On calculating the thermodynamic potential, we will use the following formula for the diamagnetic susceptibility (in SI system):

$$\chi = -\frac{\mu_0 \partial\Omega}{B\partial B}, \tag{9.3}$$

where μ_0 – vacuum permeability.

As a result, we get the formula for full diamagnetic susceptibility of electron gas as below:

$$\chi = \chi_L + \chi_{os} + \chi_{lg}, \tag{9.4}$$

where χ_L, χ_{os} and χ_{Lg} – the Landau diamagnetic susceptibility, the oscillating diamagnetic susceptibility due to dHvA effect and the Langevin diamagnetic susceptibility that does not oscillate, but is a function of a magnetic field, respectively. The first two components are essential under strong degeneracy of electron gas, the third component – under intermediate and weak degeneracy, as well as in the nondegenerate gas.

These components are defined by the following formulae [30, 31]:

$$\chi_L = -\frac{e\mu_0\mu^*}{3\pi h}k_\zeta . \tag{9.5}$$

$$\chi_{os} = \frac{2e\mu_0\mu^*}{\pi h}\sum_{l=1}^{\infty}\left\{g_l \int_{W(k_z)\le\zeta} \cos\left[\pi l \frac{\zeta - W(k_z)}{\mu^* B}\right]dk_z + \right.$$
$$\left. + f_l \int_{W(k_z)\le\zeta} \frac{\zeta - W(k_z)}{\mu^* B}\sin\left[\pi l \frac{\zeta - W(k_z)}{\mu^* B}\right]dk_z\right\} . \tag{9.6}$$

$$\chi_{Lg} = \frac{e\mu_0\mu^*}{\pi h}\sum_{l=1}^{\infty}(-1)^l h_l\left\{\int_{W(k_z)\le\zeta}\exp\left[l\frac{W(k_z)-\zeta}{kT}\right]dk_z - \int_{W(k_z)\ge\zeta}\exp\left[l\frac{\zeta - W(k_z)}{kT}\right]dk_z\right\}. \tag{9.7}$$

In these formulae:

$$g_l = \frac{(-1)^{l-1}}{(\pi l)^2}\left\{\frac{3\pi^2 lkT/\mu^* B}{\sinh(\pi^2 lkT/\mu^* B)} - \left[\frac{\pi^2 lkT/\mu^* B}{\sinh(\pi^2 lkT/\mu^* B)}\right]^2\cosh(\pi^2 lkT/\mu^* B)\right\}, \tag{9.8}$$

$$f_l = \frac{(-1)^{l-1}}{\pi l}\frac{\pi^2 lkT/\mu^* B}{\sinh(\pi^2 lkT/\mu^* B)}, \tag{9.9}$$

$$h_l = \frac{kT/\mu^* Bl}{\sinh(\mu^* Bl/kT)} - \frac{\cosh(\mu^* Bl/kT)}{\sinh^2(\mu^* Bl/kT)}. \tag{9.10}$$

Integration is performed in formulae (9.5) – (9.7) only over positive k_z values.

In the case of thin films, formulae (9.5) – (9.7) become as follows:

$$\chi_L = -\frac{e\mu_0\mu^*}{3hd}s_\zeta , \tag{9.11}$$

$$\chi_{os} = \frac{2e\mu_0\mu^*}{hd}\sum_{l=1}^{\infty}\left\{g_l\sum_{s=1}^{s_\zeta}\cos\left[\pi l\frac{\zeta - W(s)}{\mu^* B}\right] + f_l\sum_{s=s_\zeta+1}^{N}\frac{\zeta - W(s)}{\mu^* B}\sin\left[\pi l\frac{\zeta - W(s)}{\mu^* B}\right]\right\}, \tag{9.12}$$

$$\chi_{Lg} = \frac{e\mu_0\mu^*}{hd}\sum_{l=1}^{\infty}(-1)^l h_l\left\{\sum_{s=1}^{s_\zeta}\exp\left[l\frac{W(s)-\zeta}{kT}\right]-\sum_{s=s_\zeta+1}^{N}\exp\left[l\frac{\zeta-W(s)}{kT}\right]\right\}. \qquad (9.13)$$

Next we will use these formulae to analyze the influence of layered structure effects on each component of the diamagnetic susceptibility of electron gas in layered crystals and thin films on their basis.

Chapter 10

LANDAU DIAMAGNETISM AND TOTAL MAGNETIC SUSCEPTIBILITY OF ELECTRON GAS IN LAYERED CRYSTALS AND THIN FILMS ON THEIR BASIS IN A WEAK MAGNETIC FIELD

Formula (50) implies the following expression for the Landau diamagnetic susceptibility of electron gas in a real layered crystal:

$$\chi_L = -\frac{e^2\mu_0}{12\pi^2 m^* a}\arccos(1-\gamma). \quad (10.1)$$

Modulo maximum χ_L value in a layered crystal is achieved at $\gamma = 2$ and equals:

$$\chi_0 = \frac{e^2\mu_0}{12\pi m^* a}. \quad (10.2)$$

For instance, at $m^* = 0.01m_0$ and $a = 1$ nm we get $\chi_0 = 9.38\cdot 10^{-5}$.

At $\gamma << 1$, i.e. in effective mass approximation, from formula (10.1) follows the Landau formula [30]:

$$\chi_L = -\frac{e^2\mu_0}{12\pi^2 m^* a}\sqrt{2\gamma}. \quad (10.3)$$

The result by D.M.Bercha and V.T.Maslyuk is a small correction to formula (10.3), if the exact formula (10.1) takes into account the following terms of expansion with respect to small parameter γ.

In the case of a thin film, formula (10.1) goes over into:

$$\chi_L = -\frac{e^2 \mu_0}{12\pi m^* d} E\left[\frac{d \arccos(1-\gamma)}{\pi a}\right], \tag{10.4}$$

and formula (10.3) – into the following one:

$$\chi_L = -\frac{e^2 \mu_0}{12\pi m^* d} E\left(\frac{d\sqrt{2\gamma}}{\pi a}\right). \tag{10.5}$$

However, if $\gamma \geq 2$, i.e. all N subbands of the film are populated, then modulo maximum value of the Landau diamagnetic susceptibility of electron gas in a thin film coincides with its value in a bulk crystal and is defined by formula (10.2).

Total diamagnetic susceptibility χ_0 of electron gas in a weak magnetic field is obtained by passage to the limit $B \to 0$ in formulae (9.5) – (52) with regard to (53) – (55). As a result, we arrive at the following formula:

$$\chi_{0t} = -\frac{e^2 \mu_0}{12\pi^2 m^*} \int_{all} \left[1 + \exp\left(\frac{W(k_z) - \zeta}{kT}\right)\right]^{-1} dk_z . \tag{10.6}$$

Integration in formula (10.6) is performed over all positive k_z values. This formula in a certain sense is similar to the well known Peierls-Wilson formula.

In the case of a thin film, formula (64) goes over into:

$$\chi_{0t} = -\frac{e^2 \mu_0}{12\pi m^* d} \sum_{s=1}^{N} \left[1 + \exp\left(\frac{W(s) - \zeta}{kT}\right)\right]^{-1} . \tag{10.7}$$

From formulae (10.1) – (10.7) it is seen that, with a rise in temperature, the Landau diamagnetic susceptibility both in a layered crystal and a thin film must drop due to a removal of electron gas degeneracy.

However, in a bulk crystal the diamagnetic susceptibility is reduced smoothly, and in a thin film – by jumps which occur whenever the level of chemical potential crosses the boundary of any populated film subband.

However, as it follows from formulae (10.6) and (10.7), full diamagnetic susceptibility of electron gas in a weak magnetic field, both in a layered crystal and a thin film is a smooth function of temperature. The results of calculation of electron gas diamagnetic susceptibility in a weak magnetic field are given in Figures 19 – 22.

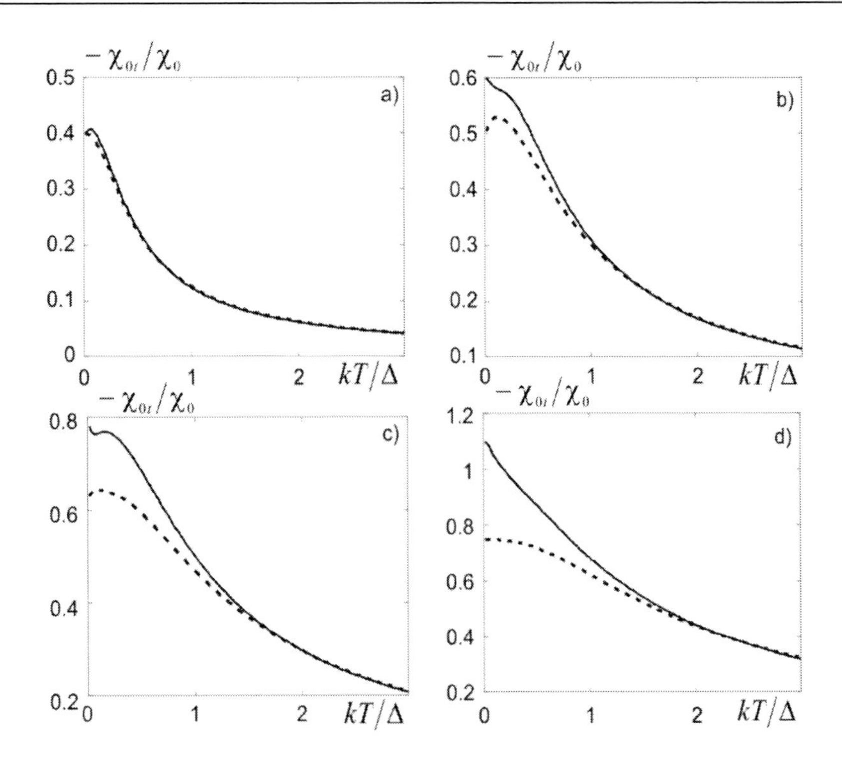

Figure 19. Temperature dependence of electron gas diamagnetic susceptibility in a weak magnetic field in a thin film at $d/a = 10$.

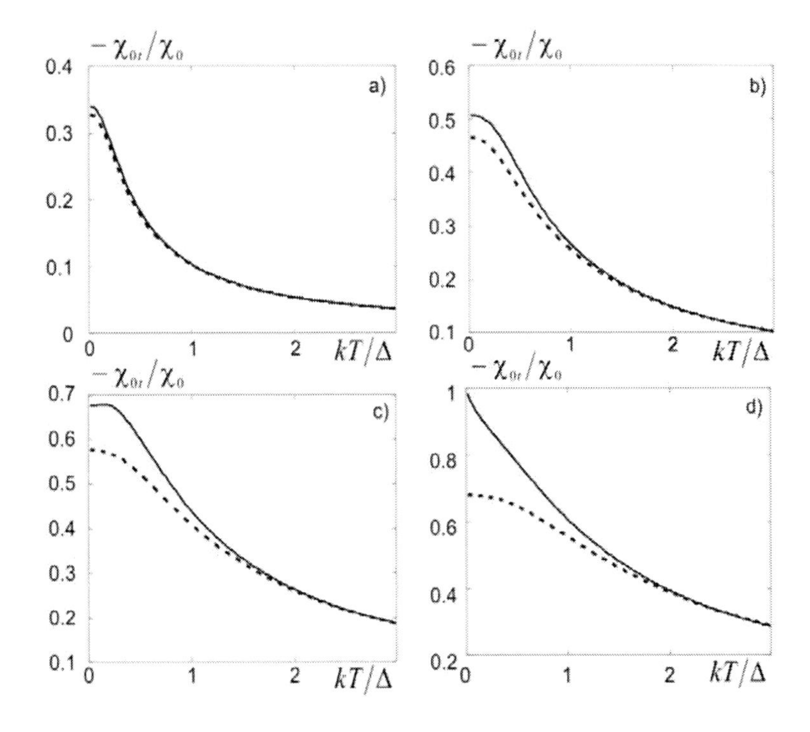

Figure 20. Temperature dependence of electron gas diamagnetic susceptibility in a weak magnetic field in a thin film at $d/a = 100$.

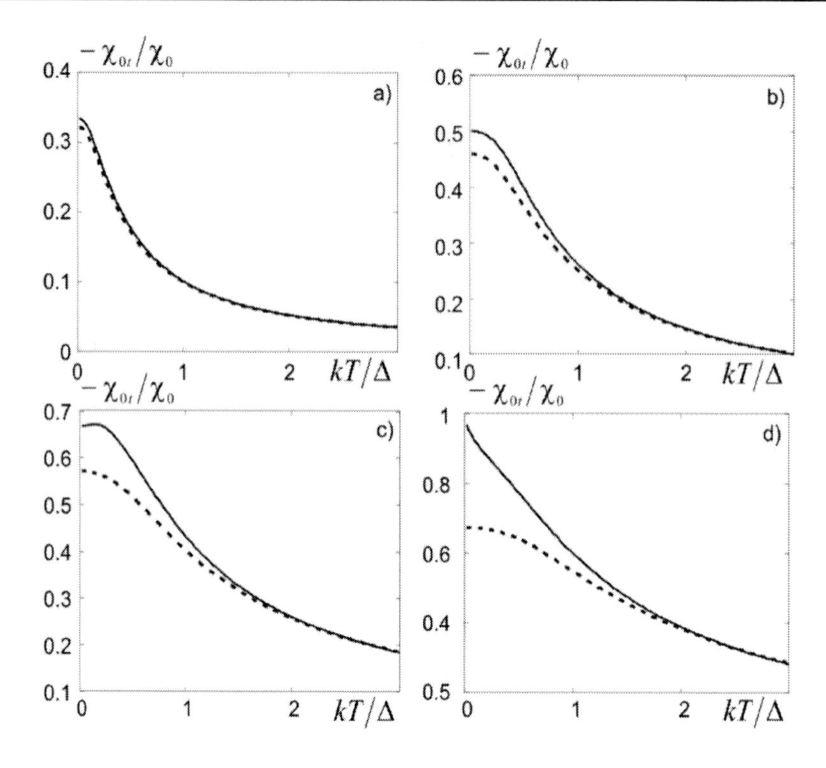

Figure 21. Temperature dependence of electron gas diamagnetic susceptibility in a weak magnetic field in a thin film at $d/a = 1000$.

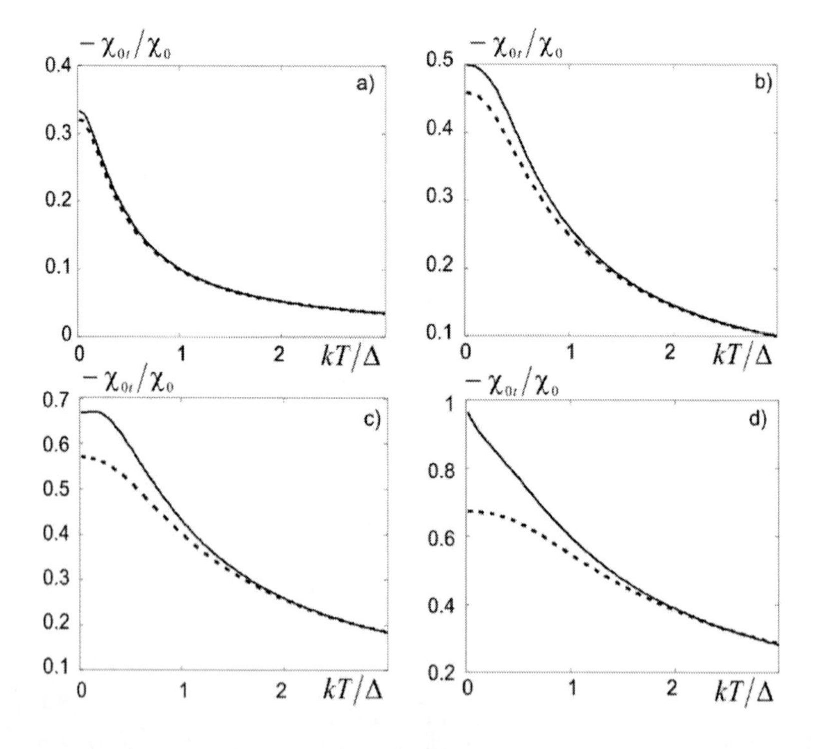

Figure 22. Temperature dependence of electron gas diamagnetic susceptibility in a weak magnetic field in a bulk crystal.

The figures demonstrate that for all considered temperatures and electron concentrations the diamagnetic susceptibility of electron gas in a real layered crystal or a thin film on its basis is higher than in effective mass approximation. This is the case because in a real layered crystal, both with and without size quantization, the density of electron states is higher. Moreover, with small film thicknesses, on the temperature dependence of full diamagnetic susceptibility of electron gas there are maxima which are smoothed with increase in the film thickness. If $d/a \geq 100$, for all electron concentrations the temperature dependences of total diamagnetic susceptibility of electron gas in thin films are almost the same as in a bulk crystal. With a growth of electron concentration at low temperatures, while electron gas degeneracy is essential, the difference in the solid and dashed curves increases. Besides, drawings "d" in all the figures corresponding to case $\zeta_0/\Delta = 2$ demonstrate that at low temperatures an extra singularity of electron density of states typical of a transient Fermi surface becomes apparent. However, for each electron concentration and film thickness, with a rise in temperature, the solid and dashed curves approach each other. This occurs because after a complete removal of electron gas degeneracy with a rise in temperature, its total diamagnetic susceptibility in a weak magnetic field does not depend on the pattern of electrons motion between the layers and is defined by the formula:

$$\chi_{0t} = -\frac{4\mu_0 \mu^{*2} n_0}{3kT}. \tag{10.8}$$

This formula corresponds to Curie's law. At first sight, it is inconsistent with a well-known theorem that a system obeying classical mechanics and classical statistics does not possess magnetic susceptibility. However, this contradiction is illusory, because the system in hand under conditions of validity of formula (10.8) really does obey *classical* statistics, but its mechanics is basically *quantum*, since it is based on the fact of existence of the Landau levels.

Chapter 11

DIAMAGNETIC SUSCEPTIBILITY OF CHARGE-ORDERED LAYERED CRYSTALS IN A WEAK MAGNETIC FIELD

Full diamagnetic susceptibility of electron gas in charge-ordered layered crystals can be defined as:

$$\chi_{0t} = -\frac{e^2\mu_0}{24\pi^2 m^* a}\int_0^{\pi}\left\{\left[1+\exp\left(\frac{-\sqrt{w^2\delta^2+\cos^2 x}-\gamma}{t}\right)\right]^{-1}+\right.$$
$$\left.+\left[1+\exp\left(\frac{\sqrt{w^2\delta^2+\cos^2 x}-\gamma}{t}\right)\right]^{-1}\right\}dx \quad . \qquad (11.1)$$

This formula is written with regard to the fact that each of minibands formed due to charge ordering contributes to total diamagnetic susceptibility of electron gas with a "weight" of ½.

The temperature dependence of this susceptibility calculated according to formula (67) with account of the temperature dependences of order parameter and chemical potential is shown in Figure 23.

The figure demonstrates that in the charge-ordered state at low temperatures the diamagnetic susceptibility of electron gas in a weak magnetic field for all values of effective interaction does not depend on temperature and order parameter and is equal to half its maximum value. This result also confirms the conclusion that interlayer charge ordering can be regarded as a second-order topological phase transition from a closed or transient FS to a strongly open one. This result also agrees, at least qualitatively, with the experimental data [23].

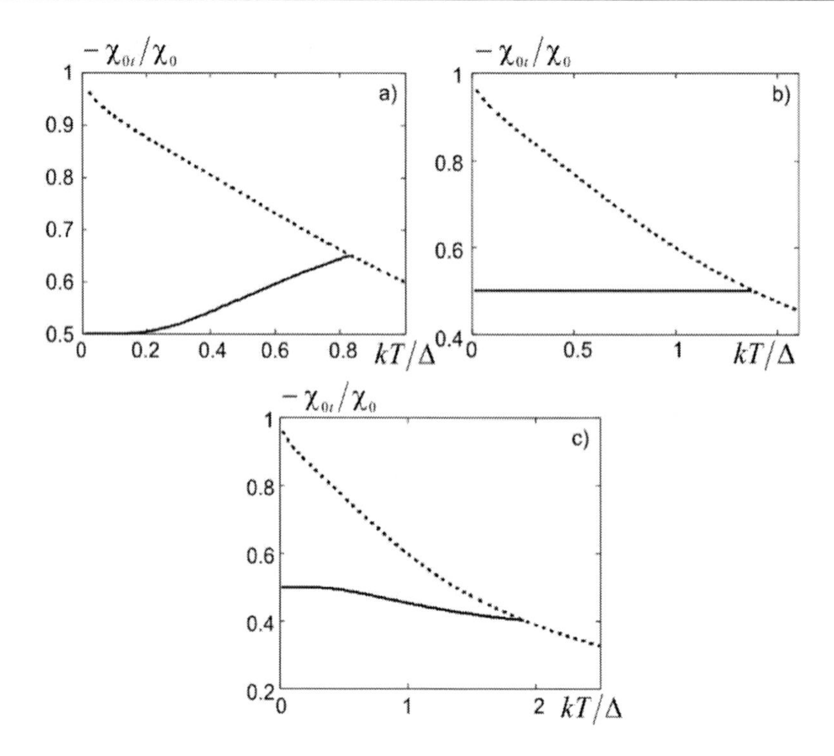

Figure 23. Temperature dependence of total diamagnetic susceptibility of electron gas in a charge ordered layered crystal in a weak magnetic field at $\zeta_{02D}/\Delta = 1$: a) $W_0/\zeta_{02D} = 1.5$; b) $W_0/\zeta_{02D} = 2$; c) $W_0/\zeta_{02D} = 2.5$. The dashed curves correspond to the disordered state.

The temperature dependence of the diamagnetic susceptibility of electron gas in a charge-ordered layered crystal for all values of effective interaction is similar to the temperature dependence of chemical potential. Namely, at a subcritical temperature of phase transition, the diamagnetic susceptibility of electron gas increases with a rise in temperature, if $W_0/\zeta_{02D} = 1.5$, does not depend on temperature, if $W_0/\zeta_{02D} = 2$ and drops with a rise in temperature, if $W_0/\zeta_{02D} = 2.5$. At a transition point, for all considered values of effective interaction, there is a kink in the temperature dependence of electron gas diamagnetic susceptibility which is typical of second-order phase transitions. At supercritical temperatures, for all values of effective interaction, the diamagnetic susceptibility of electron gas coincides with that in a disordered layered crystal and drops with a rise in temperature. Reduction of diamagnetic susceptibility in the charge-ordered state as compared to the disordered state corresponds to a reduction in electron density of states with a transition to the charge-ordered state.

Chapter 12

DE HAAS-VAN ALPHEN DIAMAGNETISM IN LAYERED CRYSTALS AND THIN FILMS ON THEIR BASIS

In the specific case, when motion of electrons in a direction normal to layers is described by strong-coupling approximation, formula (9.12) for an arbitrary film thickness acquires the form:

$$\chi_{os} = \frac{e^2 \mu_0}{2\pi m^* d} \sum_{l=1}^{\infty} \left\{ \left[g_l A_{1l} + f_l A_{2l} \right] \cos\left(\pi l \frac{\gamma - 1}{b} \right) + \left[g_l A_{3l} + f_l A_{4l} \right] \sin\left(\pi l \frac{\gamma - 1}{b} \right) \right\}. \tag{12.1}$$

In this formula:

$$A_{1l} = C_0^{\chi} J_0\left(\pi l b^{-1}\right) + 2\sum_{r=1}^{\infty} (-1)^r C_{2r}^{\chi} J_{2r}\left(\pi l b^{-1}\right). \tag{12.2}$$

$$A_{2l} = \sum_{r=0}^{\infty} (-1)^r \left[2\frac{\gamma - 1}{b} C_{2r+1}^{\chi} + b^{-1}\left(C_{2r}^{\chi} + C_{2r+2}^{\chi} \right) \right] J_{2r+1}\left(\pi l b^{-1}\right). \tag{12.3}$$

$$A_{3l} = -2\sum_{r=0}^{\infty} (-1)^r C_{2r+1}^{\chi} J_{2r+1}\left(\pi l b^{-1}\right). \tag{12.4}$$

$$A_{4l} = \left(\frac{\gamma - 1}{b} C_0^{\chi} + b^{-1} C_1^{\chi} \right) J_0\left(\pi l b^{-1}\right) + \sum_{r=1}^{\infty} (-1)^r \left[2\frac{\gamma - 1}{b} C_{2r}^{\chi} + b^{-1}\left(C_{2r+1}^{\chi} + C_{2r-1}^{\chi} \right) \right] J_{2r}\left(\pi l b^{-1}\right). \tag{12.5}$$

$$C_m^{\chi} = \cos\left\{ \frac{\pi a m}{2d} \left[E\left(\frac{d \arccos(1 - \gamma + b)}{\pi a} \right) + 1 \right] \right\} \sin\left[\frac{\pi a m}{2d} E\left(\frac{d \arccos(1 - \gamma + b)}{\pi a} \right) \right] \csc\left(\frac{\pi a m}{2d} \right). \tag{12.6}$$

In the limit $d \to \infty$, formulae (12.1) – (12.6) define the oscillating part of the diamagnetic susceptibility of electron gas in the bulk crystals. The presence of the Bessel functions is uniquely related to the nonparabolicity of electron dispersion law in a direction normal to layers. Nevertheless, in passing to the limit $d \to \infty$ and with a simultaneous fulfillment of condition $\Delta/\mu^* B >> 1$, (12.1) – (12.6) unambiguously imply the quasi-classical formula of the Lifshitz-Kosevich theory. Subtraction of b value from γ in the formula for coefficients C_m^χ logically takes into account the possibility of electron condensation in the Landau subband with the number $n = 0$ in the ultraquantum limit, and in the quasi-classical approximation, i.e. at $b << \gamma$ it is insignificant.

From (12.1) – (12.6) it is readily apparent that coefficients C_m^χ are functions of current carrier concentrations, and, on removal of electron gas degeneracy, they go to zero. Thus, these formulae explicitly take into account the fact that dHvA effect exists only in the case when one can speak of a pronounced FS of a crystal. With a traditional quasi-classical approach, it is apriori assumed that conditions of strong degeneracy and quasi-classical approximation applicability, i.e $\gamma >> b$ are fulfilled, so within this approach the extension of FS along the magnetic field direction is minor.

From (68) – (73) it is also evident that in the general case the oscillating part of the diamagnetic susceptibility of electron gas in a layered crystal comprises two sets of oscillations frequencies. In a bulk sample the frequencies of the first set are defined by the formula:

$$F_l^{(1)} = l\zeta/2\mu^* . \tag{12.7}$$

These frequencies are related to the extreme section of FS by plane $k_z = 0$.

The frequencies of the second set are defined by the formula:

$$F_l^{(2)} = l|\zeta - 2\Delta|/2\mu^* . \tag{12.8}$$

These frequencies are related to the extreme sections of FS by planes $k_z = \pm\pi/a$ only in the case of open FS. Whereas in the case of closed FS these frequencies cannot be related to any FS sections, if $\zeta/\Delta < 1$. However, if $1 \le \zeta/\Delta \le 2$, these frequencies are identified with the non-extreme sections of FS by planes $k_z = \pm\arccos(3 - 2\zeta/\Delta)/a$. The relative contribution of these frequencies to the oscillations of the diamagnetic susceptibility of electron gas depends not only on current carrier concentration, but also on the ratio of film thickness to the distance between the neighbouring layers. However, in the bulk crystals in quasi-classical approximation at $\zeta/\Delta < 2$ the oscillations with frequencies (12.8) interfere to cause a smooth or nearly smooth additive to oscillations with frequencies (12.7) described by the Lifshitz-Kosevich theory.

From this standpoint, it is of certain interest to analyze the frequency composition of dHvA oscillations for the case when FS corrugation in a bulk crystal is not described by one harmonic of the Fourier series, but is more complicated. In this case, the law of quantization of energy levels in a thin film based on such crystal is of the form:

$$W(s)=\sum_{m=1}^{\infty}\Delta_m\left(1-\cos\frac{ms\pi a}{d}\right). \tag{12.9}$$

In this case, the oscillating part of the diamagnetic susceptibility of electron gas can be represented as follows:

$$\chi_{os}=\frac{e^2\mu_0}{2\pi m^* d}\sum_{l=1}^{\infty}\sum_{M}A_{lM}\cos\pi l\alpha_M+B_{lM}\sin\pi l\alpha_M\,. \tag{12.10}$$

In this formula:

$$\alpha_M=\left(\zeta-2\sum_{m\in M}\Delta_m\right)\Big/\mu^* B, \tag{12.11}$$

$$A_{lM}=g_l\sum_{s\in S_\zeta}\operatorname{Re}\prod_{\substack{m\in M\\ m'\notin M}}F_{lm's}G_{lms}+f_l\sum_{s\in S_\zeta}\beta_s\operatorname{Im}\prod_{\substack{m\in M\\ m'\notin M}}F_{lm's}G_{lms}\,, \tag{12.12}$$

$$B_{lM}=-g_l\sum_{s\in S_\zeta}\operatorname{Im}\prod_{\substack{m\in M\\ m'\notin M}}F_{lm's}G_{lms}+f_l\sum_{s\in S_\zeta}\beta_s\operatorname{Re}\prod_{\substack{m\in M\\ m'\notin M}}F_{lm's}G_{lms}\,, \tag{12.13}$$

$$F_{lm's}=\sum_{k=0}^{\infty}(2k+1)P_k\left[\cos(sm'\pi a/d)\right]\sum_{r=0}^{k}\frac{i^{r-1}(k+r)!}{r!(k-r)!}\left(\frac{\mu^* B}{2\pi l\Delta_{m'}}\right)^{r+1}, \tag{12.14}$$

$$G_{lms}=\sum_{k=0}^{\infty}(-1)^k(2k+1)P_k\left[\cos(sm\pi a/d)\right]\sum_{r=0}^{k}\frac{(-i)^{r-1}(k+r)!}{r!(k-r)!}\left(\frac{\mu^* B}{2\pi l\Delta_m}\right)^{r+1}, \tag{12.15}$$

$$\beta_s=\left[\zeta-W(s)\right]/\mu^* B. \tag{12.16}$$

In formulae (12.10) – (12.16), M – an arbitrary subset of a plurality of harmonic numbers m, S_ζ – a plurality of numbers of populated film subbands, $P_l(x)$ – the Legendre polynomials. Hence, in sharply anisotropic layered crystals the oscillation frequencies are the same for the thin and bulk samples, and the thickness of samples governs only their relative contribution. The oscillation frequencies are defined by the formula [32]:

$$F_{lM}=l\left|\zeta-2\sum_{m\in M}\Delta_m\right|\Big/2\mu^*. \tag{12.17}$$

From formula (12.17) it is seen that any of the frequencies of the type (84) can be represented as a linear combination of a frequency for which plurality M is empty and the frequencies for which plurality M consists of one component. It is obvious that only some of these frequencies can be identified with the extreme sections of FS by planes normal to a magnetic field. The results of calculation of the oscillating part of the diamagnetic susceptibility of electron gas in layered crystals and thin films on their basis according to formulae (12.10) – (12.16) are given in Figures 24 to 27.

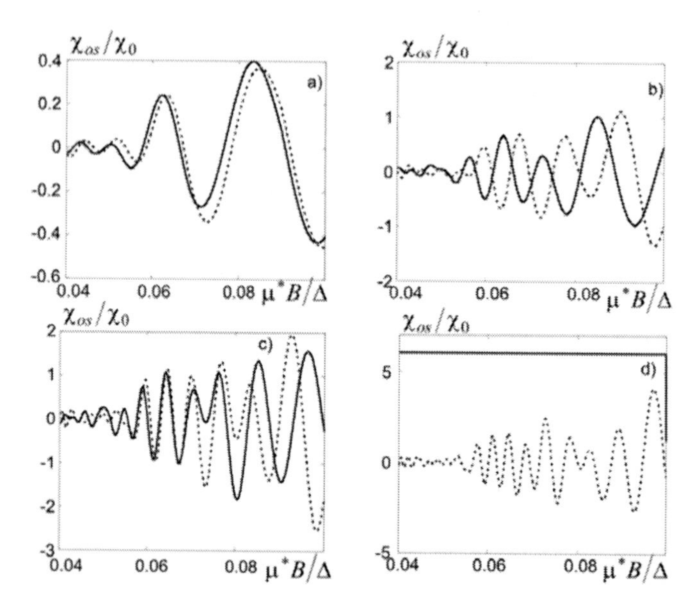

Figure 24. Field dependences of the diamagnetic susceptibility of a thin film based on a layered crystal in quasi-classical approximation at $d/a = 10$.

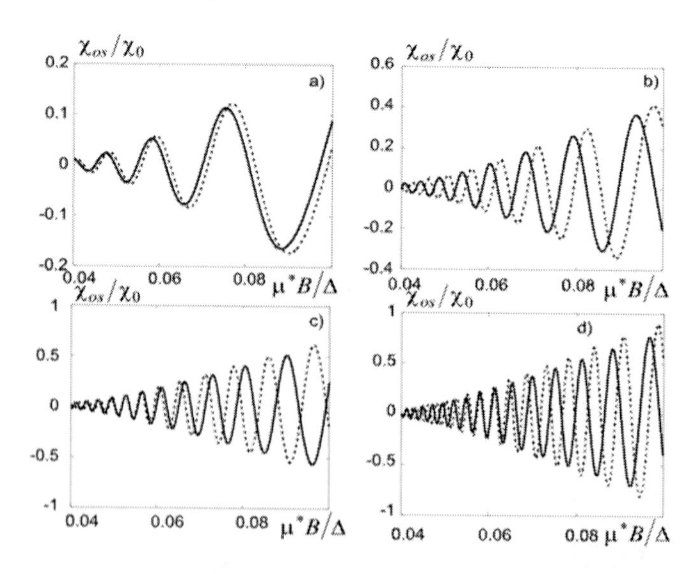

Figure 25. Field dependences of the diamagnetic susceptibility of a thin film based on a layered crystal in quasi-classical approximation at $d/a = 100$.

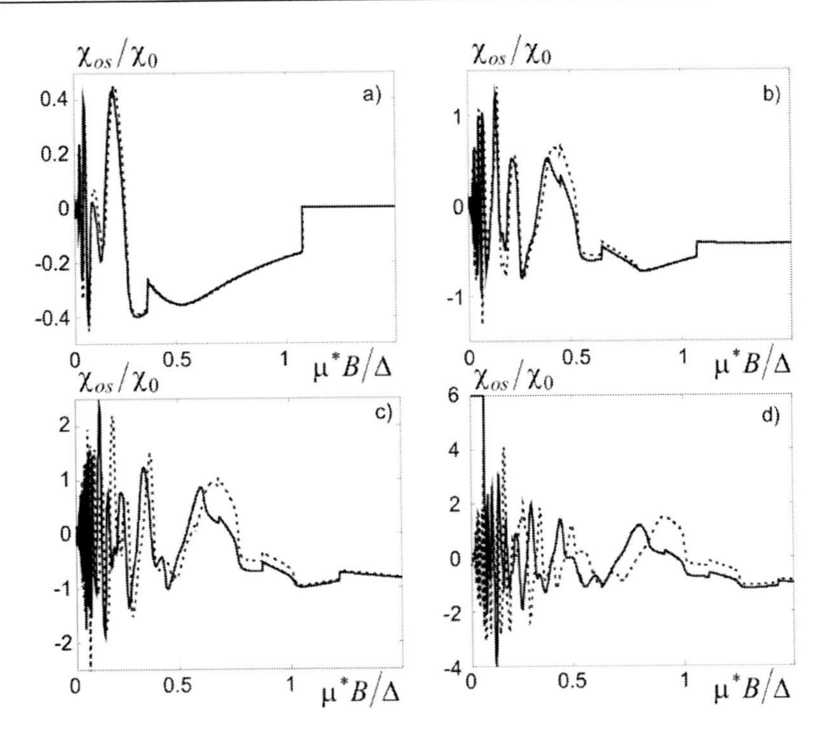

Figure 26. Field dependences of the diamagnetic susceptibility of a thin film based on a layered crystal in strong magnetic fields at $d/a = 10$.

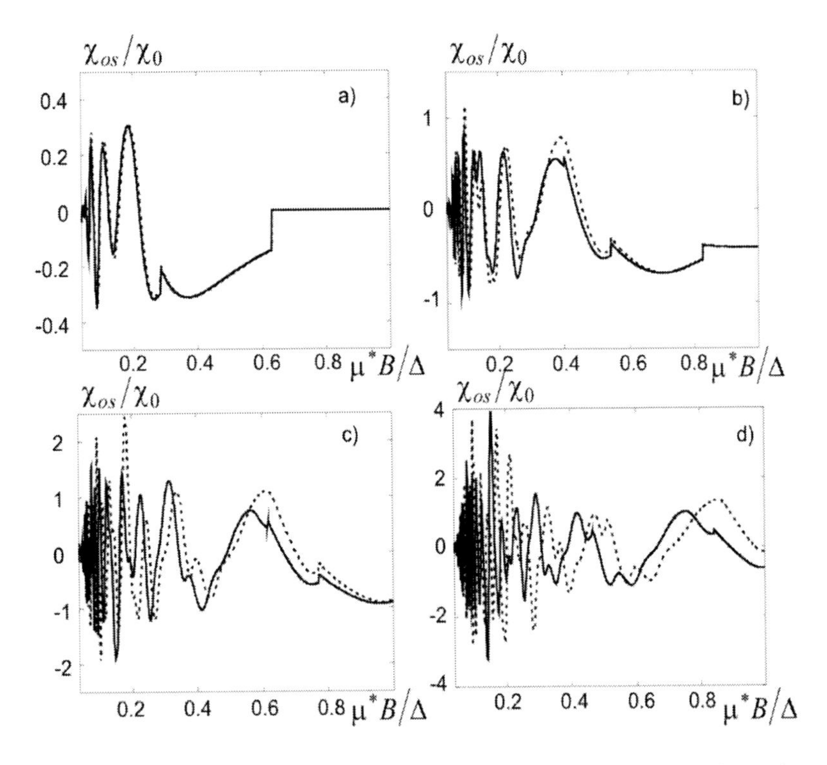

Figure 27. Field dependences of the diamagnetic susceptibility of a thin film based on a layered crystal in strong magnetic fields at $d/a = 100$.

The figures demonstrate that in the quasi-classical region of magnetic fields in the thin samples the contribution of non-quasi-classical frequencies (12.8) to dHvA oscillations is rather large. It is because the FS of a crystal in a thin sample has the shape of a system of coaxial cylinders whose cross-sectional areas decrease stepwise with increasing the number of populated film subband. By virtue of this, all cylinders contribute to oscillations, rather than only that with the largest cross-section area corresponding to the lowest populated film subband. With a growth in charge carrier concentration, the difference between the solid and dashed curves in each figure is increased, which points to a stronger manifestation of layered structure effects. At $d/a = 10$ the greatest difference between the solid and dashed curves in the quasi-classical region of magnetic fields takes place at $\zeta_0/\Delta = 2$. It is due to the fact that at $\zeta_0/\Delta = 2$ all film subbands are populated and the oscillations from the respective cylinders interfere to give a constant value as a result. However, it depends not only on charge carrier concentration, but also on the quantization law of energy levels in a film. Therefore, a constant value is obtained only in the case of a real layered crystal, when the law of quantization of energy levels in a film is cosine. Contrary to this, in effective mass approximation, when the law of quantization of energy levels in a film is quadratic, dHvA oscillations in the quasi-classical range of magnetic fields are also present at $\zeta_0/\Delta = 2$.

With a growth in sample thickness, dHvA oscillations in it become the same as in a bulk crystal, i.e. monoperiodic. Due to the above mentioned difference in the electron density of states in a layered crystal and in effective mass approximation, dHvA oscillation frequencies in a layered crystal are less than in effective mass approximation, whereby with a growth in current carrier concentration, there is not only a rise in the oscillation frequency and amplitude, but also a growth in phase delay between the solid and dashed curves. The oscillation amplitude in a layered crystal is also somewhat smaller than in effective mass approximation, as sections of FS by planes normal to a field in a layered crystal are reduced with a growth of a longitudinal quasi-pulse slower than in effective mass approximation. Hence, dHvA oscillations from the only extreme section of FS by plane $k_z = 0$ are partially damped.

This "damping" is relatively small, but measurable, and increases with charge carrier concentration. Therein lies the influence of layered structure effects in the quasi-classical range of magnetic fields.

In strong magnetic fields, the effects of layered structure both in the thin films and bulk samples become much more pronounced, and the oscillations of the diamagnetic susceptibility of electron gas lose their monochromatism not only in a thin film, but also in a bulk sample.

This happens because in a strong magnetic field, even in a bulk crystal, the FS has the form of a system of coaxial "magnetic tubes" whose cross-sectional areas change stepwise with a growth in the number of the populated Landau subband and the distinction between them is the stronger, the stronger is a magnetic field. That is why electrons from all cylinders, not only from a cylinder corresponding to plane $k_z = 0$ take part in dHvA oscillations in strong magnetic fields.

The law of quantization of energy levels in a film or the law of electron dispersion in the direction normal to layers in a bulk crystal also affect the dependence of cylinder section areas on k_z, hence, the frequency composition and oscillation amplitude. In this case, the difference between the solid and dashed curves also increases with charge carrier concentration. However, with further increase in a magnetic field, the solid and dashed curves draw together due to the fact that electrons condensate in the low Landau subband, and the difference between a real layered crystal and effective mass approximation is leveled out.

Chapter 13

De Haas-Van Alphen Diamagnetism in Charge-Ordered Layered Crystals

In the presence of charge ordering, formula (9.6) acquires the form:

$$\chi_{os} = \frac{e\mu_0\mu^*}{\pi h a}\sum_{l=1}^{\infty}\left\{ g_l \left[\int_{\gamma+\sqrt{w^2\delta^2+\cos^2 x}-b\geq 0} \cos\left[\pi l \frac{\gamma+\sqrt{w^2\delta^2+\cos^2 x}}{b}\right]dx + \right.\right.$$
$$\left. + \int_{\gamma-\sqrt{w^2\delta^2+\cos^2 x}-b\geq 0} \cos\left[\pi l \frac{\gamma-\sqrt{w^2\delta^2+\cos^2 x}}{b}\right]dx \right] +$$
$$+ f_l \left[\int_{\gamma+\sqrt{w^2\delta^2+\cos^2 x}-b\geq 0} \frac{\gamma+\sqrt{w^2\delta^2+\cos^2 x}}{b}\sin\left[\pi l \frac{\gamma+\sqrt{w^2\delta^2+\cos^2 x}}{b}\right]dx + \right.$$
$$\left.\left. \int_{\gamma-\sqrt{w^2\delta^2+\cos^2 x}-b\geq 0} \frac{\gamma-\sqrt{w^2\delta^2+\cos^2 x}}{b}\sin\left[\pi l \frac{\gamma-\sqrt{w^2\delta^2+\cos^2 x}}{b}\right]dx \right]\right\} . \quad (13.1)$$

In the determination of integration limits, charge carrier condensation in the low Landau subband was taken into account. The results of calculation of dHvA diamagnetic susceptibility by this formula are given in Figures 28 and 29.

The figures demonstrate that in the quasi-classical magnetic fields dHvA oscillations become biperiodic, and their relative contribution is reduced. The presence of additional oscillation frequencies is related to transformation of a transient FS to a closed one in transition to the charge-ordered state. As long as in the absence of a magnetic field in the charge-ordered state the level of electron gas chemical potential lies in the gap between the minibands, the FS of a layered crystal in the charge-ordered state within the first one-dimensional Brillouin zone has five extreme sections, namely three maximum sections by plane $k_z = 0$ and planes $k_z = \pm\pi/a$ and two minimum sections by planes $k_z = \pm\pi/2a$. The respective frequencies are:

$$F_l = \pi l \frac{\zeta + \sqrt{W_0^2 \delta^2 + \Delta^2}}{2\mu^*}, \tag{13.2}$$

$$F_l' = \pi l \frac{\zeta + W_0 \delta}{2\mu^*}. \tag{13.3}$$

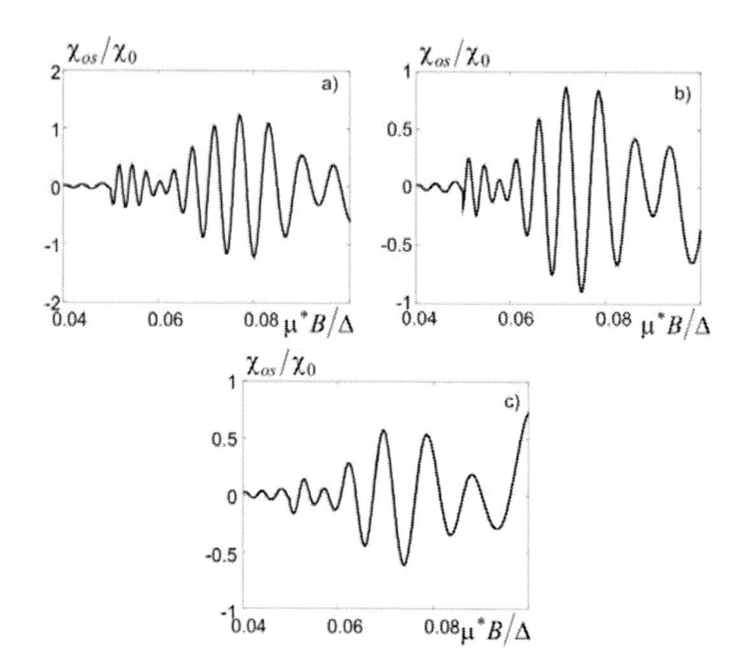

Figure 28. Field dependence of dHvA diamagnetic susceptibility in a charge-ordered layered crystal at $\zeta_{02D}/\Delta = 1, kT/\Delta = 0.03, 0.04 \le \mu^* B/\Delta \le 0.1$ and: a) $W_0/\zeta_0 = 1.5$; b) $W_0/\zeta_0 = 2$; c) $W_0/\zeta_0 = 2.5$.

With a high degree of ordering which is realized at sufficiently large values of effective attracting interaction, these frequencies are close to each other, so in the charge-ordered state, as was to be expected, dHvA oscillations have the shape of high-frequency vibrations with low-frequency beats superimposed on them. High "carrier" frequencies are approximately equal to:

$$F_{lC} = \pi l \frac{\zeta + W_0 \delta}{2\mu^*}. \tag{13.4}$$

Beat frequencies are equal to:

$$F_{lB} = \pi l \frac{\sqrt{W_0^2 \delta^2 + \Delta^2} - W_0 \delta}{4\mu^*}. \tag{13.5}$$

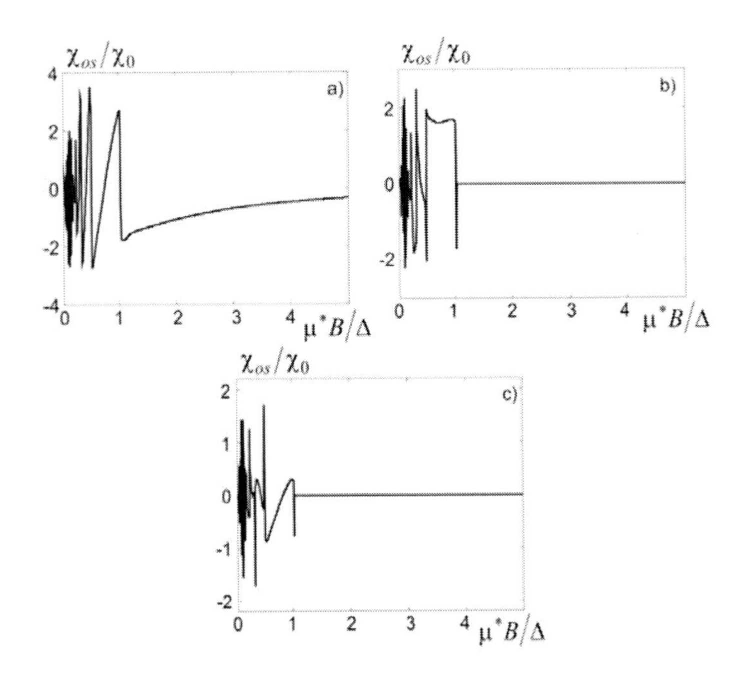

Figure 29. Field dependence of dHvA diamagnetic susceptibility in a charge-ordered layered crystal at $\zeta_{02D}/\Delta = 1, kT/\Delta = 0.03, 0 \leq \mu^* B/\Delta \leq 5$ and: a) $W_0/\zeta_0 = 1.5$; b) $W_0/\zeta_0 = 2$; c) $W_0/\zeta_0 = 2.5$.

The smallness of beat frequencies unambiguously points to narrowing of conduction minibands in the charge-ordered state, and might serve as direct experimental proof of the existence of interlayer charge ordering in layered crystals, however, the author of this book is not aware so far of such direct experiments. On the contrary, the majority of experiments have been based on layered crystals whose FS apriori are strongly open; hence, the biperiodic kind of oscillations with high "carrier" frequencies and low beat frequencies is anyway typical of them without charge ordering. In the charge-ordered state, the relative contribution of the oscillating part of diamagnetic susceptibility is less than in the absence of ordering, however, it remains comparable or even greater that the contribution of permanent part. With a growth in the value of effective interaction, the relative contribution of the oscillating part of diamagnetic susceptibility is reduced. The physical reason for this reduction is that any restriction of free motion of electrons reduces the diamagnetic susceptibility of electron gas.

In stronger magnetic fields, the relative contribution of dHvA oscillations is increased, and their shape is changed, including, among other reasons, the field dependence of chemical potential and order parameter. However, after the last oscillation maximum, at each effective interaction value, the diamagnetic susceptibility of electron gas in a charge-ordered layered crystal becomes a monotonic function of a magnetic field and tends to zero by the law $\chi \propto B^{-1}$ in conformity with relation (5.9). The physical reason for this is additional narrowing of conduction miniband due to electron condensation in the low Landau subband with the number $n = 0$ and FS squeeze along the direction of a magnetic field. On approaching the ultra-quantum limit, the oscillating part of electron gas diamagnetic susceptibility tends to zero the quicker, the larger is the effective interaction value, and hence, the narrower is conduction miniband in the absence of a magnetic field. Note that these results cannot be obtained within the quasi-classical approach for which the specific shape of FS and its extension along the direction of a magnetic field do not matter much.

Chapter 14

LANGEVIN DIAMAGNETISM IN LAYERED CRYSTALS AND THIN FILMS ON THEIR BASIS

The Langevin diamagnetism becomes apparent under the intermediate and weak degeneracy, as well as in the nondegenerate gas. If electron motion in the direction normal to layers is described by strong-coupling approximation, then formula (9.7) takes on the form [30]:

$$\chi_{Lg} = \frac{e\mu_0\mu^*}{hd}\sum_{l=1}^{\infty}(-1)^l h_l\left\{B_{1l}\exp\left[-lt^{-1}(1-\gamma)\right]+B_{2l}\cosh\left[lt^{-1}(\gamma-1)\right]+B_{3l}\sinh\left[lt^{-1}(\gamma-1)\right]\right\} \tag{14.1}$$

In this formula:

$$B_{1l} = -NI_0\left(lt^{-1}\right)-\sum_{r=0}^{\infty}I_{2r+1}\left(lt^{-1}\right), \tag{14.2}$$

$$B_{2l} = 2C_0^{\chi}I_0\left(lt^{-1}\right)-4\sum_{r=1}^{\infty}C_{2r}^{\chi}I_{2r}\left(lt^{-1}\right), \tag{14.3}$$

$$B_{3l} = 4\sum_{r=0}^{\infty}C_{2r+1}^{\chi}I_{2r+1}\left(lt^{-1}\right). \tag{14.4}$$

In so doing, $I_n(x)-$ imaginary Bessel functions of the first kind, coefficients C_m^{χ} are found by formula (12.6) on substituting $b=0$ into it. The calculation data of the Langevin diamagnetic susceptibility in real layered crystals and in thin films on their basis and in effective mass approximation are given in Figures 30 to 33. The figures demonstrate that at small film thicknesses there is a stepwise dependence of the Langevin diamagnetic susceptibility. It is because the chemical potential of the system essentially depends on magnetic field, and the Langevin diamagnetic susceptibility in a thin film changes abruptly, whenever the chemical potential level crosses the boundary of some of film subbands. If, however, the chemical potential level moves between the boundaries of film subbands, the

Langevin diamagnetic susceptibility changes smoothly. With increasing film thickness, the steps are smoothed to disappear completely in a bulk crystal. From comparison of the solid and dashed curves in the figures it follows that the value and shape of the steps are essentially dependent on the law of quantization of electron energy levels in a film, that is, on layered structure effects. In so doing, with a growth in charge carrier concentration, the difference between the solid and dashed curves is increased, pointing to a stronger manifestation of layered structure effects.

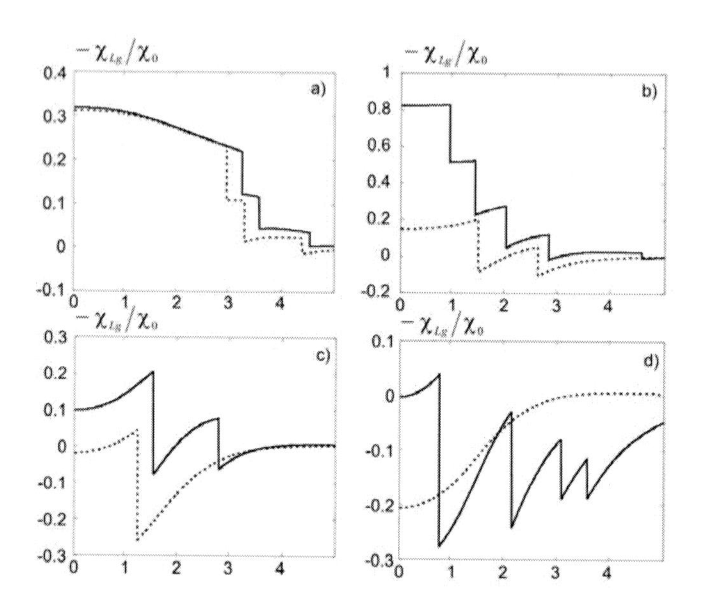

Figure 30. Field dependence of the Langevin diamagnetic susceptibility in a thin film at $d/a = 10, kT/\Delta = 1, 0 \le \mu^* B/\Delta \le 5$.

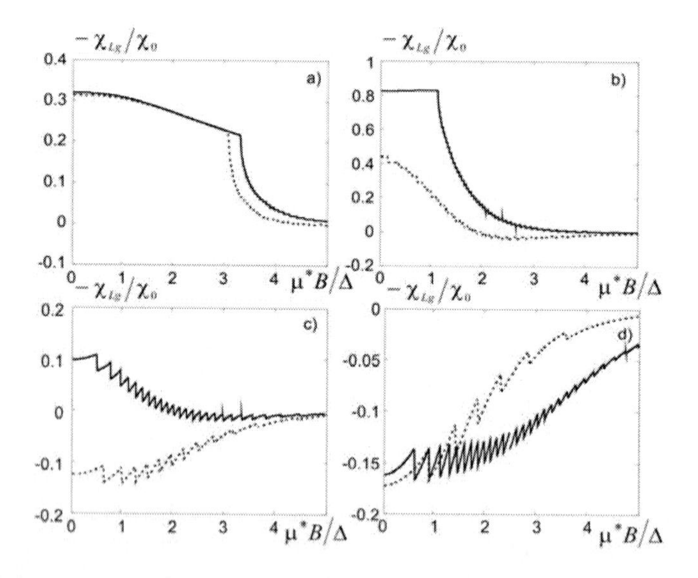

Figure 31. Field dependence of the Langevin diamagnetic susceptibility in a thin film at $d/a = 100, kT/\Delta = 1, 0 \le \mu^* B/\Delta \le 5$.

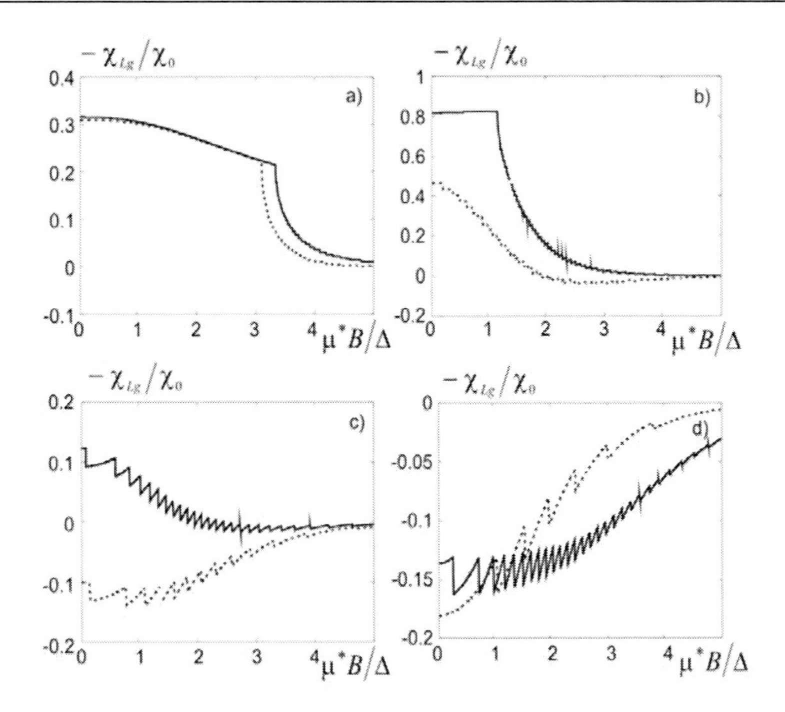

Figure 32. Field dependence of the Langevin diamagnetic susceptibility in a thin film at $d/a = 1000, kT/\Delta = 1, 0 \leq \mu^* B/\Delta \leq 5$.

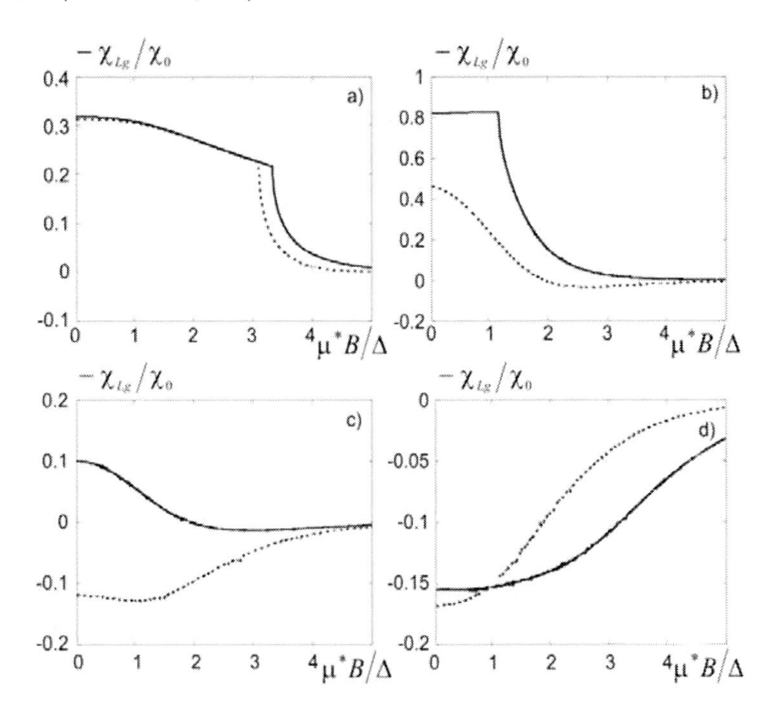

Figure 33. Field dependence of the Langevin diamagnetic susceptibility in a bulk crystal at $d/a = 10000, kT/\Delta = 1, 0 \leq \mu^* B/\Delta \leq 5$.

It should be noted that the sign and value of the Langevin diamagnetic susceptibility, as it follows from the general formula (9.13), is defined by competing contributions of charge carriers with the energy smaller than electron gas chemical potential, and with the energy larger than electron gas chemical potential. With a dominant contribution of charge carriers with the energy smaller than electron gas chemical potential, the Langevin diamagnetic susceptibility is negative. However, with a dominant contribution of charge carriers with the energy larger than electron gas chemical potential, the Langevin diamagnetic susceptibility is positive. The ratio between said contributions is affected not only by the film thickness and charge carrier concentration, but also by magnetic field value. It is largely due to magnetic field dependence of chemical potential. Therefore, inversion of the Langevin magnetic susceptibility sign is governed not only by crystal band structure, charge carrier concentration and temperature, but also by magnetic field value. However, in strong magnetic fields, in the absence of degeneracy, the magnetic moment of crystal unit volume does not depend on the type of electron motion between the layers and is fully defined by the Langevin formula:

$$M_{Lg} = \mu^* n_0 \left[\frac{kT}{\mu^* B} - \coth\left(\frac{\mu^* B}{kT} \right) \right]. \tag{14.5}$$

Therefore, in a strong magnetic field, in the absence of degeneracy, the magnetic moment of electron gas unit volume in a crystal tends to a constant value $-\mu^* n_0$. It is due to the fact that under the Landau quantization conditions the cyclotron rotation of electron in a magnetic field with induction B is equivalent to induction of magnetic moment $-\mu^*$ with its subsequent orientation along the direction of a magnetic field. It is noteworthy that other authors exclude from consideration the Langevin diamagnetic susceptibility, believing that it can be only paramagnetic, i.e. such which is governed by the orientation of magnetic moments of individual atoms, which in the absence of a magnetic field *exist and are different from zero, but not oriented.*

Chapter 15

LANGEVIN DIAMAGNETISM IN CHARGE-ORDERED LAYERED CRYSTALS

In the presence of charge ordering the Langevin diamagnetic susceptibility can be determined by the formula:

$$\chi_{Lg}=\frac{e\mu_0\mu^*}{2\pi ha}\sum_{l=1}^{\infty}(-1)^l h_l\left\{\int\limits_{\gamma+\sqrt{w^2\delta^2+\cos^2x}\ge 0}\exp\left[-l\frac{\sqrt{w^2\delta^2+\cos^2x}+\gamma}{t}\right]dx-\right.$$
$$-\int\limits_{\gamma+\sqrt{w^2\delta^2+\cos^2x}\le 0}\exp\left[l\frac{\sqrt{w^2\delta^2+\cos^2x}+\gamma}{t}\right]dx+\int\limits_{\gamma-\sqrt{w^2\delta^2+\cos^2x}\ge 0}\exp\left[-l\frac{\gamma-\sqrt{w^2\delta^2+\cos^2x}}{t}\right]dx-\quad .(15.1)$$
$$\left.-\int\limits_{\gamma-\sqrt{w^2\delta^2+\cos^2x}\le 0}\exp\left[l\frac{\gamma-\sqrt{w^2\delta^2+\cos^2x}}{t}\right]dx\right\}$$

The results of calculation of the field dependence of the Langevin diamagnetic susceptibility by this formula are given in Figure 34.

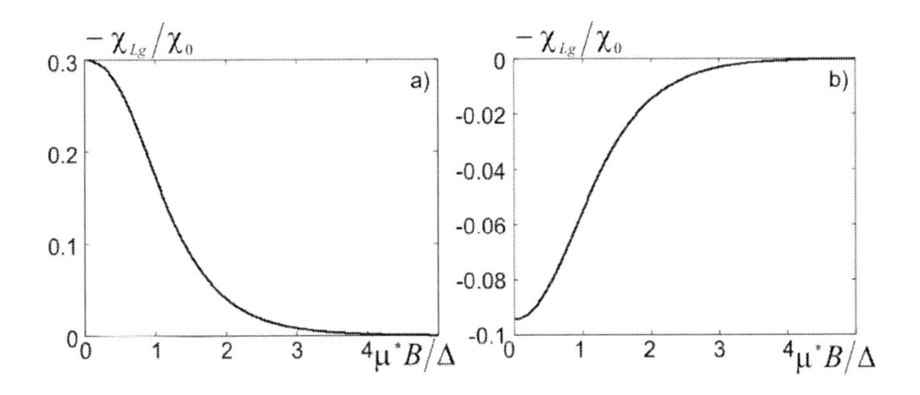

Figure 34. Field dependence of the Langevin diamagnetic susceptibility in a charge-ordered layered crystal at $\zeta_{02D}/\Delta=1, kT/\Delta=1; 0\le\mu^*B/\Delta\le 5$ and: a) $W_0/\zeta_{02D}=1.5$; b) $W_0/\zeta_{02D}=2.5$.

The figure demonstrates that the Langevin diamagnetic susceptibility in a charge-ordered crystal, decreasing in modulus with a rise in magnetic field, undergoes sign inversion depending on the value of effective interaction. Besides, as the value of effective interaction increases, the Langevin diamagnetic susceptibility also decreases in modulus. The sign inversion of the Langevin diamagnetic susceptibility in a charge-ordered crystal depending on the value of effective interaction is attributable to the following physical reasons. On the one hand, at $W_0/\zeta_{02D} = 2$, with neglect of the oscillation effects, the level of electron gas chemical potential with all magnetic field inductions lies in the middle of pseudogap between the minibands. Therefore, contributions of charge carriers with the energies above and below the chemical potential to the Langevin diamagnetic susceptibility compensate each other. Hence, it vanishes. However, if $W_0/\zeta_{02D} < 2$, the level of electron gas chemical potential lies above the middle of pseudogap between the minibands, the contribution of the upper miniband is dominant and the Langevin diamagnetic susceptibility is negative. And if $W_0/\zeta_{02D} > 2$, the level of electron gas chemical potential lies below the middle of pseudogap between the minibands, the contribution of the lower miniband becomes dominant and the Langevin diamagnetic susceptibility becomes positive. Whereas the decrease of the Langevin diamagnetic susceptibility modulus with increase in magnetic field induction is attributable, on the one hand, to complication of thermal transfer of charge carriers between the Landau levels, and, on the other hand, to increase of the order parameter, and, hence, narrowing of conduction minibands.

Chapter 16

General Formulae for a Longitudinal Electric Conductivity of a Layered Crystal in a Strong Quantizing Magnetic Field and General Conditions for the Origination of Longitudinal Galvanomagnetic Effects

When calculating a longitudinal electric conductivity of a layered crystal in a strong quantizing magnetic field normal to layers, we will proceed from the Kubo formula in the approximation of relaxation time which will be written as:

$$\sigma_{zz} = -\sum_{\alpha} g_{\alpha}\tau_{\alpha}v_{z\alpha}^{2}\frac{\partial f^{0}}{\partial \varepsilon_{\alpha}}, \qquad (16.1)$$

where $\tau_{\alpha} \equiv \tau(n,k_z)$ – relaxation time of a longitudinal quasi-momentum of current carriers, $v_{z\alpha}$ – electron velocity component in a direction normal to layers, $f^{0}(\varepsilon_{\alpha})$ – equilibrium Fermi-Dirac distribution function, the rest of designations are explained above.

Taking into account the ratio $-\partial f^{0}/\partial\varepsilon_{\alpha} = \partial f^{0}/\partial\zeta$, formula (16.1) may be written as:

$$\sigma_{zz} = \frac{\partial}{\partial\zeta}\sum_{\alpha} g_{\alpha}\tau_{\alpha}v_{z\alpha}^{2}f^{0}(\varepsilon_{\alpha}). \qquad (16.2)$$

Summation over the Landau level numbers in formula (16.2) can be performed to the end, if the relaxation time of a longitudinal quasi-pulse depends only on k_z, i.e. relaxation of a longitudinal quasi-pulse in all filled Landau subbands occurs in a similar way. Under this assumption, based on formula (16.2), with regard to Eq. (5.3) and taking into account the ratio $v_{z\alpha} = v_z(x) = 2\pi aW'(x)/h$, we immediately arrive at the following general formula for a total longitudinal conductivity of a layered crystal in a quantizing magnetic field normal to layers [26]:

$$\sigma_{zz} = \sigma_0 + \sigma_{os} + \sigma_{mr} . \tag{16.3}$$

In this formula:

$$\sigma_0 = \frac{16\pi^2 e^2 m^* a}{h^4} \int_{W(x)\le\zeta} \tau(x)|W'(x)|^2 dx , \tag{16.4}$$

$$\sigma_{os} = \frac{32\pi^2 e^2 m^* a}{h^4} \sum_{l=1}^{\infty} (-1)^l f_l^\sigma \int_{W(x)\le\zeta} \tau(x)|W'(x)|^2 \cos\left[\pi l \frac{\zeta - W(x)}{\mu^* B}\right] dx , \tag{16.5}$$

$$\begin{aligned} \sigma_{mr} = \frac{16\pi^2 e^2 m^* a}{h^4} \sum_{l=1}^{\infty} (-1)^l h_l^\sigma \Bigg\{ \int_{W(x)\le\zeta} \tau(x)|W'(x)|^2 \exp\left[l \frac{W(x)-\zeta}{kT}\right] - \\ - \int_{W(x)\ge\zeta} \tau(x)|W'(x)|^2 \exp\left[l \frac{\zeta - W(x)}{kT}\right] \Bigg\} \end{aligned} , \tag{16.6}$$

$$f_l^\sigma = \frac{\pi^2 lkT/\mu^* B}{\sinh\left(\pi^2 lkT/\mu^* B\right)} , \tag{16.7}$$

$$h_l^\sigma = \frac{\mu^* Bl/kT}{\sinh\left(\mu^* Bl/kT\right)} . \tag{16.8}$$

Thus, the first term of formula (16.3) describes the constant part of a longitudinal electric conductivity of a layered crystal, the second term – its oscillating part, i.e. Shubnikov-de Haas effect (ShdH), the third term – the monotonic part of magnetoresistance. These formulae explicitly take into account the dependence of FS extension along the direction of a magnetic field on the concentration of current carriers and temperature and the fact that ShdH effect is only possible under the existence of a pronounced FS. Transition to the limit $B \to 0$ yields the following formula for a total longitudinal electric conductivity of a layered crystal:

$$\sigma_{zz} = \frac{16\pi^2 e^2 m^* a}{h^4} \int_{all} \tau(x)|W'(x)|^2 \left\{1 + \exp\left[\frac{W(x)-\zeta}{kT}\right]\right\}^{-1} dx . \tag{16.9}$$

In the absence of degeneracy, when electron gas chemical potential ζ is modulo larger negative value, hence, condition $W(x) > \zeta$ is always satisfied, formula (16.3) with regard to (5.3) goes into the following:

$$\sigma_{zz} = \frac{4\pi^2 e^2 a^2 n_0}{kTh^2} \frac{\int\limits_{all} \tau(x) |W'(x)|^2 \exp[-W(x)/kT] dx}{\int\limits_{all} \exp[-W(x)/kT] dx}. \tag{16.10}$$

Thus it is seen that in the nondegenerate gas, i.e. when electron gas obeys *classical statistics,* no longitudinal galvanomagnetic effects arise, if relaxation time does not depend on a magnetic field. Therefore, they can arise either in the degenerate electron gas with any kind of relaxation time, or in the nondegenerate electron gas on condition that relaxation time is magnetic field dependent.

Chapter 17

LONGITUDINAL ELECTRIC CONDUCTIVITY OF LAYERED CRYSTALS IN A QUANTIZING MAGNETIC FIELD UNDER STRONG DEGENERACY CONDITIONS AND IN THE APPROXIMATION OF CONSTANT RELAXATION TIME

In a strong quantizing magnetic field under strong degeneracy conditions in formula (16.3) the first two terms are essential, and the third one is exponentially small. In a real layered crystal in the approximation of constant relaxation time, i.e. at $\tau(x) \equiv \tau_0$, the first two components are given by [2]:

$$\sigma_0 = \frac{8\pi^2 e^2 m^* a \tau_0 \Delta^2}{h^4}\left(C_0^\sigma - C_2^\sigma\right), \tag{17.1}$$

$$\begin{aligned}\sigma_{os} = \frac{16\pi^2 e^2 m^* a \tau_0 \Delta^2}{h^4}\sum_{l=1}^{\infty}(-1)^l f_l^\sigma \Bigg\{\Bigg[\cos\left(\pi l\frac{\gamma-1}{b}\right)\left(C_0^\sigma - C_2^\sigma\right)J_0\left(\pi l b^{-1}\right)+ \\ +\sum_{r=1}^{\infty}(-1)^r\left(2C_{2r}^\sigma - C_{2r+2}^\sigma - C_{2r-2}^\sigma\right)J_{2r}\left(\pi l b^{-1}\right)\Bigg]- \\ -\sin\left(\pi l\frac{\gamma-1}{b}\right)\sum_{r=0}^{\infty}(-1)^r\left(2C_{2r+1}^\sigma - C_{2r+3}^\sigma - C_{|2r-1|}^\sigma\right)J_{2r+1}\left(\pi l b^{-1}\right)\Bigg\}\end{aligned} \tag{17.2}$$

In these formulae, $C_0^\sigma = \arccos(1-\gamma+b)$ and $C_m^\sigma = \sin mC_0^\sigma / m$ at $m \neq 0$. Dependence of C_m^σ on b explicitly takes into account the fact of electron condensation in the lower Landau subband in the ultra-quantum limit under strong degeneracy of electron gas.

For the open FS, formulae (17.1) and (17.2) acquire the form:

$$\sigma_0 = \frac{8\pi^3 e^2 m^* a \tau_0 \Delta^2}{h^4}, \tag{17.3}$$

$$\sigma_{os} = \frac{16\pi^3 e^2 m^* a\tau_0\Delta^2}{h^4}\sum_{l=1}^{\infty}(-1)^l f_l^{\sigma}\cos\left(\pi l\frac{\gamma-1}{b}\right)\left[J_0\left(\pi l b^{-1}\right)+J_2\left(\pi l b^{-1}\right)\right]. \qquad (17.4)$$

In the effective mass approximation, formulae (17.1) and (17.2) become:

$$\sigma_0 = \frac{16\pi^2 e^2 m^* a\tau_0\Delta^2}{3h^4}\left[2(\gamma-b)\right]^{3/2}. \qquad (17.5)$$

$$\sigma_{os} = \frac{32\pi^{1/2}e^2 m^* a\tau_0\Delta^{1/2}\left(\mu^* B\right)^{3/2}}{h^4}\sum_{l=1}^{\infty}(-1)^l l^{-3/2} f_l^{\sigma}\left\{\sin\left(\pi l\gamma b^{-1}\right)C\left[\sqrt{2l\left(\gamma b^{-1}-1\right)}\right]-\right. \qquad (17.6)$$
$$\left.-\cos\left(\pi l\gamma b^{-1}\right)S\left[\sqrt{2l\left(\gamma b^{-1}-1\right)}\right]\right\}$$

When passing in formulae (17.2) and (17.6) to the asymptotic limits $b^{-1} >> 1$, $\gamma b^{-1} >> 1$, we obtain the identical result:

$$\sigma_{os} = \frac{16\sqrt{2}\pi^{1/2}e^2 m^* a\tau_0\Delta^{1/2}\left(\mu^* B\right)^{3/2}}{h^4}\sum_{l=1}^{\infty}(-1)^l l^{-3/2}\sin\left(\pi l\gamma b^{-1}-\pi/4\right). \qquad (17.7)$$

And this is the formula of Lifshits-Kosevich theory for FS with the only extreme, namely maximum section of FS by plane $k_z = 0$. This is not surprising, as long as for the quasi-classical approximation the specific law of electron dispersion along the direction of a magnetic field and the FS extension in this direction are insignificant. Thus, at constant Fermi energy, in this approximation it is impossible to identify the influence of layered structure effects on the longitudinal conductivity of crystal.

The results of calculation of longitudinal electric conductivity for a real layered crystal and in the effective mass approximation at constant relaxation time are given in Figures 35 and 36.

The figures demonstrate that in the quasi-classical approximation the layered structure effects become apparent as a phase delay of conductivity oscillations of a real layered crystal and as a relative contribution of oscillating conductivity part as compared to the effective mass approximation. With a growth in electron concentration, the oscillation frequency of longitudinal electric conductivity increases. The phase difference between the dashed and solid curves and the difference in relative oscillation contributions increase, too. This is due to a larger density of electron states in a real layered crystal as compared to the effective mass approximation and a lower rate of change in FS section by a plane normal to field, with a growth of k_z in a real layered crystal as compared to the effective mass approximation. Thus, with increasing electron concentration, the layered structure effects become more pronounced. In the quasi-classical region of a magnetic field, the relative contribution of oscillating part to total conductivity for crystals with closed and transient FS in the investigated range of electron concentrations does not exceed 0.15 – 1%.

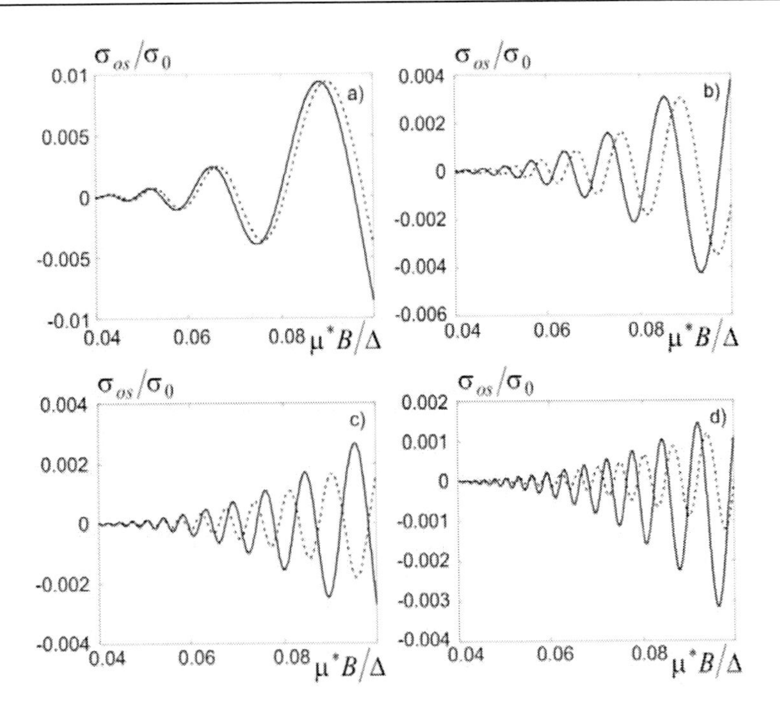

Figure 35. Field dependence of longitudinal electric conductivity of a layered crystal at $\tau(x) \equiv \tau_0$, $kT/\Delta = 0.03$, $0.04 \le \mu^* B/\Delta \le 0.1$.

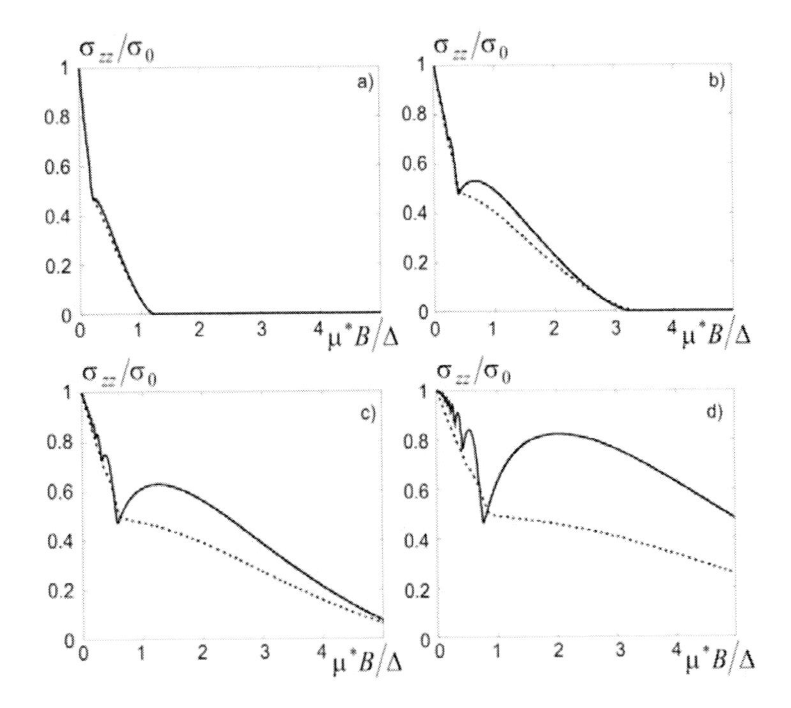

Figure 36. Field dependence of longitudinal electric conductivity of a layered crystal at $\tau(x) \equiv \tau_0$, $kT/\Delta = 0.03$, $0 \le \mu^* B/\Delta \le 5$.

In strong magnetic fields, the relative contribution of oscillations to total conductivity at first increases, but conductivity in general is reduced. However, as the field dependence of electron gas chemical potential starts to be manifested, on the field dependence of total longitudinal electric conductivity there appears a kink following which this electric conductivity starts to grow and reaches a maximum. This maximum is typical of only a real layered crystal and not characteristic of the effective mass approximation, i.e. it is a unique attribute of a layered structure. With increasing electron concentration, this maximum is increased and in the investigated range of electron concentrations for crystals with closed and transient FS it makes 0.5 to 0.8 from the value of full electric conductivity of crystal in the absence of a magnetic field.

Now we turn to the analysis of almost linear parts of the field dependences of total longitudinal electric conductivity depicted in Figures 35a-d that become apparent in no longer quasi-classical, but still relatively weak quantizing magnetic fields. It is exactly the presence of these parts that accounts for the longitudinal Kapitsa effect that consists in a linear dependence of crystal magnetoresistance on a magnetic field induction [34]. Expanding expression (17.1) in a Taylor series and assuming coefficients C_m^σ at $b=0$, we obtain the following formula for relative magnetoresistance of a real layered crystal:

$$\frac{\Delta\rho(B)}{\rho(0)}=\frac{2\sin C_0^\sigma}{C_0^\sigma-C_2^\sigma}\frac{\mu^* B}{\Delta}. \tag{17.8}$$

In the effective mass approximation, this formula takes the form:

$$\frac{\Delta\rho(B)}{\rho(0)}=\frac{1.5}{\gamma}\frac{\mu^* B}{\Delta}. \tag{17.9}$$

When using formula (17.9) it should be borne in mind that, with the same electron concentration, γ in the effective mass approximation is somewhat larger than γ for a real layered crystal. Therefore, for instance, at $\mu^* B/\Delta = 0.1$ and $\gamma_0 = 0.5$, 1, 1.5 and 2, respectively, for the ratio $\Delta\rho(B)/\rho(0)$ we obtain 0.282, 0.127, 0.0685 and 0 for a real layered crystal and 0.295, 0.144, 0.0936 and 0.0672, respectively, in the effective mass approximation. Hence it is seen that the layered crystal structure manifests itself in reduced Kapitsa coefficient as compared to the effective mass approximation. It should be noted that the longitudinal Kapitsa effect cannot be explained in the framework of conventional approaches. Within their limits, only the transverse Kapitsa effect can be explained. The physical reason for this effect is that in a strong magnetic field the role of electron mean free paths is adopted by their cyclotron orbital radii [35].

With further increase in a magnetic field, when inequality $\mu^* B/\Delta >> 1$ becomes valid, the longitudinal conductivity of a layered crystal turns to zero, not identically, however, but according to certain asymptotic laws which will be discussed below.

Now we estimate the value σ_0 of layered crystal electric conductivity in the absence of a magnetic field. First, we assume that electron scattering on the ionized impurities is predominant. Scattering cross section, hence, the length l_0 of electron mean free path, will be considered to be energy independent. To consider electron scattering on impurities as isotropic and suppose relaxation time to be constant, we will replace the real FS of a layered crystal by the equivalent sphere on condition that electron concentration and chemical potential of electron gas will remain the same as in a real layered crystal. If k_0 is the equivalent sphere radius, and m_{es}^* – effective mass of electron on it, then the relaxation time at electron scattering on impurities for a real layered crystal can be defined as:

$$\tau_0 = \frac{2\pi l_0 m_{es}^*}{hk_0} = \frac{hk_0 l_0}{4\pi\gamma_0\Delta}. \tag{17.10}$$

In the effective mass approximation, the relaxation time is defined as:

$$\tau_0 = \frac{hk_0 l_0}{4\pi\gamma_{0em}\Delta}. \tag{17.11}$$

The equivalent sphere radius k_0 is defined as:

$$k_0 = \sqrt[3]{\frac{12\pi^2 m^* \Delta f(\gamma_0)}{ah^2}}. \tag{17.12}$$

This radius is the same both for a real layered crystal and in the effective mass approximation, since electron concentration is assumed to be constant. In so doing, function $f(\gamma_0)$ is defined as:

$$f(\gamma_0) = (\gamma_0 - 1)\arccos(1-\gamma_0) + \sqrt{2\gamma_0 - \gamma_0^2}. \tag{17.13}$$

Substituting (17.10) and (17.11) with regard to (17.12) and (17.13) into (17.1) and (17.5) for $b = 0$, we obtain the following expressions for σ_0:

$$\sigma_0 = \frac{2\pi e^2 m^* a^2 \Delta N}{h^3\gamma_0}\sqrt[3]{\frac{12\pi^2 m^* \Delta f(\gamma_0)}{ah^2}}\{\arccos(1-\gamma_0) - 0.5\sin[2\arccos(1-\gamma_0)]\}, \tag{17.14}$$

which is valid for a real layered crystal, and

$$\sigma_0 = \frac{4\pi e^2 m^* a^2 \Delta N}{3h^3\gamma_{0em}}\sqrt[3]{\frac{12\pi^2 m^* \Delta f(\gamma_0)}{ah^2}}(2\gamma_{0em})^{3/2}, \tag{17.15}$$

which is valid in the effective mass approximation. In these formulae, an additional designation $N \equiv l_0/a$ is introduced. For instance, for a real layered crystal at $m^* = 0.01m_0$, $\Delta = 0.01$ eV, $a = 1$ nm, $N = 1000$ and $\gamma_0 = 0.5;1;1.5;2$, respectively, we obtain $\sigma_0 = 1.089 \cdot 10^3; 1.991 \cdot 10^3; 2.651 \cdot 10^3; 2.916 \cdot 10^3$ S/m, respectively. The latter conductivity value in the absence of a magnetic field at selected problem parameters is maximum possible for a layered crystal and with further increase in electron concentration, i.e. at topological transition from a closed or transient FS to an open one, it increases. Whereas in the effective mass approximation with the same problem parameters, taking into account the difference between γ_0 and γ_{0em}, we obtain $\sigma_0 = 1.193 \cdot 10^3; 2.437 \cdot 10^3; 3.756 \cdot 10^3; 5.228 \cdot 10^3$ S/m, respectively. Concerning the selection of N value it should be noted that with the other selected problem parameters this is a minimum value of relative mean free path whereby one can ignore the influence of Dingle factor on SdH oscillations over the entire considered range of quantizing magnetic fields. Thus, this value corresponds to sufficiently pure and perfect crystals. From comparison of σ_0 values calculated at different γ_0 values for a real layered crystal and in the effective mass approximation it follows that with relatively low γ_0 their difference, as could be expected, is small, but it drastically grows with γ_0 increase. The values calculated for a real layered crystal are always lower than in the effective mass approximation. This is also quite clear, since any restriction of free electron motion reduces conductivity. In the effective mass approximation, maximum electric conductivity value in a direction normal to layers does not exist, but this approximation becomes *basically non-applicable* in the case of transient and closed FS not least because these FS have no tangential planes normal to k_z axis and to a magnetic field.

Consider now the case of electron scattering on the deformation potential of acoustic phonons. In this case, using the relaxation time from [36], we obtain the following formula for electric conductivity of a real layered crystal in the absence of a magnetic field:

$$\sigma_0 = \frac{32e^2hm^*a\rho s_0^6\Delta^3\{\arccos(1-\gamma_0) - 0.5\sin[2\arccos(1-\gamma_0)]\}\gamma_0}{31\pi^2\Gamma(5)\zeta(5)\Xi^2(kT)^5}\sqrt[3]{\frac{12\pi^2 f(\gamma_0)m^*\Delta}{ah^2}}, \tag{17.16}$$

and in the effective mass approximation:

$$\sigma_0 = \frac{64e^2hm^*a\rho s_0^6\Delta^3(2\gamma_{0em})^{3/2}\gamma_{0em}}{93\pi^2\Gamma(5)\zeta(5)\Xi^2(kT)^5}\sqrt[3]{\frac{12\pi^2 f(\gamma_0)m^*\Delta}{ah^2}}. \tag{17.17}$$

In these formulae, $\Gamma(x)$ is gamma-function, $\zeta(x)$ is Riemann zeta function, ρ is crystal density, s_0 is sound velocity in it, Ξ is deformation potential of acoustic phonons. So, for instance, at $\rho = 5000$ kg/m^3, $s_0 = 5000$ m/s, $\Xi = 10$ eV, $kT/\Delta = 0.03$ and previously specified other problem parameters for a real layered crystal at $\gamma_0 = 0.5;1;1.5;2$, respectively, we will obtain $\sigma_0 = 1.079 \cdot 10^4; 7.888 \cdot 10^4; 2.364 \cdot 10^5; 4.621 \cdot 10^5$ S/m. In the effective mass approximation with the same problem parameters we will obtain $\sigma_0 = 1.225 \cdot 10^4; 1.044 \cdot 10^5; 3.825 \cdot 10^5; 1.031 \cdot 10^6$ S/m, respectively.

Finally, we will establish the asymptotic laws of decay in longitudinal electric conductivity in the ultra-quantum limit. First, we will do it for the case of electron scattering on the ionized impurities without regard to magnetic field effect on this scattering. Let us use relationship (5.9) and substitute it into general formula (16.5), since exactly σ_{os} is the principal term in the ultra-quantum limit. In so doing, factor $(-1)^l$ is compensated, and cosine factor can be replaced by unity. We will expand the integrand as a series in x, being restricted to a principal, i.e. quadratic term.

Moreover, we will take into account that, as shown by numerical analysis, at $kT/\mu^* B << 1$ the following relation is valid:

$$\sum_{l=1}^{\infty} f_l^{\sigma} = \frac{2.467 \mu^* B}{\pi^2 kT} . \tag{17.18}$$

Finally, performing integration with respect to x in the range from 0 to $f(\gamma_0)\Delta(2\mu^* B)^{-1}$ and combining numerical multipliers into one, we obtain the following asymptotic formula for a longitudinal electric conductivity:

$$\sigma_{zz} = 1.285 \frac{e^2 m^* a^2 \Delta^4 f^3(\gamma_0)}{\gamma_0 h^3 (\mu^* B)^2 kT} \sqrt[3]{\frac{f(\gamma_0) m^* \Delta}{a h^2}} N . \tag{17.19}$$

This formula is valid for a real layered crystal. In the effective mass approximation, it should be replaced by the following one:

$$\sigma_{zz} = 1.285 \frac{e^2 m^* a^2 \Delta^4 f^3(\gamma_0)}{\gamma_{0em} h^3 (\mu^* B)^2 kT} \sqrt[3]{\frac{f(\gamma_0) m^* \Delta}{a h^2}} N . \tag{17.20}$$

This law is somewhat difficult for understanding, because it implies that longitudinal electric conductivity of crystal at $T = 0$ tends to infinity, rather than to a finite value, as could be expected at electron scattering on the ionized impurities. However, at real finite low temperatures this law does not produce physically incorrect results. Indeed, at previously specified problem parameters and $B = 60\,\mathrm{T}$ for a real layered crystal we obtain at $\gamma_0 = 0.5;1;1.5;2$ $\sigma_{zz} = 0.082;1.454;8.427;33.011$ S/m, respectively. In the effective mass approximation we obtain $\sigma_{zz} = 0.08;1.398;7.885;29.593$ S/m, respectively. And it is 2 to 4 orders of magnitude lower than in the absence of a magnetic field. Therefore, we obtain an asymptotic law of the type $\rho_{zz} \propto TB^2$.

Let us also consider the possibility to obtain for longitudinal electric conductivity at electron scattering on the ionized impurities a different law which does not contain temperature. To do this, we will take into account the influence of a magnetic field on electron scattering and use the results of work [37] from which it follows that for an open FS in the ultra-quantum limit the electric conductivity formula coincides with formula (105) for σ_0, in this case with substitution of constant relaxation time τ_0 by the time which is a function of magnetic field, namely:

$$\tau(B) = \tau_0\left(1 - \cos\frac{\pi\zeta}{\mu^* B}\right). \tag{17.21}$$

But this result has been obtained by solving a kinetic equation for *each* of Landau subbands that are completely or partially filled. Therefore, it can be also extended for the case of closed FS, by substituting into (17.21) instead of ζ its value (5.9). Then $\tau(B) = 2\tau_0$. Hence, using formula (16.4), acting in the same way as in the previous case and considering the Landau subband with the number $n = 0$ as partially filled, for the longitudinal electric conductivity of a real layered crystal we obtain the following asymptotic expression:

$$\sigma_{zz} = \frac{\pi e^2 m^* a^2 \Delta^4 f^3(\gamma_0)}{3h^3\gamma_0(\mu^* B)^3}\sqrt[3]{\frac{12\pi^2 f(\gamma_0) m^*\Delta}{ah^2}}N. \tag{17.22}$$

In the effective mass approximation formula (17.22) goes into:

$$\sigma_{zz} = \frac{\pi e^2 m^* a^2 \Delta^4 f^3(\gamma_0)}{3h^3\gamma_{0em}(\mu^* B)^3}\sqrt[3]{\frac{12\pi^2 f(\gamma_0) m^*\Delta}{ah^2}}N. \tag{17.23}$$

Thus, we obtain the law $\rho_{zz} \propto B^3$.

However, if the Landau subband with the number $n = 0$ is considered as completely filled, formulae (17.22) and (17.23) will take the form:

$$\sigma_{zz} = \frac{\pi^2 e^2 m^* a^2 \Delta^5 f^4(\gamma_0)}{64h^3\gamma_0(\mu^* B)^4}\sqrt[3]{\frac{12\pi^2 f(\gamma_0) m^*\Delta}{ah^2}}N, \tag{17.24}$$

$$\sigma_{zz} = \frac{\pi^2 e^2 m^* a^2 \Delta^5 f^4(\gamma_0)}{64h^3\gamma_{0em}(\mu^* B)^4}\sqrt[3]{\frac{12\pi^2 f(\gamma_0) m^*\Delta}{ah^2}}N. \tag{17.25}$$

Therefore, in this case we obtain the law $\rho_{zz} \propto B^4$. With the previously specified problem parameters we obtain from formula (17.22) for a real layered crystal by the law $\rho_{zz} \propto B^3$ and $\gamma_0 = 0.5;1;1.5;2$ the values of $\sigma_{zz} = 2.819 \cdot 10^{-4};5.018 \cdot 10^{-3};0.029;0.114$ S/m, respectively, and in the effective mass approximation from formula (17.23) the values of $\sigma_{zz} = 2.769 \cdot 10^{-4};4.825 \cdot 10^{-3};0.027;0.102$ S/m, respectively. From formula (17.24) for a real layered crystal under the law $\rho_{zz} \propto B^4$ and the same problem parameters for a real layered crystal we obtain $\sigma_{zz} = 4.087 \cdot 10^{-7};2.124 \cdot 10^{-5};2.356 \cdot 10^{-4};1.515 \cdot 10^{-3}$ S/m, respectively. Whereas, in the effective mass approximation, from formula (17.25) we get $\sigma_{zz} = 4.014 \cdot 10^{-7};2.043 \cdot 10^{-5};2.204 \cdot 10^{-4};1.358 \cdot 10^{-3}$ S/m, respectively. Therefore, under restrictions of the law $\rho_{zz} \propto B^3$ the longitudinal electric conductivity of a layered crystal is reduced by 4 to 6 orders of magnitude and under restrictions of the law $\rho_{zz} \propto B^4$ – by 6 to 9 orders of magnitude as compared to longitudinal electric conductivity in the absence of a magnetic field.

Let us now turn to the case of electron scattering on the deformation potential of acoustic phonons. Acting in the same way as before, we will obtain the following asymptotic laws of decay in the longitudinal electric conductivity of a real layered crystal in the ultra-quantum limit when considering the only Landau subband as partially filled:

$$\sigma_{zz} = 8.598 \cdot 10^{-4} \frac{e^2 h m^* a \rho s_0^6 \Delta^6 f^3(\gamma_0)\gamma_0}{\Xi^2 (kT)^6 (\mu^* B)^2} \sqrt[3]{\frac{f(\gamma_0) m^* \Delta}{a h^2}}. \tag{17.26}$$

In the effective mass approximation, this formula acquires the form:

$$\sigma_{zz} = 8.598 \cdot 10^{-4} \frac{e^2 h m^* a \rho s_0^6 \Delta^6 f^3(\gamma_0)\gamma_{0em}}{\Xi^2 (kT)^6 (\mu^* B)^2} \sqrt[3]{\frac{f(\gamma_0) m^* \Delta}{a h^2}}. \tag{17.27}$$

Whereas when considering the only Landau subband as completely filled, formulae (17.26) and (17.27) take the form:

$$\sigma_{zz} = 1.266 \cdot 10^{-4} \frac{e^2 h m^* a \rho s_0^6 \Delta^7 f^4(\gamma_0)\gamma_0}{\Xi^2 (kT)^6 (\mu^* B)^3} \sqrt[3]{\frac{f(\gamma_0) m^* \Delta}{a h^2}}. \tag{17.28}$$

$$\sigma_{zz} = 1.266 \cdot 10^{-4} \frac{e^2 h m^* a \rho s_0^6 \Delta^7 f^4(\gamma_0)\gamma_{0em}}{\Xi^2 (kT)^6 (\mu^* B)^3} \sqrt[3]{\frac{f(\gamma_0) m^* \Delta}{a h^2}}. \tag{17.29}$$

Thus, we obtain the asymptotic laws $\rho_{zz} \propto T^6 B^2$ and $\rho_{zz} \propto T^6 B^3$, respectively. Therefore, with the same problem parameters and $\gamma_0 = 0.5;1;1.5;2$, from formula (17.26) we obtain $\sigma_{zz} = 0.809;57.631;751.57;5.234 \cdot 10^3$ S/m, respectively, and from formula (17.27) $\sigma_{zz} = 0.824;59.937;803.177;5.839 \cdot 10^3$ S/m, respectively. In turn, from formula (17.28) we obtain $\sigma_{zz} = 1.173 \cdot 10^{-3};0.244;6.086;69.603$ S/m, respectively, and from formula (17.29) $\sigma_{zz} = 1.194 \cdot 10^{-3};0.254;6.504;77.462$ S/m, respectively. Thus, at electron scattering on acoustic phonons under restrictions of the law $\rho_{zz} \propto T^6 B^2$ the longitudinal electric conductivity of crystal is reduced by 2 to 4 orders of magnitude as compared to electric conductivity in the absence of a magnetic field, and under restrictions of the law $\rho_{zz} \propto T^6 B^3$ – by 4 to 6 orders of magnitude as compared to electric conductivity in the absence of a magnetic field. It must be emphasized that these results also cannot be obtained in the framework of traditional approaches, where specific shape of FS and its extension along the direction of a magnetic field are of marginal importance.

Comparing the results obtained in this section to the results reported by previous authors, we note that in the case of constant relaxation time, or, equally, constant Dingle temperature, formula (17.3) for the case of open FS is identical to formula obtained in Ref. [34], and to an accuracy of correction exponentially small at low temperatures coincides with the formula obtained in Ref. [33]. And this means that, at least ignoring magnetic field effect on electron scattering, formulae (17.4) and (17.19) must be valid.

Chapter 18

LONGITUDINAL ELECTRIC CONDUCTIVITY OF LAYERED CRYSTALS IN A QUANTIZING MAGNETIC FIELD UNDER STRONG DEGENERACY CONDITIONS AND IN THE APPROXIMATION OF CONSTANT ELECTRON MEAN FREE PATH

In the calculation of longitudinal electric conductivity of a layered crystal for this case, we will consider that electron mean free path l is defined by cross-section of its scattering on the ionized impurities calculated with the energy equal to the Fermi energy; hence it does not depend on quantum numbers. In this case the relaxation time can be written as:

$$\tau(x) = \frac{l}{v_f(x)}. \tag{18.1}$$

In so doing, electron velocity on FS $v_f(x)$ in the case of a real layered crystal will be defined as:

$$v_f(x) = \sqrt{2\frac{[\zeta - \Delta + \Delta\cos x]}{m^*} + \frac{4\pi^2 a^2 \Delta^2 \sin^2 x}{h^2}}. \tag{18.2}$$

Substituting (18.1) to general formulae for electric conductivity components (16.4) and (16.5) and passing from integration with respect to dimensionless longitudinal quasi-pulse x to integration with respect to energy, we obtain the following expressions for these components:

$$\sigma_0 = \frac{4e^2 N}{\pi a A h}\int_0^{\gamma - b}\sqrt{\frac{2y - y^2}{A(\gamma - y) + 2y - y^2}}dy, \tag{18.3}$$

$$\sigma_{os} = \frac{8e^2N}{\pi aAh}\sum_{l=1}^{\infty}(-1)^l f_l^{\sigma}\int_0^{\gamma-b}\sqrt{\frac{2y-y^2}{A(\gamma-y)+2y-y^2}}\cos\left[\pi lb^{-1}(\gamma-y)\right]dy. \tag{18.4}$$

In these formulae, $A = h^2/(2\pi^2 m^* \Delta a^2)$ is a duplicated ratio of longitudinal effective mass of current carriers to transverse effective mass.

In the effective mass approximation, formulae (18.3) and (18.4) acquire the form:

$$\sigma_0 = \frac{4e^2N}{\pi aAh}\int_0^{\gamma-b}\sqrt{\frac{2y}{A(\gamma-y)+2y}}dy, \tag{18.5}$$

$$\sigma_{os} = \frac{8e^2N}{\pi aAh}\sum_{l=1}^{\infty}(-1)^l f_l^{\sigma}\int_0^{\gamma-b}\sqrt{\frac{2y}{A(\gamma-y)+2y}}\cos\left[\pi lb^{-1}(\gamma-y)\right]dy. \tag{18.6}$$

The effect of magnetic field on electron scattering in these formulae is taken into account in terms of γ dependence thereupon. If, however, the effect of magnetic field on scattering is disregarded, then under the signs of square roots in formulae (18.3) – (18.6) one should insert the value γ determined in the absence of a magnetic field.

The results of calculation of longitudinal electric conductivity of a layered crystal by formulae (18.3) – (18.6) are represented in Figures 37 – 51.

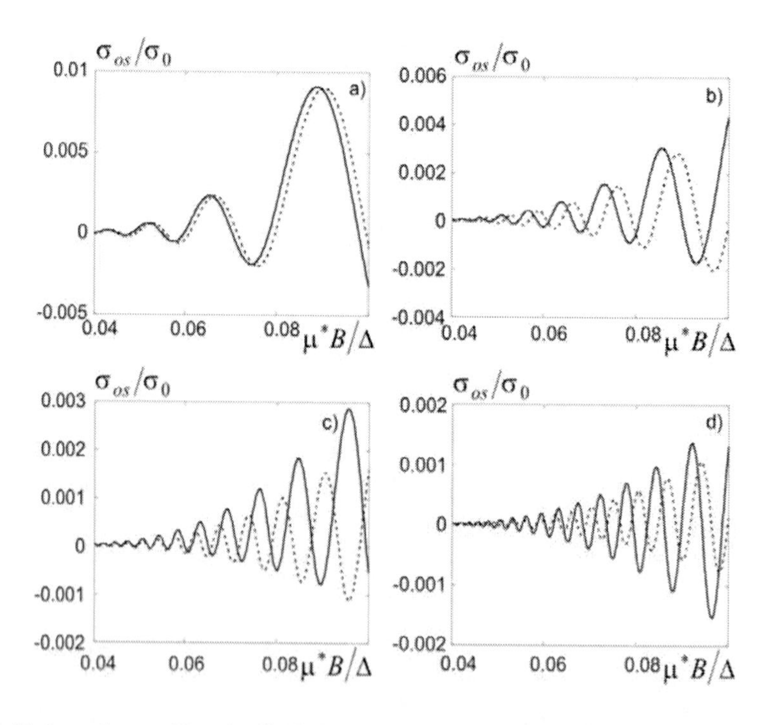

Figure 37. Field dependence of longitudinal electric conductivity of a layered crystal in the model $\tau(x) \propto v_f^{-1}(x)$ at $A = 5, kT/\Delta = 0.03, 0.04 \le \mu^* B/\Delta \le 0.1$.

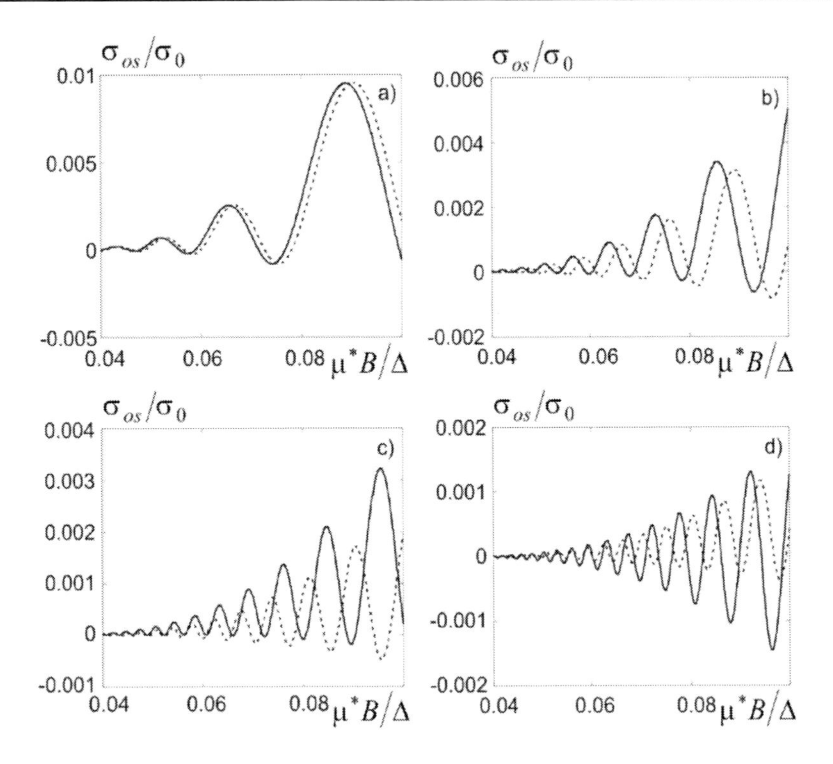

Figure 38. Field dependence of longitudinal electric conductivity of a layered crystal in the model $\tau(x) \propto v_f^{-1}(x)$ at $A = 10, kT/\Delta = 0.03, 0.04 \le \mu^* B/\Delta \le 0.1$.

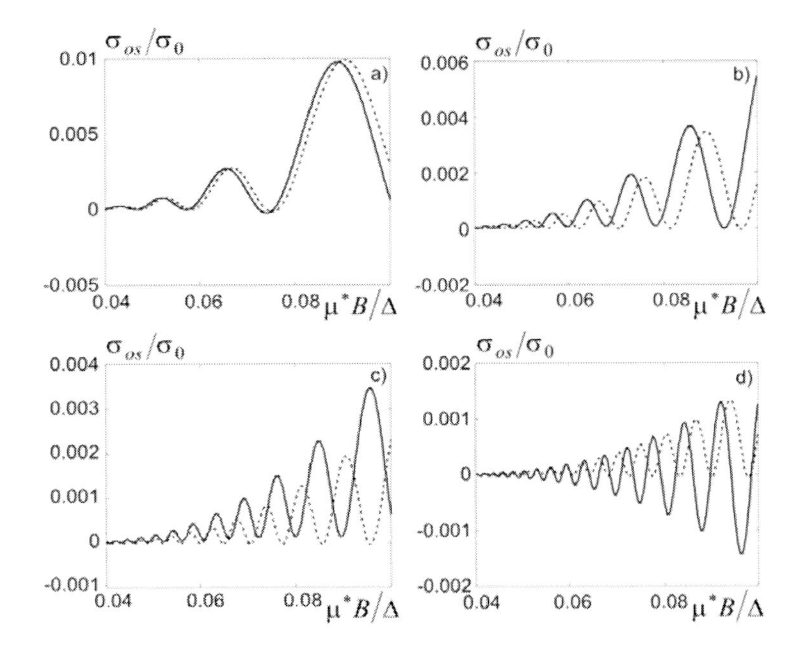

Figure 39. Field dependence of longitudinal electric conductivity of a layered crystal in the model $\tau(x) \propto v_f^{-1}(x)$ at $\Lambda = 15, kT/\Delta = 0.03, 0.04 \le \mu^* B/\Lambda \le 0.1$.

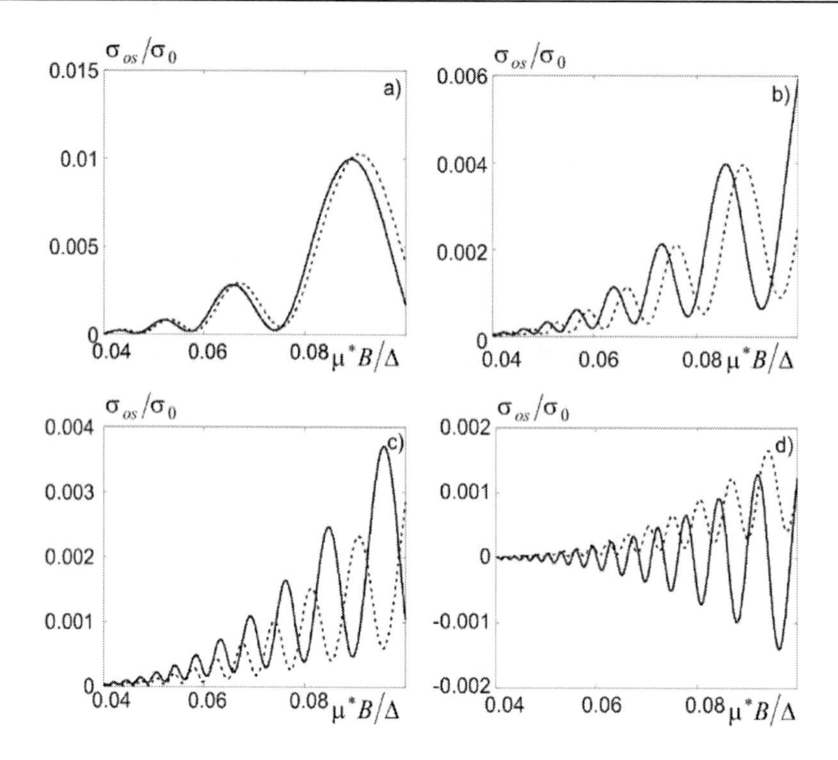

Figure 40. Field dependence of longitudinal electric conductivity of a layered crystal in the model $\tau(x) \propto v_f^{-1}(x)$ at $A = 25, kT/\Delta = 0.03, 0.04 \leq \mu^* B/\Delta \leq 0.1$.

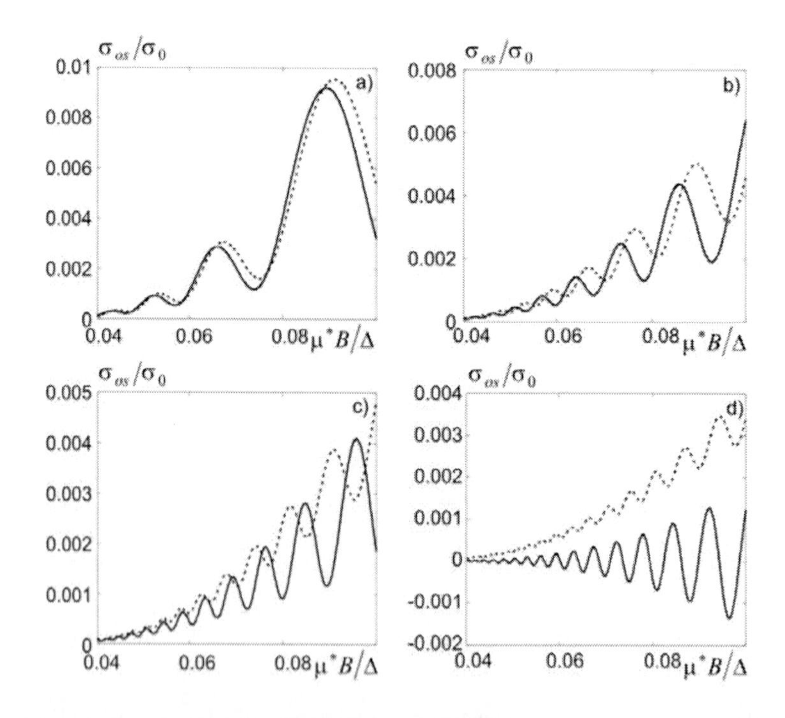

Figure 41. Field dependence of longitudinal electric conductivity of a layered crystal in the model $\tau(x) \propto v_f^{-1}(x)$ at $A = 10^5, kT/\Delta = 0.03, 0.04 \leq \mu^* B/\Delta \leq 0.1$.

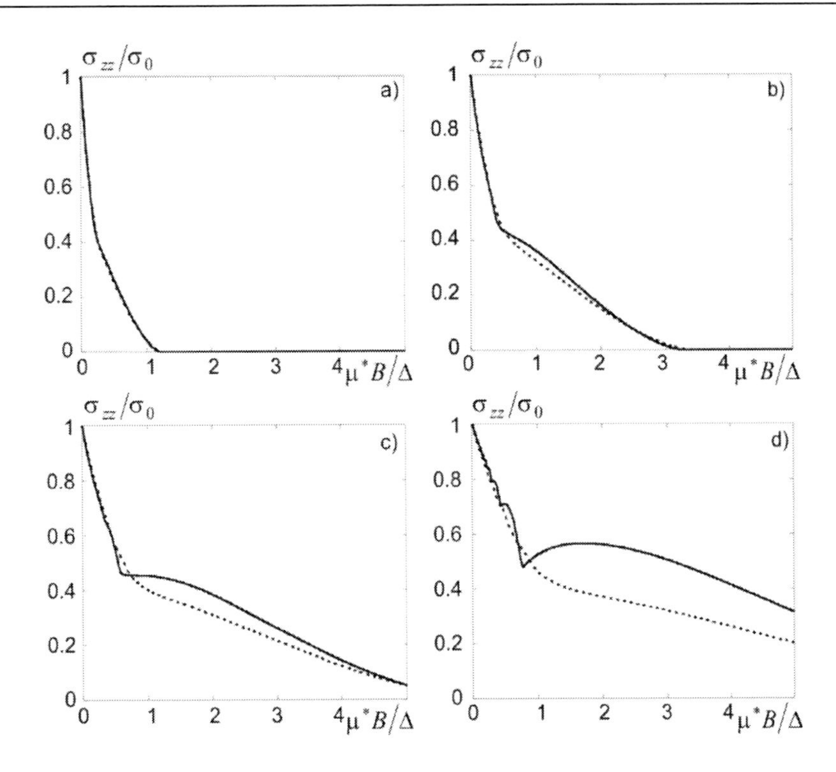

Figure 42. Field dependence of longitudinal electric conductivity of a layered crystal in the model $\tau(x) \propto v_f^{-1}(x)$ at $A = 5, kT/\Delta = 0.03, 0 \le \mu^* B/\Delta \le 5$.

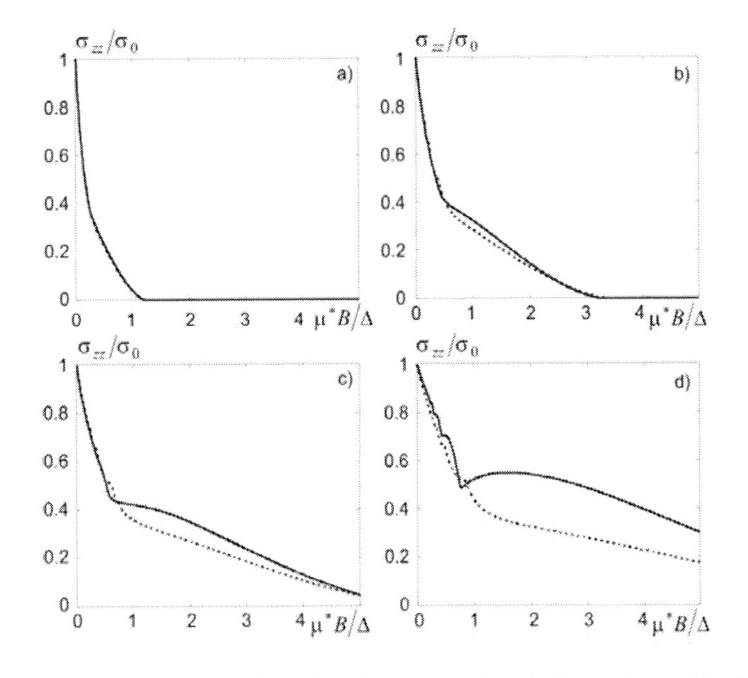

Figure 43. Field dependence of longitudinal electric conductivity of a layered crystal in the model $\tau(x) \propto v_f^{-1}(x)$ at $A = 10, kT/\Delta = 0.03, 0 \le \mu^* B/\Delta \le 5$.

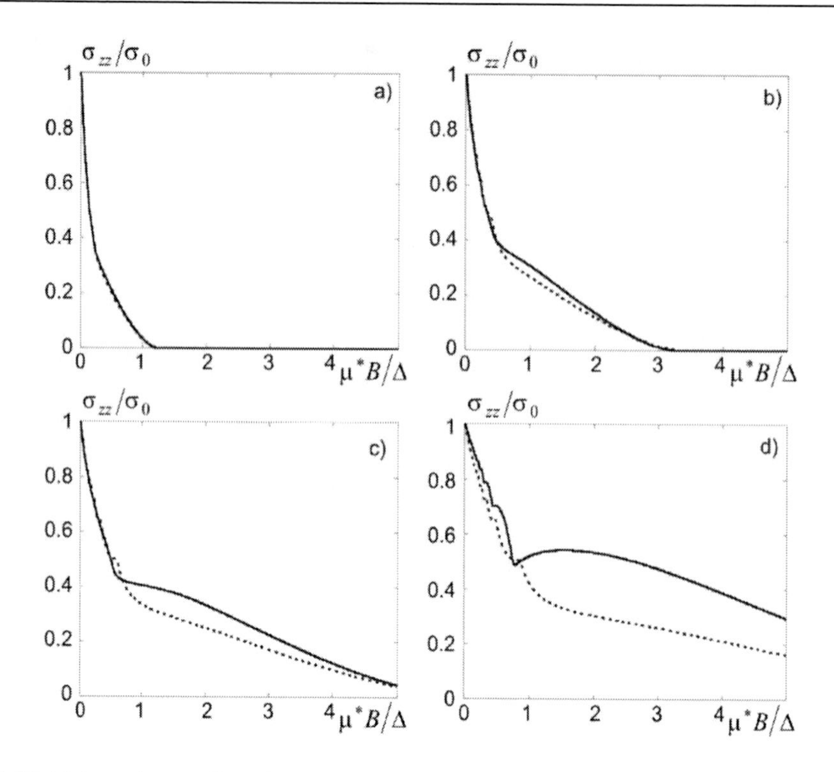

Figure 44. Field dependence of longitudinal electric conductivity of a layered crystal in the model $\tau(x) \propto v_f^{-1}(x)$ at $A = 15, kT/\Delta = 0.03, 0 \le \mu^* B/\Delta \le 5$.

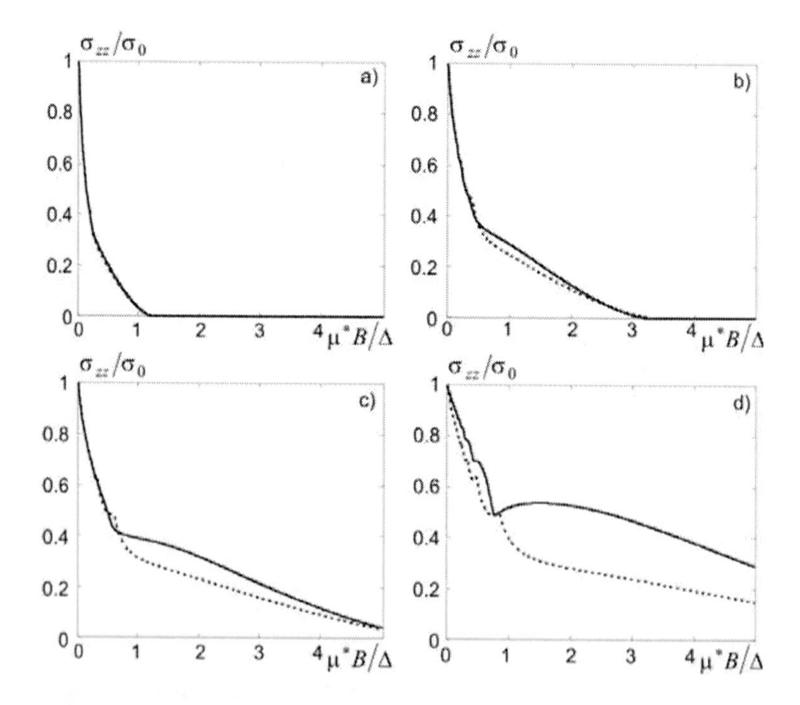

Figure 45. Field dependence of longitudinal electric conductivity of a layered crystal in the model $\tau(x) \propto v_f^{-1}(x)$ at $A = 25, kT/\Delta = 0.03, 0 \le \mu^* B/\Delta \le 5$.

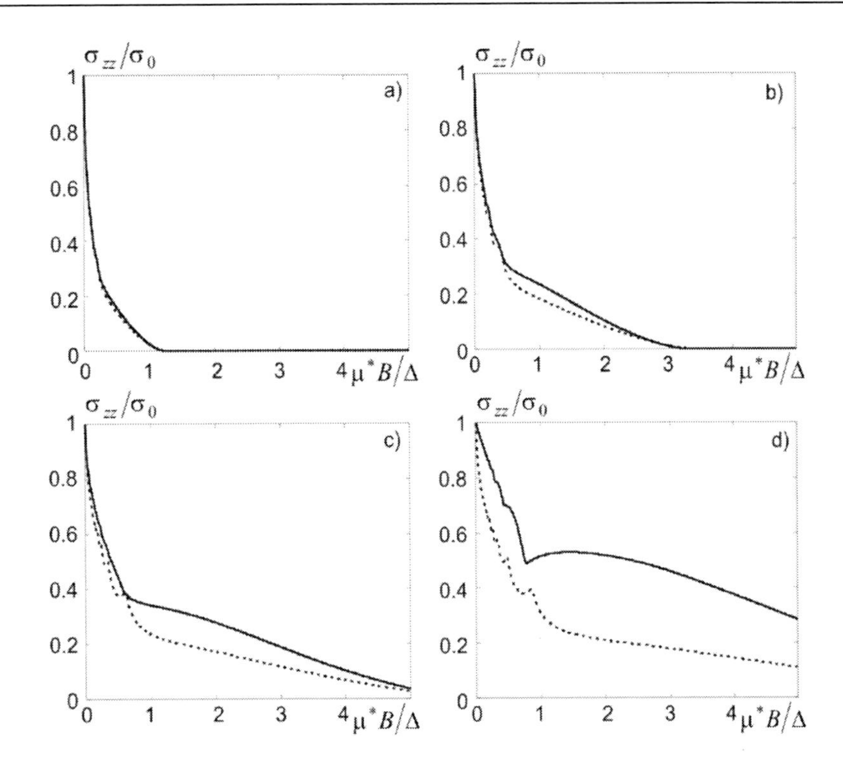

Figure 46. Field dependence of longitudinal electric conductivity of a layered crystal in the model $\tau(x) \propto v_f^{-1}(x)$ at $A = 10^5, kT/\Delta = 0.03, 0 \le \mu^* B/\Delta \le 5$.

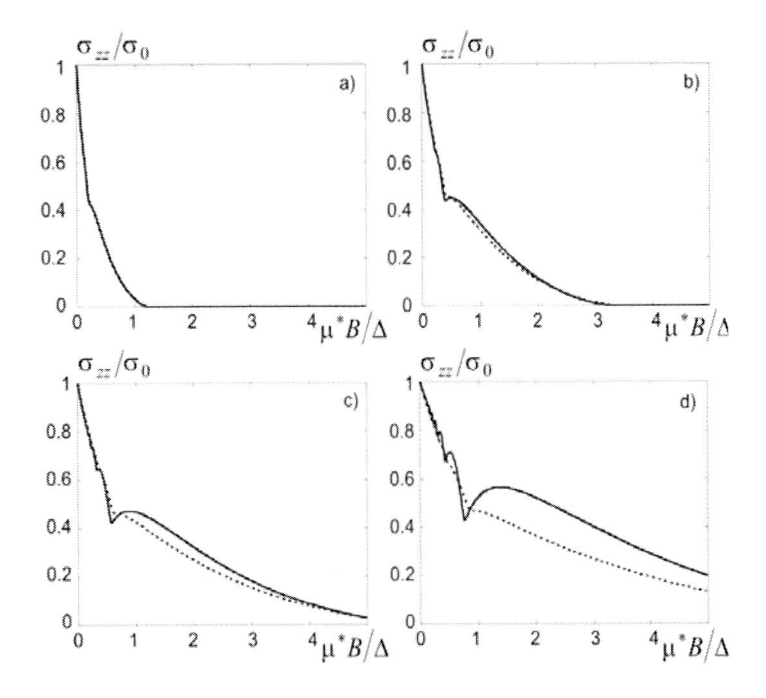

Figure 47. Field dependence of longitudinal electric conductivity of a layered crystal in the model $\tau(x) \propto v_f^{-1}(x, B)$ at $A = 5, kT/\Delta = 0.03, 0 \le \mu^* B/\Delta \le 5$.

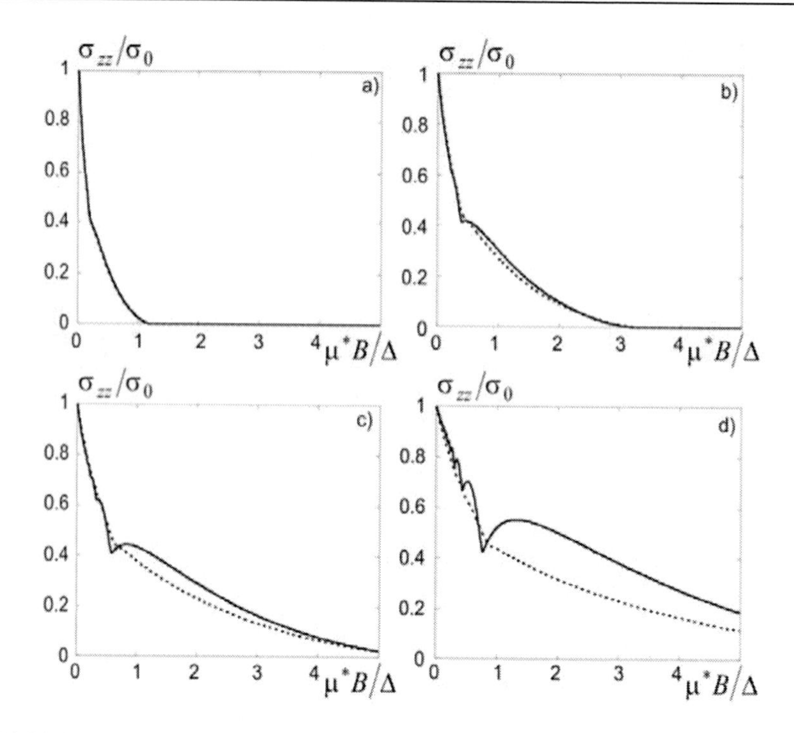

Figure 48. Field dependence of longitudinal electric conductivity of a layered crystal in the model $\tau(x) \propto v_f^{-1}(x, B)$ at $A = 10, kT/\Delta = 0.03, 0 \le \mu^* B/\Delta \le 5$.

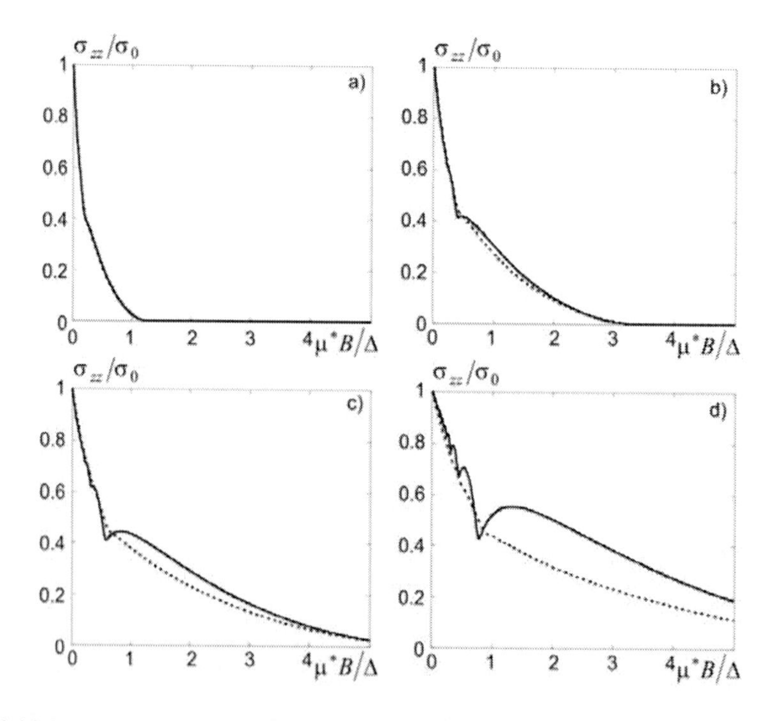

Figure 49. Field dependence of longitudinal electric conductivity of a layered crystal in the model $\tau(x) \propto v_f^{-1}(x, B)$ at $A = 15, kT/\Delta = 0.03, 0 \le \mu^* B/\Delta \le 5$.

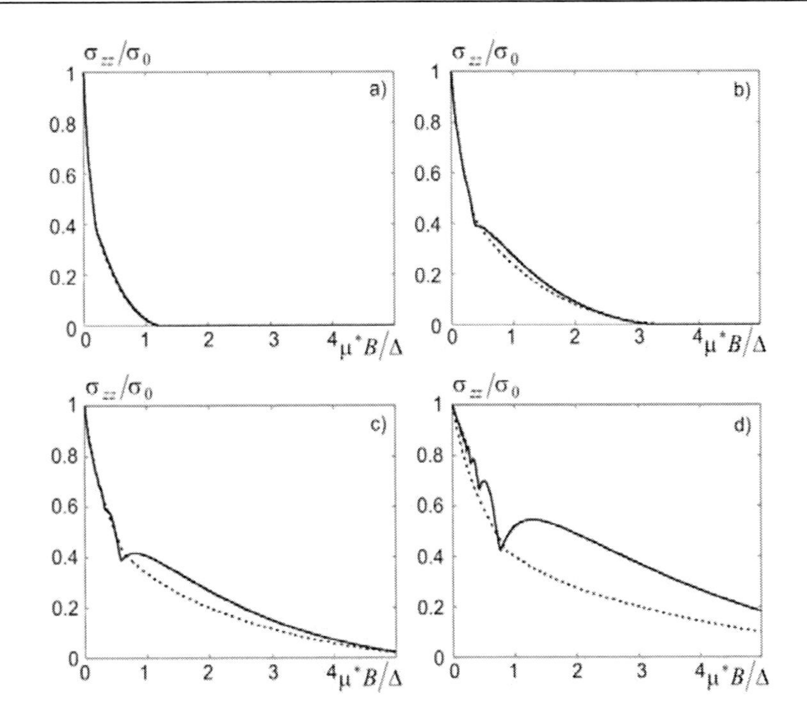

Figure 50. Field dependence of longitudinal electric conductivity of a layered crystal in the model $\tau(x) \propto v_f^{-1}(x,B)$ at $A = 25, kT/\Delta = 0.03, 0 \le \mu^* B/\Delta \le 5$.

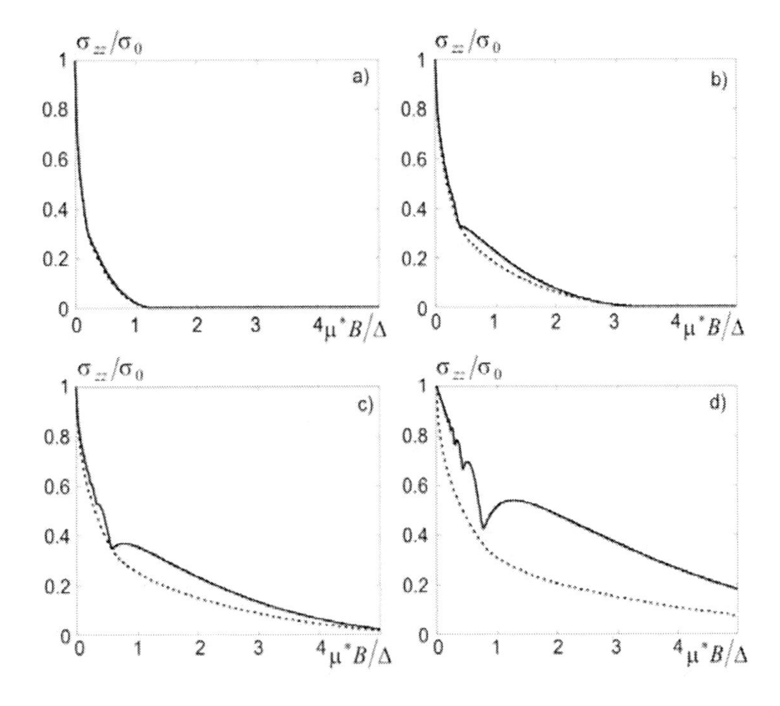

Figure 51. Field dependence of longitudinal electric conductivity of a layered crystal in the model $\tau(x) \propto v_f^{-1}(x,B)$ at $A = 10^5, kT/\Delta = 0.03, 0 \le \mu^* B/\Delta \le 5$.

The figures demonstrate that in quasi-classical magnetic fields the relative contribution of SdH oscillations to total longitudinal conductivity, as in the case of constant relaxation time, amounts to 0.2 – 1%, and with a growth in electron concentration, hence, the degree of degeneracy, it is reduced. As in the case of constant relaxation time, the layered structure effects become apparent in phase delay of longitudinal conductivity oscillations and increase of their relative contribution as compared to the effective mass approximation. With increasing current carrier concentration, just as with increasing anisotropy parameter A, the difference between the solid and dashed curves grows. The greatest difference between the solid and dashed curves occurs at $A = 10^5$ and $\zeta_0/\Delta = 2$. Exactly this case can be realized in dichalcogenides of transient metals and intercalated graphite compounds.

In stronger magnetic fields, without regard to the effect of magnetic field on scattering, there is an optimal range of magnetic field inductions wherein the layered structure effects are most pronounced. These effects imply that longitudinal conductivity of a real layered crystal with a growth in magnetic field induction within the limits of optimal range of magnetic fields is reduced slower than in the effective mass approximation. Without regard to the effect of magnetic field on scattering, electric conductivity maximum in the optimal range of magnetic fields occurs only for $\zeta_0/\Delta = 2$ and is about 0.5 from the electric conductivity value in the absence of a magnetic field. If, however, we take into account the effect of magnetic field on electron scattering on ionized impurities, it turns out that electric conductivity maximum in a real layered crystal for all values of A is more pronounced, and not only at $\zeta_0/\Delta = 2$, i.e. for a transient FS, but also at $\zeta_0/\Delta < 2$, i.e. for closed FS. In this case it is also of the order of 0.4÷0.6 from electric conductivity value in the absence of a magnetic field. Moreover, layered structure effects consist in a more drastic manifestation of oscillations in the range of magnetic fields $\mu^* B/\Delta \le 1$. In so doing, the oscillations, as could be expected, become more pronounced with a growth of anisotropy parameter A.

Consider now the longitudinal Kapitsa effect in this model. Expanding expression (136) as a Taylor series in b, for a real layered crystal we obtain:

$$\Delta\rho(B)\Big/\rho_0 = \left[\int_0^{\gamma_0} \sqrt{\frac{2y - y^2}{A(\gamma_0 - y) + 2y - y^2}}\,dy\right]^{-1} \frac{\mu^* B}{\Delta}, \tag{18.7}$$

Quite similarly, from expression (138) in the effective mass approximation we get:

$$\Delta\rho(B)\Big/\rho_0 = \left[\int_0^{\gamma_{0em}} \sqrt{\frac{2y}{A(\gamma_{0em} - y) + 2y}}\,dy\right]^{-1} \frac{\mu^* B}{\Delta}. \tag{18.8}$$

For instance, at $A = 5$ and $\mu^* B/\Delta = 0.1$, we have at $\gamma_0 = 0.5;1;1.5;2$, respectively, for a real layered crystal $\Delta\rho(B)/\rho_0 = 38.5;20.2;14.6;13.2\%$, respectively, and in the effective mass approximation $-36.4;17.8;11.57;8.31\%$, respectively. At $A = 10$, for a real layered crystal we have $\Delta\rho(B)/\rho_0 = 47.5;25.3;18.6;17.7\%$, respectively, and in the effective mass approximation $- 44.4;21.8;14.1;10.14\%$, respectively. At $A = 15$, for a real layered crystal we have $\Delta\rho(B)/\rho_0 = 54.5;29.3;21.8;21.3\%$, respectively, and in the effective mass approximation – $50.6;24.8;16.1;11.6\%$, respectively. At $A = 25$, for a real layered crystal we have $\Delta\rho(B)/\rho_0 = 65.6;35.6;26.9;27.1\%$, respectively, and in the effective mass approximation – $60.5;29.6;19.2;13.8\%$, respectively. At $A = 10^5$, for a real layered crystal we have $\Delta\rho(B)/\rho_0 = 3.2 \cdot 10^3;1.82 \cdot 10^3;1.47 \cdot 10^3;1.68 \cdot 10^3\%$, respectively, and in the effective mass approximation – $2.82 \cdot 10^3;1.37 \cdot 10^3;0.89 \cdot 10^3;0.64 \cdot 10^3\%$, respectively. The results $A = 10^5$, absurd on the face of it, testify that with large effective mass anisotropy degrees there is a fast growth of longitudinal magnetoresistance, and the Kapitsa linear law in this case is valid only in weak quantizing fields. Therefore, in a model of constant mean free path the Kapitsa coefficient for a real layered crystal is larger than in the effective mass approximation. In this model, the Kapitsa coefficient decreases with increasing electron concentration and increases with increasing anisotropy parameter. With increasing electron concentration, as could be expected, the difference between the Kapitsa coefficients calculated for a real layered crystal and in the effective mass approximation increases as well.

We now estimate in the model of electron relaxation time under consideration the longitudinal conductivity of crystal in the absence of a magnetic field. With this purpose, a dimensionless parameter $\omega_c\tau$, where ω_c – cyclotron frequency, will be written as follows:

$$\omega_c\tau = \frac{N}{\sqrt{A\gamma_0}}\frac{\mu^* B}{\Delta} . \tag{18.9}$$

Taking into account that neglect of the Dingle factor is considered to be valid, if over the entire range of magnetic fields under study $\omega_c\tau \geq 25$, substituting $\mu^* B/\Delta = 0.04$ into (18.9) yields $N \geq 625\sqrt{A\gamma_0}$. Therefore, for instance, at $A = 5$ and $\gamma_0 = 0.5;1;1.5;2$, respectively, based on formulae (18.3) and (18.5) for a real layered crystal we have $\sigma_0 = 2.523 \cdot 10^6;6.787 \cdot 10^6;1.156 \cdot 10^7;1.476 \cdot 10^7$ S/m, respectively, and in the effective mass approximation $2.691 \cdot 10^6;7.858 \cdot 10^6;1.503 \cdot 10^7;2.469 \cdot 10^7$ S/m, respectively. At $A = 10$, for a real layered crystal we have $1.447 \cdot 10^6;3.835 \cdot 10^6;6.391 \cdot 10^6;7.749 \cdot 10^6$ S/m, respectively, and in the effective mass approximation $1.56 \cdot 10^6;4.555 \cdot 10^6;1.503 \cdot 10^7;2.469 \cdot 10^7$ S/m, respectively. At $A = 15$, for a real layered crystal we have $1.03 \cdot 10^6;2.708 \cdot 10^6;4.457 \cdot 10^6;5.258 \cdot 10^6$ S/m, respectively, and in the effective mass approximation $1.118 \cdot 10^6;3.264 \cdot 10^6;6.245 \cdot 10^6;1.025 \cdot 10^7$ S/m, respectively. At $A = 25$, for a real layered crystal we have $6.626 \cdot 10^5;1.725 \cdot 10^6;2.798 \cdot 10^6;3.202 \cdot 10^6$ S/m, respectively, and in the effective mass approximation $7.245 \cdot 10^5;2.116 \cdot 10^6;4.048 \cdot 10^6;6.647 \cdot 10^6$ S/m, respectively. At $A = 10^5$, for a real layered crystal we have $215.057; 534.143; 808.693; 819.381$ S/m, respectively, and in the effective mass approximation $245.755; 719.506; 1.378 \cdot 10^3; 2.263 \cdot 10^3$ S/m, respectively. Thus, we see that within the assumed relaxation time model and the mean free path ratio, the longitudinal electric conductivity of a layered crystal drops with increasing anisotropy parameter A and increases with increasing electron concentration, while crystal FS is closed or transient. In this case, as in the model of constant relaxation time, with the same electron concentration the electric conductivity of a real layered crystal is lower than in the effective mass approximation. In so doing, as could be expected, the longitudinal electric conductivities, calculated in the absence of a magnetic field, for a real layered crystal and in the effective mass approximation vary the greater, the larger is electron concentration. However, it should be borne in mind that σ_0 values obtained within the model of constant mean free path with selected problem parameters should be considered as *minimum required* in order that neglect of the Dingle factor, or, more precisely, its equating to unity be valid.

In conclusion, let us establish the asymptotic laws for tending to zero of longitudinal electric conductivity calculated in the approximation of constant mean free path in the ultra-quantum limit.

We begin with the case when the effect of magnetic field on electron scattering is disregarded. Taking into account that in the ultra-quantum limit the integration variable y is low as compared to γ_0 and proceeding as in the case of constant relaxation time, for a real layered crystal we get:

$$\sigma_{zz} = 3.448 \cdot 10^{-6} \frac{h^5 e^2 N f^3(\gamma_0)}{a^7 A^4 m^{*3} kT \sqrt{A\gamma_0} (\mu^* B)^2}, \tag{18.10}$$

and in the effective mass approximation we obtain:

$$\sigma_{zz} = 3.448 \cdot 10^{-6} \frac{h^5 e^2 N f^3(\gamma_0)}{a^7 A^4 m^{*3} kT \sqrt{A\gamma_{0em}} (\mu^* B)^2}. \tag{18.11}$$

If, however, we take into account the effect of magnetic field on scattering, within the approach set forth in [35] the relaxation time should be written as below:

$$\tau(x) = \tau = \frac{2Na\sqrt{m^*}}{\sqrt{2\mu^* B}}, \tag{18.12}$$

and the electric conductivity should be calculated by formula (16.4). In this case, considering the Landau subband with the number $n = 0$ as partially filled, using the ratio (5.9) we get:

$$\sigma_{zz} = \frac{16 h^{5/2} f^3(\gamma_0) N}{3\sqrt{2}\pi^{9/2} A^5 a^8 e^{3/2} B^{7/2}}. \tag{18.13}$$

If the subband with the number $n = 0$ is considered as completely filled, we obtain:

$$\sigma_{zz} = \frac{h^{7/2} f^4(\gamma_0) N}{2\sqrt{2}\pi^{9/2} A^6 a^{10} e^{5/2} B^{9/2}}. \tag{18.14}$$

Therefore, in the approximation of constant mean free path without regard to the effect of magnetic field on scattering we have the law of magnetoresistance change with a magnetic field of the form $\rho_{zz} \propto TB^2$, and with regard to the effect of magnetic field on scattering – the law of the form $\rho_{zz} \propto B^{7/2}$, if the Landau subband with the number $n = 0$ is considered as partially filled, and the law of the form $\rho_{zz} \propto B^{9/2}$, if the Landau subband with the number $n = 0$ is considered as completely filled.

Note that the above laws are valid, if anisotropy parameter A satisfies the condition $A >> 2h/(\pi a^2 eB)$. At $a = 1\,\text{nm}$, $B = 60\,\text{T}$, from the above considered values of A this condition is satisfied only by $A = 10^5$. According to formulae (18.10) – (18.11), within the assumed model ratio for N with the specified crystal parameters, $T = 3\,\text{K}$ and $\gamma_0 = 0.5;1;1.5;2$, respectively, we get $\sigma_{zz} = 2.936\cdot10^{-8}, 7.312\cdot10^{-7}, 5.121\cdot10^{-6}, 2.267\cdot10^{-5}$ S/m, respectively, i.e. reduction of longitudinal electric conductivity by 8 – 9 orders of magnitude as compared to electric conductivity in the absence of a magnetic field. From formula (18.13) with the same problem parameters for a real layered crystal we obtain $\sigma_{zz} = 1.296\cdot10^{-12}; 4.563\cdot10^{-11}; 3.914\cdot10^{-10}; 2.001\cdot10^{-9}$ S/m, respectively, and in the effective mass approximation $1.307\cdot10^{-12}; 4.654\cdot10^{-11}; 4.046\cdot10^{-10}; 2.113\cdot10^{-9}$ S/m, respectively, i.e. reduction of electric conductivity by 11 – 15 orders of magnitude as compared to its value in the absence of magnetic field. From formula (18.14) we obtain for a real layered crystal with the same problem parameters $\sigma_{zz} = 2.87\cdot10^{-27}; 2.95\cdot10^{-25}; 4.85\cdot10^{-24}; 4.07\cdot10^{-23}$ S/m, respectively, and in the effective mass approximation $2.90\cdot10^{-27}; 3.01\cdot10^{-25}; 5.01\cdot10^{-24}; 4.30\cdot10^{-23}$ S/m, respectively. Therefore, reduction of longitudinal electric conductivity in this case is 25 – 29 orders of magnitude as compared to its value in the absence of a magnetic field.

Chapter 19

SHUBNIKOV-DE HAAS EFFECT IN LAYERED CRYSTALS WITH TWO-SHEETED FERMI SURFACES IN QUASI-CLASSICAL APPROXIMATION

Let us now consider the Shubnikov-de Haas effect in a crystal whose FS consists of two coaxial sheets with a common axis k_z – open and closed. The band parameters a, m^* and Δ will be assumed identical for both sheets. The sheets will be distinguished by their respective Fermi energies ζ_1 and ζ_2. For a closed sheet we will assume $\zeta_1 < 2\Delta$, and for an open sheet $-\zeta_2 >> \Delta$. Then, considering electron mean free path at scattering on ionized impurities dependent on the Fermi energy by the power law $l \propto \zeta^r$ and denoting $l_1/a = N$, we obtain the following formulae defining components of longitudinal electric conductivity of a crystal with such FS at low temperatures [38]:

$$\sigma_0 = \frac{4e^2 N}{\pi a A h}\left[\int_0^{\gamma_1}\sqrt{\frac{2y-y^2}{A(\gamma_1 - y)+2y-y^2}}dy + \left(\frac{\gamma_2}{\gamma_1}\right)^r \int_0^2 \sqrt{\frac{2y-y^2}{A(\gamma_2 - y)+2y-y^2}}dy\right], \tag{19.1}$$

$$\begin{aligned}\sigma_{os} = \frac{8e^2 N}{\pi a A h}\sum_{l=1}^{\infty}(-1)^l f_l^{\sigma}\Bigg\{&\int_0^{\gamma_1}\sqrt{\frac{2y-y^2}{A(\gamma_1 - y)+2y-y^2}}\cos\left[\pi l b^{-1}(\gamma_1 - y)\right]dy + \\ &+\left(\frac{\gamma_2}{\gamma_1}\right)^r \int_0^2 \sqrt{\frac{2y-y^2}{A(\gamma_2 - y)+2y-y^2}}\cos\left[\pi l b^{-1}(\gamma_2 - y)\right]dy\Bigg\}\end{aligned} \tag{19.2}$$

In the general case, integrals appearing in formulae (148) and (149) can be determined only numerically. However, if in formula (149) we take as relaxation times their values on the extreme sections of corresponding FS sheets, and calculate the remaining integrals by saddle-point method, then for σ_{os} we obtain the following quasi-classical formula:

$$\sigma_{os}^{(cc)} = \frac{4\sqrt{2}e^2 N b^{3/2}}{\pi a A^{3/2} h} \sum_{l=1}^{\infty} (-1)^l l^{-3/2} f_l^{\sigma} \left\{ \frac{\sin(\pi l b^{-1}\gamma_1 - \pi/4)}{\sqrt{\gamma_1}} + \left(\frac{\gamma_2}{\gamma_1}\right)^r \left[\frac{\sin(\pi l b^{-1}\gamma_2 - \pi/4)}{\sqrt{\gamma_2}} - \right.\right.$$
$$\left.\left. - \frac{\sin(\pi l b^{-1}(\gamma_2 - 2) - \pi/4)}{\sqrt{\gamma_2 - 2}} \right] \right\} \tag{19.3}$$

According to general concepts of quantum mechanics, parameter r assumes the values from 0 to 4 [39].

The results of numerical calculation of a relative contribution of the oscillating part of electric conductivity in the range of magnetic fields $0.04 \le \mu^* B/\Delta \le 0.1$ at $kT/\Delta = 0.03, A = 15, \gamma_1 = 1, \gamma_2 = 130$ and $r = 0.1;0.2;0.3;0.4$, respectively, are represented in Figures 51, 52.

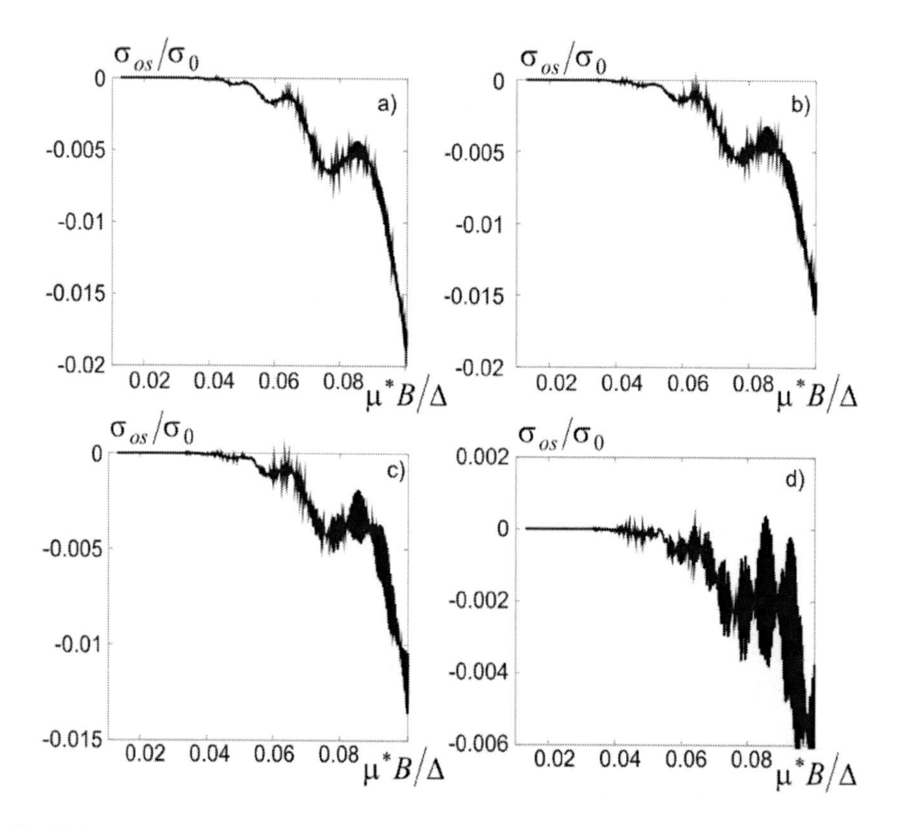

Figure 52. Field dependence of the oscillating part of longitudinal electric conductivity of a layered crystal with a two-sheeted FS according to shrewd formula at $\gamma_1 = 1, \gamma_2 = 130, A = 15, kT/\Delta = 0.03, 0.04 \le \mu^* B/\Delta \le 0.1$ and a) $r = 0.1$; b) $r = 0.2$; c) $r = 0.3$; d) $r = 0.4$.

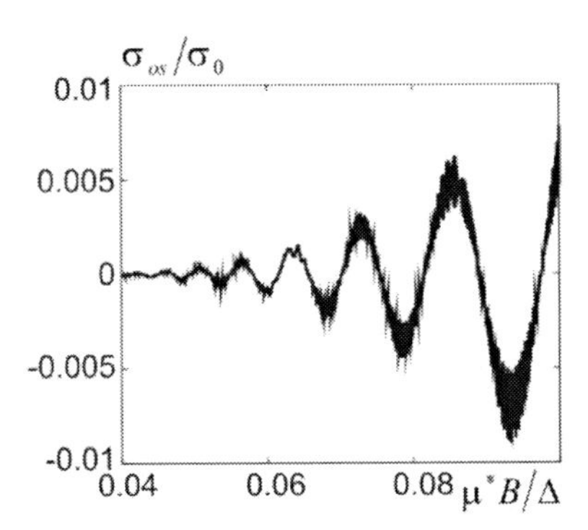

Figure 53. Field dependence of the oscillating part of longitudinal electric conductivity of a layered crystal with a two-sheeted FS according to the Lifshitz-Kosevich theory at $\gamma_1 = 1, \gamma_2 = 130, A = 15, kT/\Delta = 0.03, 0.04 \leq \mu^* B/\Delta \leq 0.1$ and $r = 0.1$.

The figures demonstrate that the contribution of high frequencies related to an open sheet of FS increases with increasing *r*. The relative contribution of the oscillating part of conductivity and the amplitude of fundamental oscillations are measured by tenths of a percent. If scattering section is weakly dependent on energy, then high frequencies related to an open sheet of FS manifest themselves as a "fringe" on the fundamental oscillations related to a closed sheet of FS. With a strong energy dependence of scattering section, on the contrary, the fundamental oscillations are related to an open sheet of FS. Oscillation frequencies in the inverse magnetic field for the two-sheeted FS under consideration are defined by the following formulae:

$$F_l^{(1)} = 2l\zeta_1/\mu^*, \tag{19.4}$$

$$F_l^{(2)} = 2l\zeta_2/\mu^*, \tag{19.5}$$

$$F_l^{(3)} = 2l(\zeta_2 - 2\Delta)/2\mu^*. \tag{19.6}$$

The first of the above frequency sets is related to the only extreme, namely maximum section of a closed sheet of FS by plane $k_{z1} = 0$. The second set is related to one extreme, namely maximum section of an open sheet of FS by plane $k_{z2} = 0$. The third set is related to two extreme, namely minimum sections of an open sheet of FS by planes $k_z = \pm\pi/a$. The proximity of frequencies $F_l^{(2)}$ and $F_l^{(3)}$ is due to beats in the "fringe" caused by $\zeta_2 >> \Delta$. Their contribution is the larger the stronger is energy dependence of scattering section. Yet, since layered structure effects are also apparent in the form of a closed FS sheet, alongside with the above frequency sets there exists a set described by the formula:

$$F_l^{(4)} = 2l|\zeta_1 - 2\Delta|/\mu^*. \tag{19.7}$$

These frequencies are not related to the extreme section of a closed sheet of FS, hence they interfere such that cause only additive smooth contribution into the fundamental oscillations. These frequencies have been discussed above while on the subject of diamagnetic susceptibility oscillations.

Comparison of numerical calculation results to experimental data for a layered organic conductor reported in works [9, 40] shows that experimentally observed dependences of the oscillating part of conductivity on magnetic field induction are rather close to those shown in Figure 51a,b in their type, the order of relative contribution and the oscillation range. Therefore, in this case we most probably deal with a two-sheeted FS consisting of strongly open and closed sheets, and with a weak energy dependence of scattering section. However, in other cases dependences are possible that are considerably closer to those depicted, say, in Figure 51c, d from which it can be inferred that here we deal either with FS consisting of the only strongly open sheet which is considered in the majority of conventional approaches, or with a two-sheeted FS with a strong energy dependence of scattering section.

As an example, Figure 52 depicts the field dependence of a relative contribution of the oscillating conductivity part determined in a conventional quasi-classical approximation with the use of formula (19.3). The figure shows that this dependence is dramatically different from that obtained as a result of a rigorous numerical calculation. This discrepancy is mainly due to the fact that formula (19.3), in fact, corresponds to approximation of constant relaxation time which is true for a strongly open sheet of FS under consideration, but too rough for its closed sheet, by virtue of which for the case of such FS the conventional quasi-classical approach requires modification.

In conclusion, using formula (19.1), we will estimate the constant part of conductivity σ_0 . Taking into account that the Dingle factor in our conditions has a weak effect on the oscillating part of conductivity, for $N \geq 5000$, assuming $a = 1$ nm, we get at $A = 15$, $\gamma_1 = 1$ and $\gamma_2 = 130$, $r = 0.1;0.2;0.3;0.4$ and for $N = 5000$ we get the values of $\sigma_0 = 6.118 \cdot 10^6;6.715 \cdot 10^6;7.686 \cdot 10^6;9.266 \cdot 10^6$ S/m, respectively. It should be noted that the results of this paragraph cannot be obtained within a conventional quasi-classical approach, for which an explicit law of charge carrier dispersion and the finite extent of FS sheets along the direction of a magnetic field are inessential.

Chapter 20

LONGITUDINAL ELECTRIC CONDUCTIVITY OF LAYERED CRYSTALS IN A QUANTIZING MAGNETIC FIELD WITH RELAXATION TIME PROPORTIONAL TO TOTAL ELECTRON VELOCITY

When describing charge carrier scattering on the deformation potential of acoustic phonons, use is very often made of relaxation time model within the framework of which this time is proportional to total electron velocity. In the context of given model this time is:

$$\tau(x) = \tau_{ap}\sqrt{A(\gamma - 1 + \cos x) + \sin^2 x}\,, \tag{20.1}$$

where τ_{ap} – temperature and magnetic field dependent nominal time of electron scattering on acoustic phonons, different for various models of transport factor. So, in conformity with general formulae (16.4) and (16.5), we obtain the following expressions for components of longitudinal electric conductivity of a layered crystal in a quantizing magnetic field under strong degeneracy conditions:

$$\sigma_0 = \frac{16\pi^2 e^2 m^* a \tau_{ap} \Delta^2}{h^4} \int_0^{\gamma - b} \sqrt{\left[A(\gamma - y) + 2y - y^2\right]\left(2y - y^2\right)}\,dy\,, \tag{20.2}$$

$$\sigma_{os} = \frac{32\pi^2 e^2 m^* a \tau_{ap} \Delta^2}{h^4} \sum_{l=1}^{\infty} (-1)^l f_l^{\sigma} \int_0^{\gamma - b} \sqrt{\left[A(\gamma - y) + 2y - y^2\right]\left(2y - y^2\right)} \cos\left[\pi l b^{-1}(\gamma - y)\right] dy\,. \tag{20.3}$$

In the effective mass approximation these formulae acquire the form:

$$\sigma_0 = \frac{16\pi^2 e^2 m^* a \tau_{ap} \Delta^2}{h^4} \int_0^{\gamma - b} \sqrt{2y\left[A(\gamma - y) + 2y\right]}\,dy\,, \tag{20.4}$$

$$\sigma_{os} = \frac{32\pi^2 e^2 m^* a \tau_{ap} \Delta^2}{h^4} \sum_{l=1}^{\infty} (-1)^l f_l^{\sigma} \int_0^{\gamma - b} \sqrt{2y[A(\gamma - y) + 2y]} \cos[\pi l b^{-1} (\gamma - y)] dy . \quad (20.5)$$

The magnetic field dependence of relaxation time in these formulae is taken into account through the dependence of γ thereupon. If, however, the magnetic field dependence of relaxation time is ignored, then under the signs of square roots in formulae (156) – (159) one should substitute the value γ determined in the absence of a magnetic field. The results of calculations of the field dependences of longitudinal electric conductivity of a layered crystal by the formulae (20.2) – (20.5) are represented in Figures 54 – 68.

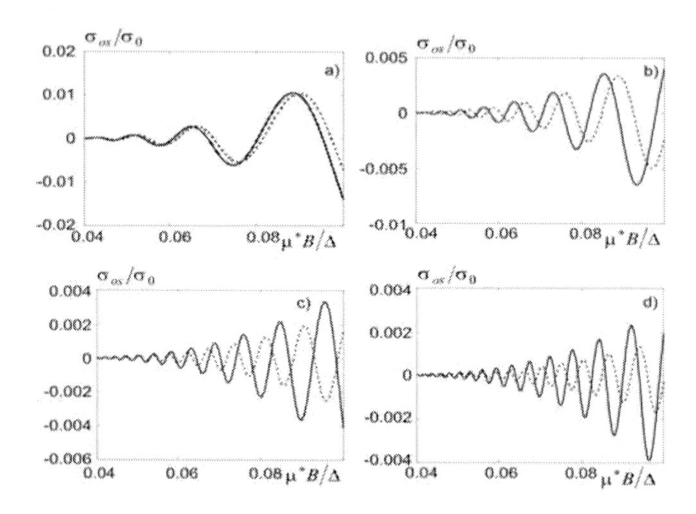

Figure 54. Field dependences of the oscillating part of longitudinal electric conductivity in the model $\tau \propto v_f(x)$ at $0.04 \le \mu^* B/\Delta \le 0.1, kT/\Delta = 0.03, A = 5$.

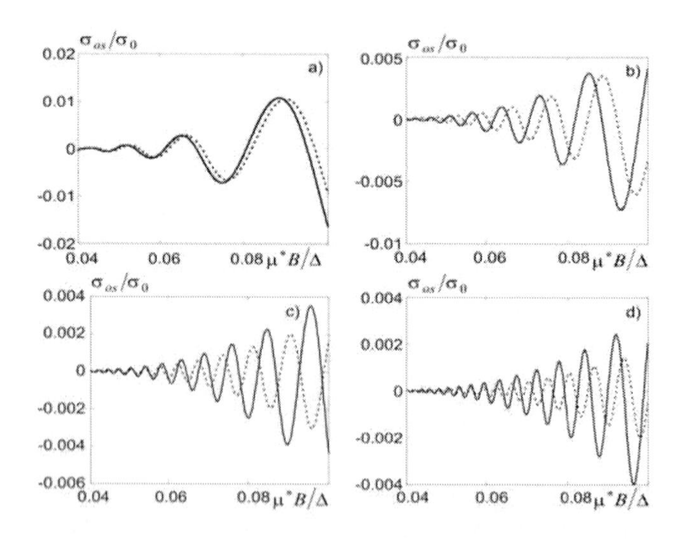

Figure 55. Field dependences of the oscillating part of longitudinal electric conductivity in the model $\tau \propto v_f(x)$ при $0.04 \le \mu^* B/\Delta \le 0.1, kT/\Delta = 0.03, A = 10$.

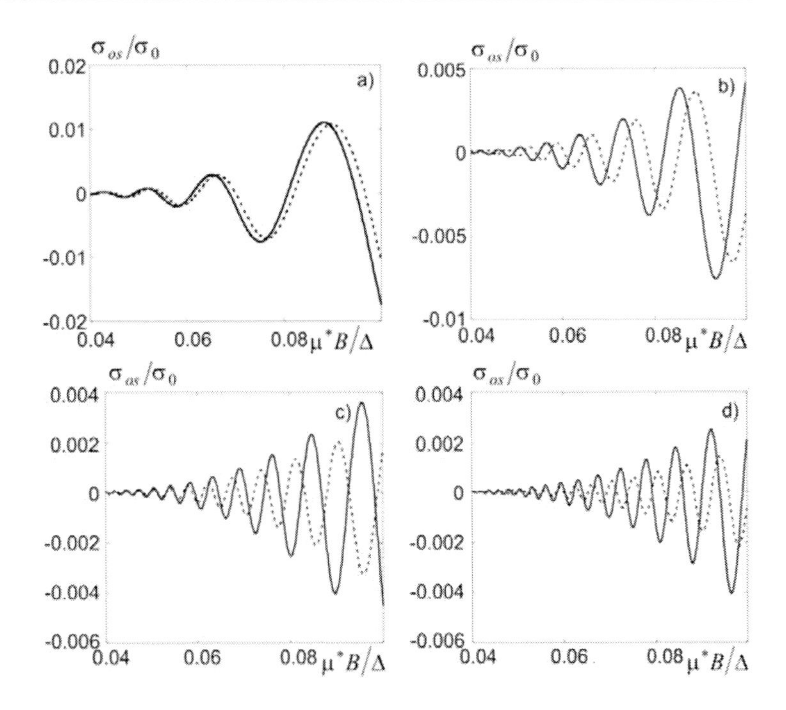

Figure 56. Field dependences of the oscillating part of longitudinal electric conductivity in the model $\tau \propto v_f(x)$ at $0.04 \leq \mu^* B/\Delta \leq 0.1, kT/\Delta = 0.03, A = 15$.

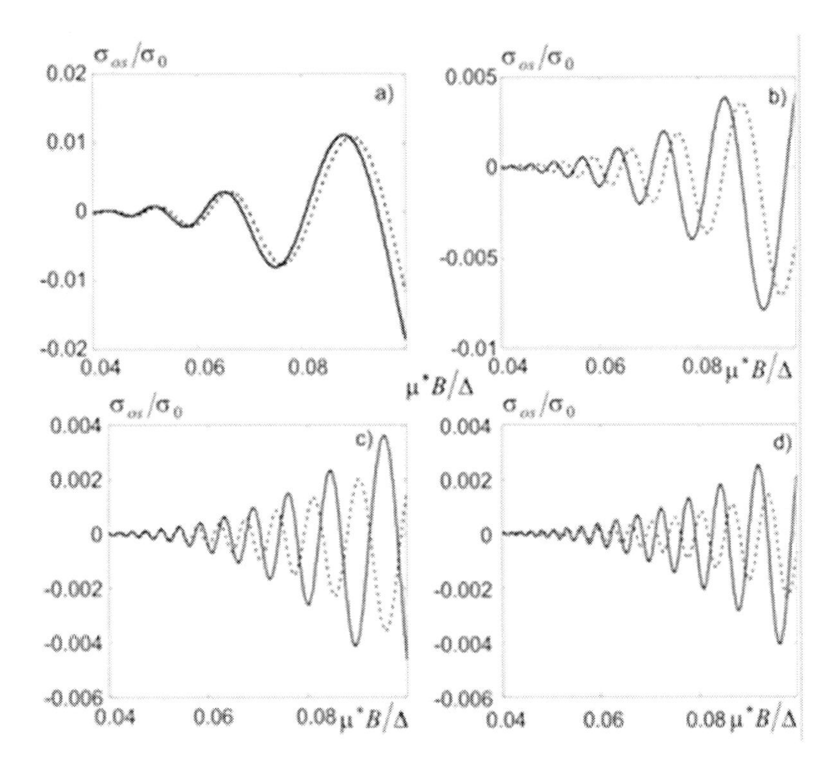

Figure 57. Field dependences of the oscillating part of longitudinal electric conductivity in the model $\tau \propto v_f(x)$ at $0.04 \leq \mu^* B/\Delta \leq 0.1, kT/\Delta = 0.03, A = 25$.

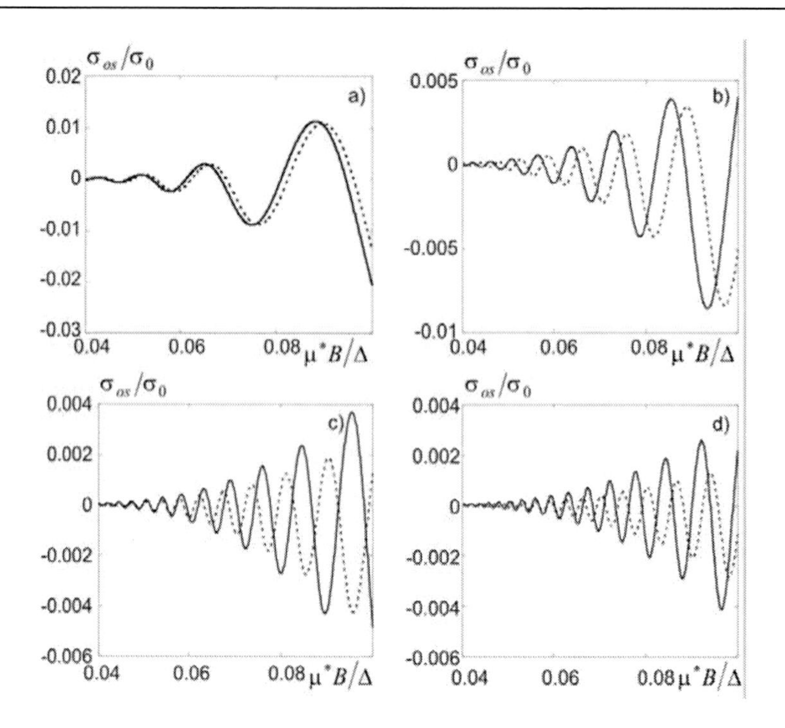

Figure 58. Field dependences of the oscillating part of longitudinal electric conductivity in the model $\tau \propto v_f(x)$ at $0.04 \le \mu^* B/\Delta \le 0.1, kT/\Delta = 0.03, A = 10^5$.

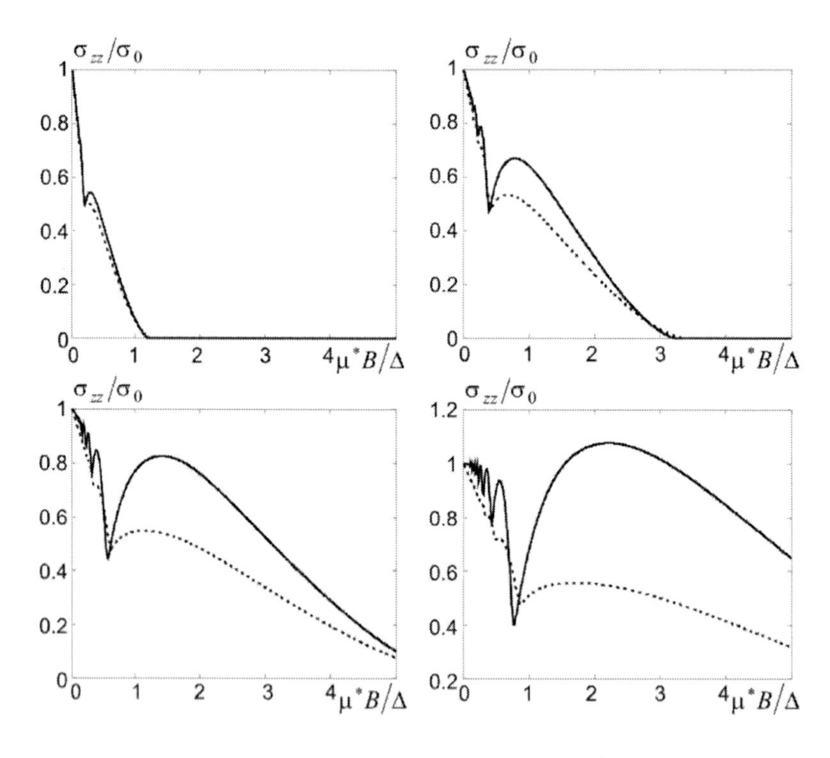

Figure 59. Field dependences of the total longitudinal electric conductivity in the model $\tau \propto v_f(x)$ at $0 \le \mu^* B/\Delta \le 5, kT/\Delta = 0.03, A = 5$.

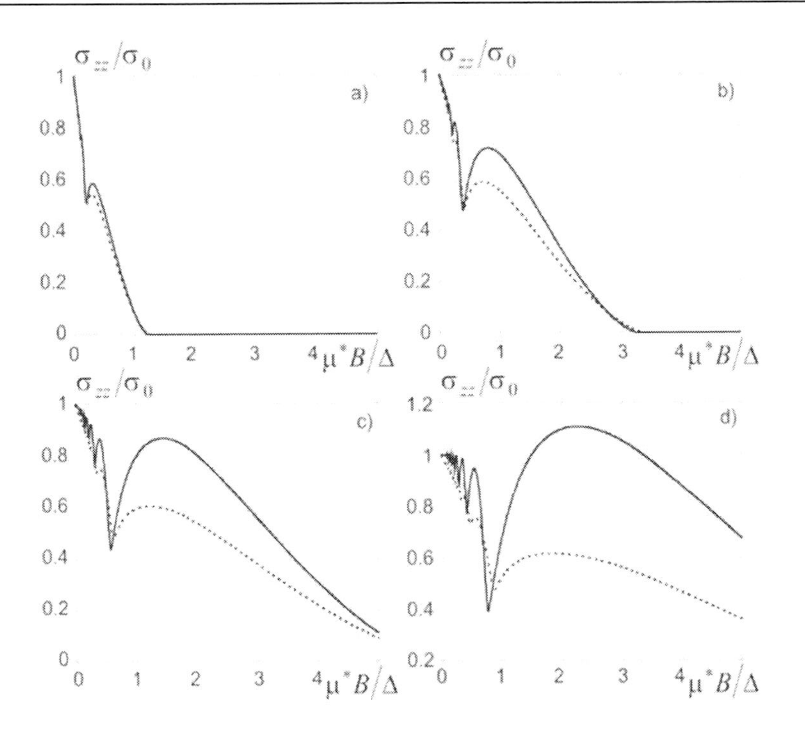

Figure 60. Field dependences of the total longitudinal electric conductivity in the model $\tau \propto v_f(x)$ at $0 \le \mu^* B/\Delta \le 5, kT/\Delta = 0.03, A = 10$.

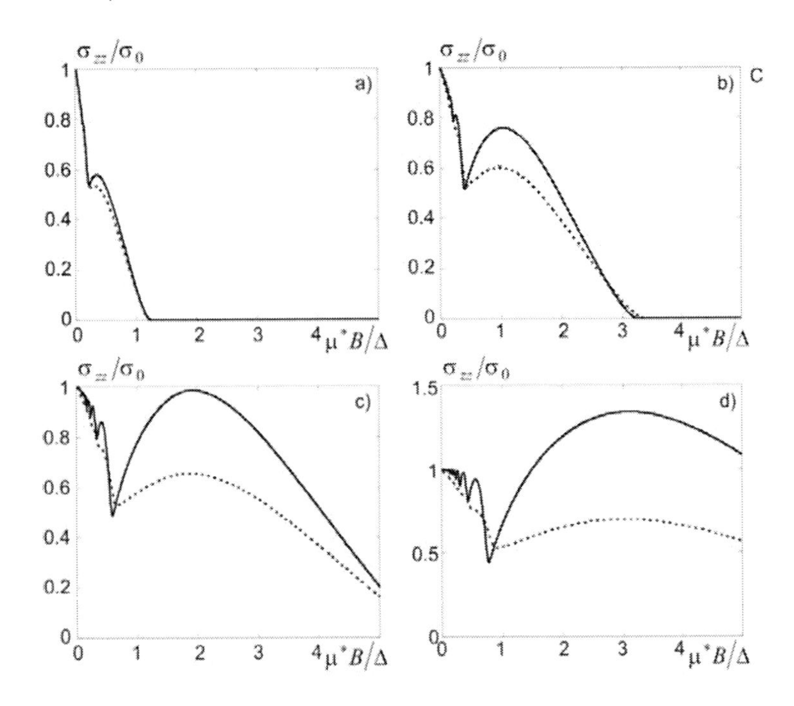

Figure 61. Field dependences of the total longitudinal electric conductivity in the model $\tau \propto v_f(x)$ at $0 \le \mu^* B/\Delta \le 5, kT/\Delta = 0.03, A = 15$.

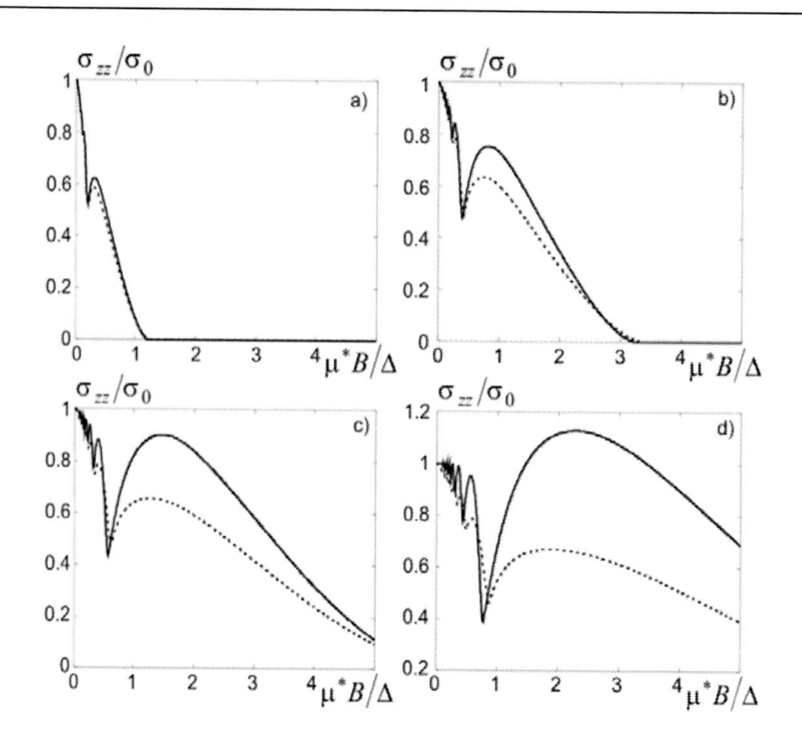

Figure 62. Field dependences of the total longitudinal electric conductivity in the model $\tau \propto v_f(x)$ at $0 \le \mu^* B/\Delta \le 5, kT/\Delta = 0.03, A = 25$.

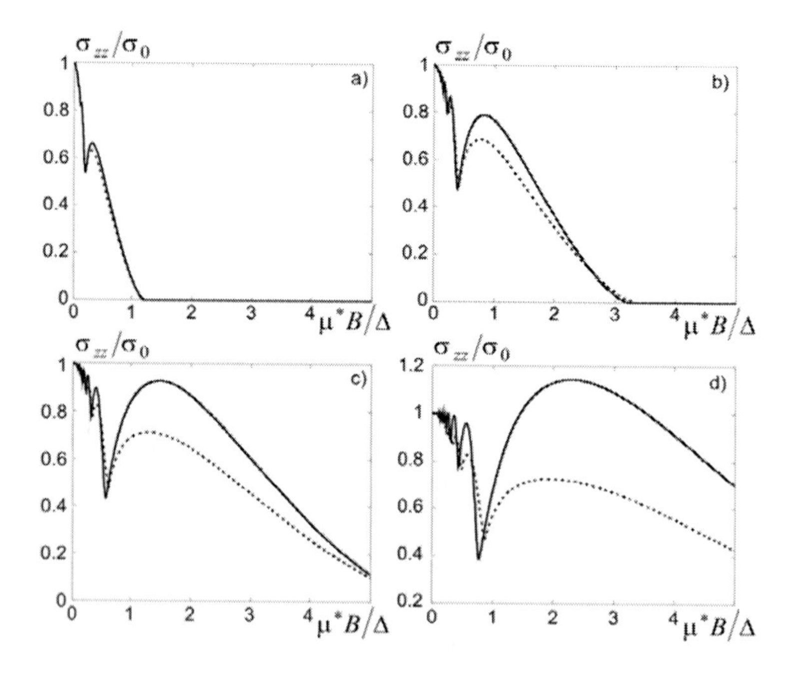

Figure 63. Field dependences of the total longitudinal electric conductivity in the model $\tau \propto v_f(x)$ at $0 \le \mu^* B/\Delta \le 5, kT/\Delta = 0.03, A = 10^5$.

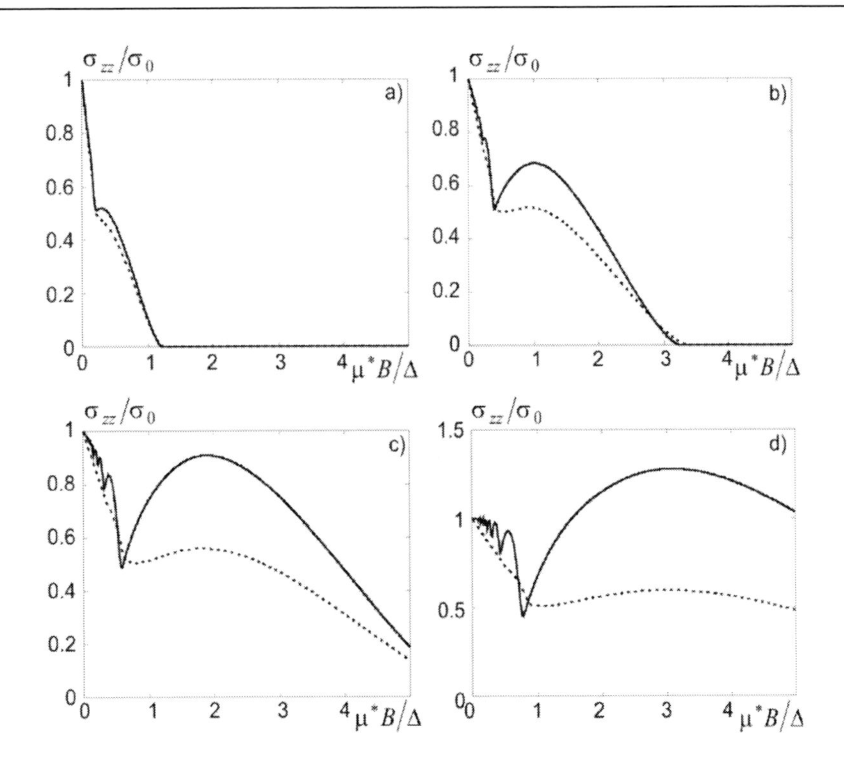

Figure 64. Field dependences of the total longitudinal electric conductivity in the model $\tau \propto v_f(x, B)$ at $0 \leq \mu^* B/\Delta \leq 5, kT/\Delta = 0.03, A = 5$.

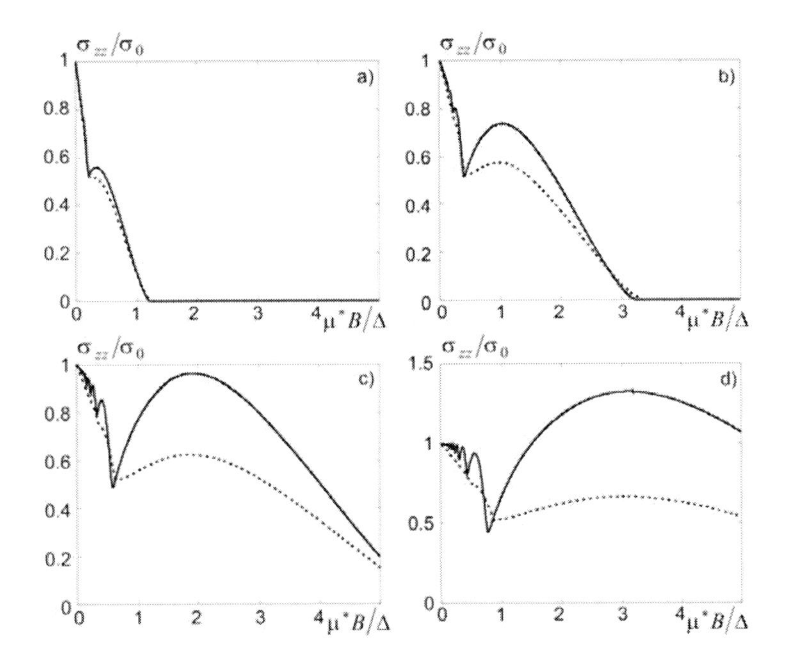

Figure 65. Field dependences of the total longitudinal electric conductivity in the model $\tau \propto v_f(x, B)$ at $0 \leq \mu^* B/\Delta \leq 5, kT/\Delta = 0.03, A = 10$.

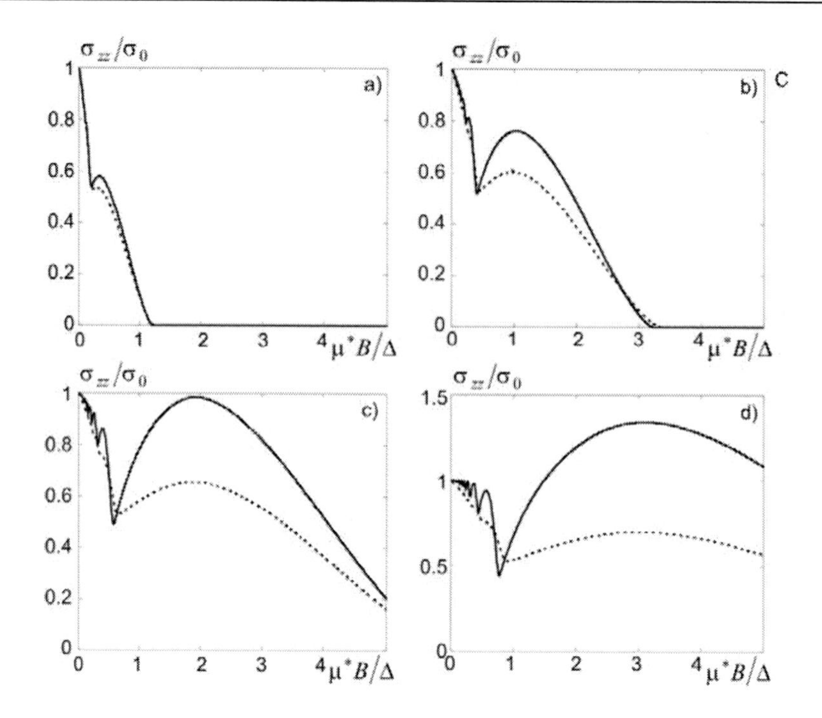

Figure 66. Field dependences of the total longitudinal electric conductivity in the model $\tau \propto v_f(x,B)$ at $0 \le \mu^* B/\Delta \le 5, kT/\Delta = 0.03, A = 15$.

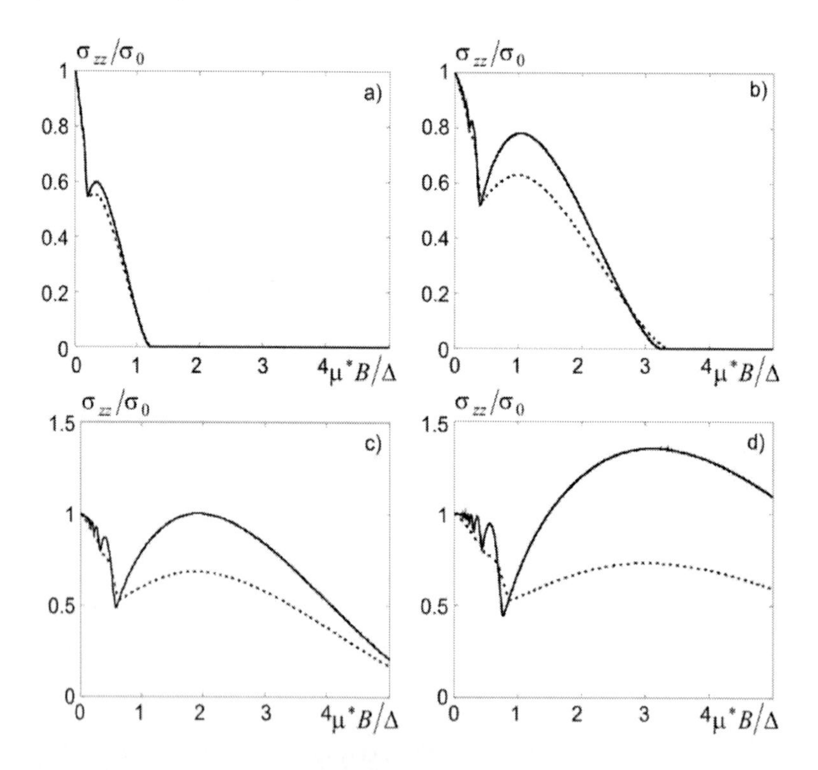

Figure 67. Field dependences of the total longitudinal electric conductivity in the model $\tau \propto v_f(x,B)$ at $0 \le \mu^* B/\Delta \le 5, kT/\Delta = 0.03, A = 25$.

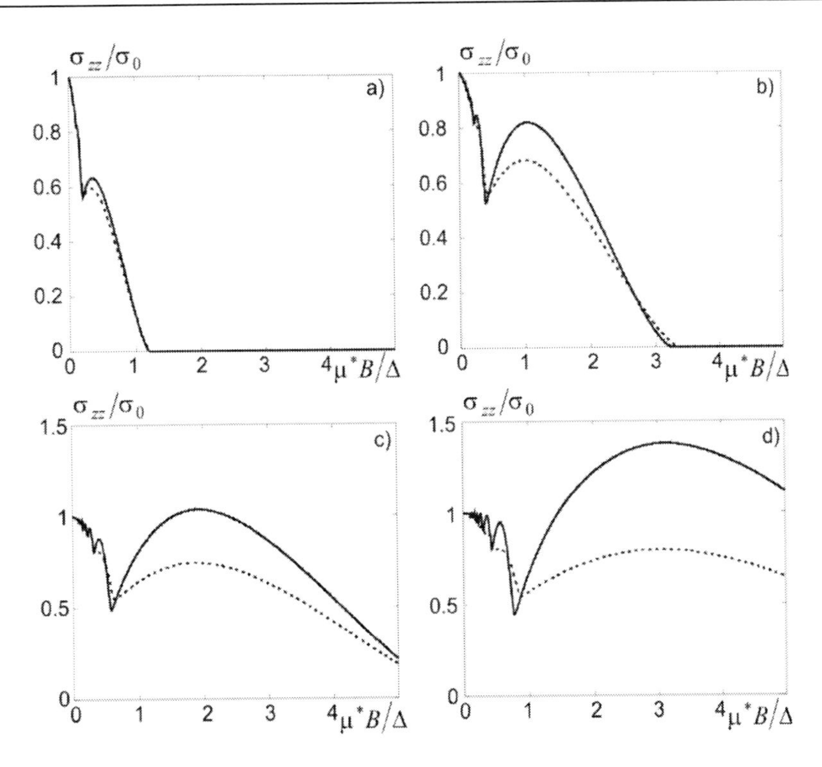

Figure 68. Field dependences of the total longitudinal electric conductivity in the model $\tau \propto v_f(x,B)$ at $0 \le \mu^* B/\Delta \le 5, kT/\Delta = 0.03, A = 10^5$.

The figures demonstrate that unlike the case of constant mean free path approximation, the oscillations of longitudinal electric conductivity of a layered crystal in quasi-classical magnetic fields with relaxation time proportional to total electron velocity on FS are of a variable polarity. The relative contribution of oscillations in this case is 0.4-1%. As before, the layered structure effects become apparent in phase delay of SdH oscillations and greater value of their relative contribution. The greatest difference between the solid and dashed curves takes place in the case of a transient FS, i.e. at $\zeta_0/\Delta = 2$.

In stronger magnetic fields the layered structure effects manifest themselves in a drastically nonmonotonous magnetic field dependence of total longitudinal electric conductivity, this nonmonotonicity being much more intense than in the case of constant mean free path model. As long as the model $\tau \propto v_f(x)$ corresponds to prevalence of charge carrier scattering on the deformation potential of acoustic phonons, the degree of similarity of really observed field dependences of total longitudinal electric conductivity to those shown in Figures 53-67 might serve a criterion of purity and perfection of crystals. In the model $\tau \propto v_f(x)$, local minimum and maximum of total longitudinal electric conductivity take place both for a real layered crystal and in the effective mass approximation, but in the case of a real layered crystal they are more pronounced. The maximum of total longitudinal electric conductivity in the models $\tau \propto v_f(x)$ and $\tau \propto v_f(x,B)$ in the case of a real layered crystal is from 0.65 to 1.25 of electric conductivity value in the absence of a magnetic field. However, in the effective mass approximation this maximum is from 0.6 to 0.75 of electric conductivity value in the absence of a magnetic field. Account of the magnetic field dependence of relaxation time through chemical potential dependence thereupon leads to some changes in the shape of the curves reflecting field dependence of total longitudinal electric conductivity on a magnetic field and to some increase in the relative value of local maximum of total longitudinal electric conductivity. The increase in anisotropy parameter A in the model of relaxation time in proportion to total velocity of electron on FS, produces a relatively weak effect on the character of field dependence of total longitudinal electric conductivity of a layered crystal, though affects considerably its absolute value.

Consider now the Kapitsa effect within the framework of the model $\tau \propto v_f(x)$. Expanding the expressions (20.2) and (20.4) as a Taylor series, we obtain that in the case of a real layered crystal

$$\frac{\Delta\rho(B)}{\rho} = \frac{2\gamma_0 - \gamma_0^2}{\int_0^{\gamma_0} \sqrt{\left[A(\gamma_0 - y) + 2y - y^2\right]\left(2y - y^2\right)}dy} \frac{\mu^* B}{\Delta}. \tag{20.6}$$

In the effective mass approximation, formula (160) takes on the form:

$$\frac{\Delta\rho(B)}{\rho} = \frac{2\gamma_{0em}}{\int_0^{\gamma_0} \sqrt{\left[A(\gamma_{0em} - y) + 2y\right]2y}dy} \frac{\mu^* B}{\Delta}. \tag{20.7}$$

Therefore, for instance, at $A = 5, \mu^* B/\Delta = 0.1$ and $\gamma_0 = 0.5;1;1.5;2$, respectively, the value $\Delta\rho(B)/\rho$ is 11.8; 5.7; 2.4 and 0%, respectively, for a real layered crystal and 15.1; 10.6; 8.5 and 7.2%, respectively, in the effective mass approximation. With $A = 10$ and other problem parameters specified above, $\Delta\rho(B)/\rho$ is 9; 4.3; 1.8 and 0%, respectively, for a real layered crystal and 11.7; 8.2; 6.6 and 5.6%, respectively, in the effective mass approximation. With $A = 15$ and other problem parameters specified above, $\Delta\rho(B)/\rho$ is 7.6; 3.6.3; 1.5 and 0%, respectively, for a real layered crystal and 9.9; 6.9; 5.6 and 4.7%, respectively, in the effective mass approximation. With $A = 25$ and other problem parameters specified above, $\Delta\rho(B)/\rho$ is 6; 2.9; 1.2 and 0%, respectively, for a real layered crystal and 7.9; 5.5; 4.5 and 3.8%, respectively, in the effective mass approximation. With $A = 10^5$ and other problem parameters specified above, $\Delta\rho(B)/\rho$ is 0.101; 0.047; 0.019 and 0%, respectively, for a real layered crystal and 0.133; 0.093; 0.075 and 0.064%, respectively, in the effective mass approximation. Hence, it is seen that the Kapitsa coefficient relating the magnetoresistance to magnetic field induction decreases both with increasing the degree of filling a narrow conduction miniband γ_0, and with growing anisotropy parameter A. In so doing, in the effective mass approximation the Kapitsa coefficient turns to be larger than for a real layered crystal. Moreover, for a real layered crystal in the case of a transient FS, i.e. at $\gamma_0 = 2$, the

Kapitsa coefficient turns to zero, while in the effective mass approximation this does not occur. Thus, layered structure effects with the open FS become apparent in the reduction of the Kapitsa coefficient. In the models $\tau \propto v_f(x)$ and $\tau \propto v_f(x, B)$ used for the description of charge carrier scattering on the deformation potential of acoustic phonons the Kapitsa coefficient turns out to be considerably higher than in the model of constant mean free path used for the description of charge carrier scattering on ionized impurities. And as long as it is exactly electron-phonon interaction that is responsible for crystal transition to superconducting state, the degree of intensity of the longitudinal Kapitsa effect, in a sense, on the one hand, can serve the indicator of crystal purity and perfection, on the other hand – the feasibility of superconducting transition in it. As to the transverse Kapitsa effect, it has no similar relation to superconductivity.

Let us now estimate the longitudinal conductivity σ_0 in the absence of a magnetic field for some specific cases. As a rule, it is considered that in sharply anisotropic layered crystals small phonon wave vectors form no small angle with a long axis of FS directed along k_z [36]. So, the relaxation time is strongly anisotropic, by virtue of which

$$\tau_{ap} = \frac{h^3 \rho s^4 \Delta a}{7\pi^2 \Xi^2 (kT)^3 \Gamma(3)\zeta(3)}. \qquad (20.8)$$

Therefore, taking into account the relation between A and Δ, in the case of a real layered crystal we finally get:

$$\sigma_0 = \frac{2e^2 \rho s^4 h^5}{7\Xi^2 (kT)^3 \Gamma(3)\zeta(3)\pi^6 A^3 m^{*2} a^4} \int_0^{\gamma_0} \sqrt{\left[A(\gamma_0 - y) + 2y - y^2\right]\left(2y - y^2\right)}\, dy. \qquad (20.9)$$

In the effective mass approximation formula (163) takes on the form:

$$\sigma_0 = \frac{2e^2 \rho s^4 h^5}{7\Xi^2 (kT)^3 \Gamma(3)\zeta(3)\pi^6 A^3 m^{*2} a^4} \int_0^{\gamma_{0em}} \sqrt{2\left[A(\gamma_{0em} - y) + 2y\right] y}\, dy. \qquad (20.10)$$

Therefore, for instance, at $\rho = 5000$ kg/m^3, $s = 5000$ m/s, $\Xi = 10$ eV, $T = 3$ K, $m^* = 0.01 m_0$, $a = 1$ nm, $A = 5$ and $\gamma_0 = 0.5;1;1.5;2$, respectively, the value σ_0 is $2.512 \cdot 10^{11}$; $8.795 \cdot 10^{11}$; $1.689 \cdot 10^{12}$ and $2.458 \cdot 10^{12}$ S/m, respectively, for a real layered crystal, and $2.923 \cdot 10^{11}$; $1.220 \cdot 10^{12}$; $2.898 \cdot 10^{12}$ and $5.613 \cdot 10^{12}$ S/m, respectively, in the effective mass approximation. With $A = 10$ and other problem parameters specified above, σ_0 is $4.010 \cdot 10^{10}$; $1.430 \cdot 10^{11}$; $2.806 \cdot 10^{11}$ and $4.185 \cdot 10^{11}$S/m, respectively, for a real layered crystal and $4.594 \cdot 10^{10}$; $1.918 \cdot 10^{11}$; $4.554 \cdot 10^{11}$ and $8/823 \cdot 10^{11}$S/m, respectively, in the effective mass approximation. With $A = 15$ and other problem parameters specified above, σ_0 is $1.396 \cdot 10^{10}$; $5.020 \cdot 10^{10}$; $9.949 \cdot 10^{10}$ and $1.499 \cdot 10^{11}$S/m, respectively, for a real layered crystal and $1.587 \cdot 10^{10}$; $6.626 \cdot 10^{10}$; $1.574 \cdot 10^{11}$ and $3.049 \cdot 10^{11}$S/m, respectively, in the effective mass approximation. With $A = 25$ and other problem parameters specified above, σ_0 is $3.749 \cdot 10^9$; $1.360 \cdot 10^{10}$; $2.719 \cdot 10^{10}$ and $4.134 \cdot 10^{10}$S/m, respectively, for a real layered crystal and $4.231 \cdot 10^9$; $1.766 \cdot 10^{10}$; $4.194 \cdot 10^{10}$ and $8.125 \cdot 10^{10}$S/m, respectively, in the effective mass approximation. With $A = 10^5$ and other problem parameters specified above, σ_0 is 3.457; 12.764; 25.995 and 40.175S/m, respectively, for a real layered crystal and 3.833; 15.999; 38.001 and 73.616S/m, respectively, in the effective mass approximation.

Again we see that with increasing the degree of filling a narrow conductivity miniband, the longitudinal electric conductivity in the absence of a magnetic field increases, and in the effective mass approximation it is larger than for a real layered crystal. The difference between the longitudinal conductivities in the absence of a magnetic field determined for a real layered crystal and in the effective mass approximation, as could be expected, increases with increasing the degree of filling a narrow conductivity miniband. As to its absolute values at helium temperatures, they correspond to sufficiently pure and perfect crystals, in which at these temperatures charge carrier scattering on acoustic phonons is dominant. However, with high degrees of effective mass anisotropy even in this case the absolute values of longitudinal conductivity are low.

Let us now establish the asymptotic laws that are followed by longitudinal electric conductivity of a layered crystal within the framework of given relaxation time model. In the ultra-quantum limit within the framework of model $\tau \propto v_f(x, B)$ the relaxation time with a strong anisotropy of FS is given by the expression:

$$\tau = \frac{h^4 \rho s^4}{14\pi^3 \Xi^2 (kT)^3 m^* \Gamma(3)\zeta(3)} \sqrt{2\mu^* B m^*}\,, \tag{20.11}$$

i.e. it is constant with respect to quantum numbers. Hence, formula (99) implies the following asymptotic law of longitudinal electric conductivity variation with a magnetic field in the ultra-quantum limit:

$$\sigma_{zz} = \frac{1.488 \cdot 10^{-9} e^2 \rho s^4 h^{10} f^3(\gamma_0)}{(\mu^* B)^{3/2} \Xi^2 (kT)^4 A^5 m^{*9/2} a^9}\,. \tag{20.12}$$

At $B = 60\,\text{T}$ and $a = 1\,\text{nm}$ this law is valid only if $A \geq 44$. Hence, with $A = 10^5$, $\gamma_0 = 0.5;1;1.5;2$, $B = 60\,\text{T}$ and other problem parameters specified above, we get a reduction of longitudinal electric conductivity by 5 to 7 orders of magnitude as compared to its value in the absence of a magnetic field. So, we see that this law does not involve physically incorrect consequences and the only logical question would concern drift approximation applicability with low electric conductivity values. However, elucidation of this problem is beyond the scope of this book. Therefore, in this model of relaxation time in the ultra-quantum limit at low temperatures we obtain the law $\rho_{zz} \propto T^4 B^{3/2}$.

Let us now consider the longitudinal conductivity of a layered crystal with scattering on acoustic phonons in the framework of the same approach which was used in [37] when considering scattering on the potential of random charged impurities. In this case the relaxation time for a closed FS in the ultra-quantum limit should be considered equal to double value of (165). Hence, based on the general formula (16.4) and considering the Landau subband with the number $n=0$ as partially filled, for a longitudinal crystal conductivity we get:

$$\sigma_{zz} = \frac{5.951 \cdot 10^{-9} e^2 \rho s^4 h^{10} f^3(\gamma_0)}{(kT)^3 \Xi^2 (\mu^* B)^{5/2} A^5 m^{*9/2} a^9}. \tag{20.13}$$

In this case, with $A=10^5$, $\gamma_0 = 0.5;1;1.5;2$, $B = 60$ T and other problem parameters specified above, we get a reduction of longitudinal electric conductivity by 8 to 9 orders of magnitude as compared to its value in the absence of a magnetic field. In this case we get the law of the form $\rho_{zz} \propto T^3 B^{5/2}$.

If, however, we consider the Landau subband with the number $n=0$ as totaly filled, for crystal longitudinal conductivity we get:

$$\sigma_{zz} = \frac{4.44 \cdot 10^{-11} e^2 \rho s^4 h^{12} f^4(\gamma_0)}{\Xi^2 (kT)^3 m^{*11/2} a^{11} A^6 (\mu^* B)^{7/2}}. \tag{20.14}$$

In this case, with $A=10^5$, $\gamma_0 = 0.5;1;1.5;2$, $B = 60$ T and other problem parameters specified above we get a reduction of longitudinal electric conductivity by 12 to 14 orders of magnitude as compared to its value in the absence of a magnetic field. In this case we get the law of the form $\rho_{zz} \propto T^3 B^{7/2}$.

Chapter 21

LONGITUDINAL CONDUCTIVITY OF A LAYERED CRYSTAL IN A QUANTIZING MAGNETIC FIELD IN A MODEL OF RELAXATION TIME PROPORTIONAL TO LONGITUDINAL VELOCITY

We now consider longitudinal conductivity of a layered crystal in a model of relaxation time proportional to electron longitudinal velocity modulus. In this model, for a real layered crystal the relaxation time is

$$\tau(x)=C\left|v_z(x)\right|=\frac{2\pi Ca}{h}\Delta\left|\sin x\right|. \tag{21.1}$$

In this formula, C is a constant with respect to quantum numbers whose explicit expressions are given below.

Therefore, formulae (16.4) and (16.5) in this model in the case of a closed FS for a real layered crystal acquire the form [26, 41-43]:

$$\sigma_0=\frac{32\pi^3e^2m^*a^2C\Delta^3}{h^5}\left[(\gamma-b)^2-\frac{(\gamma-b)^3}{3}\right], \tag{21.2}$$

$$\begin{aligned}\sigma_{os}=\frac{128\pi^3e^2m^*a^2C\Delta^3}{h^5}\sum_{l=1}^{\infty}(-1)^l f_l^\sigma(\pi l)^{-3}\left[-\pi b^2l\gamma+\pi b^3l+b^3\sin\left(\pi lb^{-1}\gamma\right)-\right.\\ \left.-\pi b^2l\cos\left(\pi lb^{-1}\gamma\right)+(-1)^l\pi b^2l\right]\end{aligned}. \tag{21.3}$$

In the effective mass approximation formulae (20.2) and (20.3) take the form [26, 41, 42]:

$$\sigma_0=\frac{32\pi^3e^2m^*a^2C\Delta^3}{h^5}(\gamma-b)^2, \tag{21.4}$$

$$\sigma_{os} = \frac{128\pi^3 e^2 m^* a^2 C\Delta^3 b^2}{h^5} \sum_{l=1}^{\infty} (-1)^l f_l^{\sigma} (\pi l)^{-2} \left[(-1)^l - \cos\left(\pi l b^{-1} \gamma\right) \right]. \quad (21.5)$$

According to traditional Lifshits-Kosevich theory, only a trigonometric term is retained in formula (21.5).

The results of longitudinal electric conductivity calculations by formulae (21.2)–(21.5) are given in Figures 69 – 70.

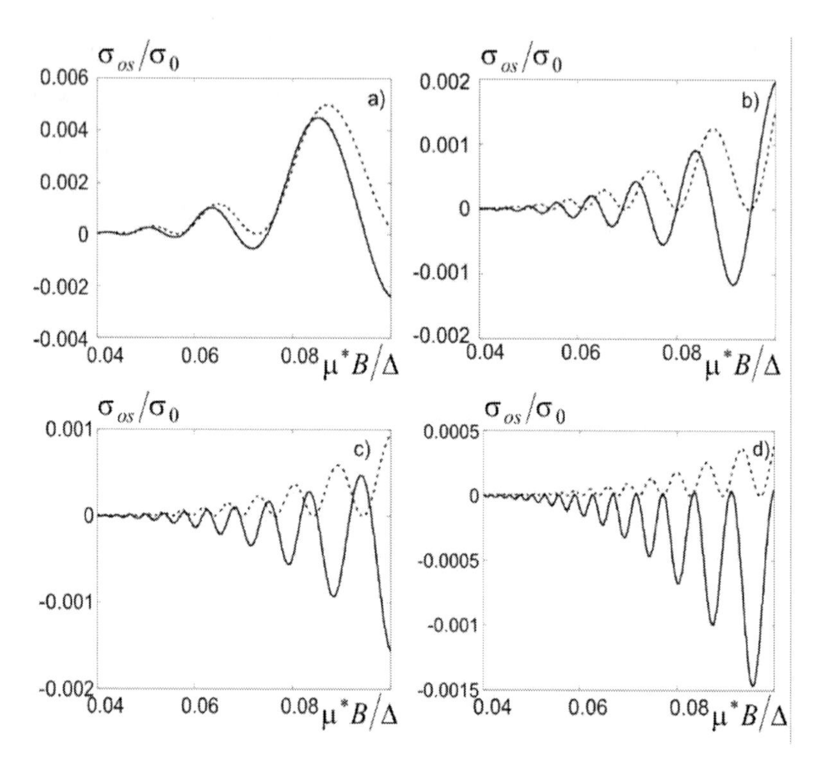

Figure 69. Field dependence of the oscillating part of longitudinal conductivity of a layered crystal in the model $\tau \propto v_z(x)$ at $kT/\Delta = 0.03$ and $0.04 \le \mu^* B/\Delta \le 0.1$.

The figures demonstrate that in quasi-classical magnetic fields the oscillations of longitudinal electric conductivity are monoperiodic, and their frequency conforms to the only extreme, namely, minimum section of FS by plane $k_z = 0$. Like in other relaxation time models discussed above, the layered structure effects in quasi-classical magnetic fields become apparent in phase delay of oscillations and their increased relative contribution. With increase in the ratio ζ_0/Δ from 0.5 to 2, the relative contribution of oscillations to full longitudinal electric conductivity of a real layered crystal falls from 0.6 to 0.15%, and in the effective mass approximation – from 0.5 to 0.04%. As could be expected, with increasing ratio ζ_0/Δ, the oscillation frequencies of electric conductivity and the difference in solid and dashed curves increase. The greatest difference occurs at ζ_0/Δ=2, which is due to the absence of planes tangential to FS at $k_z = \pm\pi/a$ in the case of a real layered crystal.

Moreover, the figures, just as formulae (20.3) and (20.5), show that even in quasi-classical magnetic fields in the model of relaxation time under consideration, like in the models considered above, of essential importance is a monotonous addition which cannot be correctly taken into account within the framework of conventional Lifshits-Kosevich theory for which finite extension of FS along the direction of a magnetic field and its corrugation pattern are of minor importance.

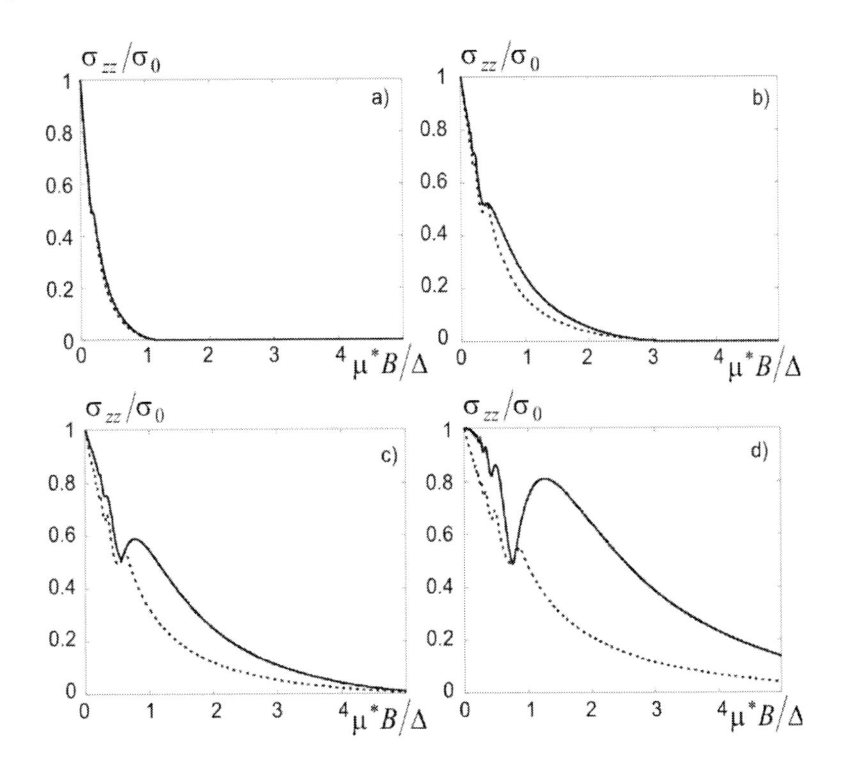

Figure 70. Field dependence of the oscillating part of longitudinal electric conductivity of a layered crystal in the model $\tau \propto v_z(x)$ at $kT/\Delta = 0.03$ and $0 \leq \mu^* B/\Delta \leq 5$.

In stronger magnetic fields there is an optimal range of magnetic field inductions, wherein the layered structure effects are most pronounced and manifest themselves in maximum longitudinal electric conductivity and its weaker decrease with growing magnetic field induction. The above maximum in model $\tau \propto v_z(x)$ also exists in the effective mass approximation, but with ζ_0/Δ equal to 0.5 and 1 it is more pronounced, and at ζ_0/Δ equal to 1.5 and 2 – considerably weaker than for a real layered crystal. Under the above values of ζ_0/Δ the maximum of full longitudinal electric conductivity for a real layered crystal achieves 0.5 – 0.8 of its value in the absence of a magnetic field, whereas in the effective mass approximation it is 0.5–0.55 of the value of full longitudinal electric conductivity in the absence of a magnetic field.

Let us now consider within the framework of this relaxation time model the longitudinal Kapitsa effect. Expanding expressions (20.2) and (20.4) as a Taylor series in b, for a real layered crystal we obtain:

$$\Delta\rho(B)\Big/\rho_0 = \frac{2\gamma_0 - \gamma_0^2}{\gamma_0^2 - (\gamma_0^3/3)}\frac{\mu^* B}{\Delta}, \tag{21.6}$$

and in the effective mass approximation we get:

$$\Delta\rho(B)\Big/\rho_0 = \frac{2}{\gamma_{0em}}\frac{\mu^* B}{\Delta}. \tag{21.7}$$

Therefore, for instance, at $\mu^* B/\Delta = 0.1$ and $\gamma_0 = 0.5;1;1.5;2$ for a real layered crystal we have a relative magnetoresistance equal to 36, 15, 6.67 and 0%, respectively, and in the effective mass approximation 32.3, 19.2, 12.5, 9%, respectively.

We now estimate crystal electric conductivity σ_0 in the absence of a magnetic field. If charge carrier scattering takes place on the deformation potential of acoustic phonons, then, by analogy with formula (20.8) it can be assumed:

$$C = \frac{h^4 \rho s^4}{14\pi^3 \Xi^2 (kT)^3 \Gamma(3)\zeta(3)}. \tag{21.8}$$

Hence, for a real layered crystal we obtain:

$$\sigma_0 = \frac{16 e^2 m^* a^2 \Delta^3 \rho s^4}{7h\Xi^2 (kT)^3 \Gamma(3)\zeta(3)}\left[\gamma_0^2 - \frac{\gamma_0^3}{3}\right], \tag{21.9}$$

and in the effective mass approximation we have:

$$\sigma_0 = \frac{16 e^2 m^* a^2 \Delta^3 \rho s^4 \gamma_{0em}^2}{7h\Xi^2 (kT)^3 \Gamma(3)\zeta(3)}. \tag{21.10}$$

Therefore, for instance, at $m = 0.01m_0$, $a = 1\text{nm}$, $\Delta = 0.01\text{eV}$, $\rho = 5000\text{kg/m}^3$, $s = 5000\text{m/s}$, $\Xi = 10\text{eV}$ and $\zeta_0/\Delta = 0.5;1;1.5;2$, respectively, we get that the values of σ_0 for a real layered crystal make $4.904 \cdot 10^3$, $1.569 \cdot 10^4$, $2.648 \cdot 10^4$, $3.139 \cdot 10^4$S/m, respectively, and in the effective mass approximation – $6.100 \cdot 10^3$, $2.546 \cdot 10^4$, $6.048 \cdot 10^4$, $1.172 \cdot 10^5$ S/m, respectively.

Let us consider another model expression for coefficient C which is based on the Bloch-Gruneisen law [36]. Taking into account that we determine the relaxation time of a longitudinal quasi-pulse, this coefficient may be written as:

$$C = \frac{2h^6 k_f^2 \rho s^6}{31\pi^6 \Gamma(5)\zeta(5)(kT)^5 \Xi^2}. \tag{21.11}$$

In this formula, k_f – extension of the FS along the direction of a magnetic field. Therefore, longitudinal electric conductivity of a real layered crystal in the absence of a magnetic field in this model will make:

$$\sigma_0 = \frac{64 e^2 h m^* \rho s^6 \Delta^3}{31\pi^3 \Gamma(5)\zeta(5)(kT)^5 \Xi^2}\left(\gamma_0^2 - \frac{\gamma_0^3}{3}\right)\arccos^2(1-\gamma_0). \tag{21.12}$$

In the effective mass approximation, formula (180) acquires the form:

$$\sigma_0 = \frac{128 e^2 h m^* \rho s^6 \Delta^3 \gamma_{0em}^3}{31\pi^3 \Gamma(5)\zeta(5)(kT)^5 \Xi^2}. \tag{21.13}$$

Therefore, with the same problem parameters and $\zeta_0/\Delta = 0.5;1;1.5;2$, respectively, we obtain that the values of σ_0 in this model of coefficient C for a real layered crystal are $9.703\cdot10^4$, $6.986\cdot10^5$, $2.096\cdot10^6$, $5.589\cdot10^6$S/m, respectively, and in the effective mass approximation – $1.120\cdot10^5$, $9.556\cdot10^5$, $3.498\cdot10^5$, $9.431\cdot10^6$S/m, respectively.

We now consider scattering on a short-range potential of ionized impurities. In so doing, it should be borne in mind that electrically active impurities possess a short-range potential only at sufficiently low temperatures. According to [36], coefficient C at scattering of charge carriers on such potential in a strong quantizing magnetic field is determined by the expression:

$$C = \left(\frac{\varepsilon\varepsilon_0}{Ze^2 a_D^2}\right)^2 \frac{h^3}{8\pi^3 N_i eB}, \tag{21.14}$$

In this formula, ε – relative dielectric constant of crystal, ε_0 – dielectric constant of vacuum, Z – impurity charge, a_D – Debye shielding length, N_i – bulk concentration of impurities, the rest of designations are explained above or commonly accepted. As long as neither shielding length nor concentration of scattering centres is known for certain, we proceed as follows. Introduce the mean free path of charge carriers in the absence of a magnetic field in conformity with the ratio:

$$l \equiv Na = \frac{1}{N_i \sigma_{tr}}, \tag{21.15}$$

and transport cross-section σ_{tr} of charge carrier scattering on the short-range impurity potential in conformity with the ratio:

$$\sigma_{tr} = 128\pi^5 \left(\frac{Ze^2 m_{es}^* a_D^2}{h^2 \varepsilon\varepsilon_0} \right)^2 . \tag{21.16}$$

Then for coefficient C in a quantizing magnetic field we obtain the following expression:

$$C = \frac{h^3 Na}{4\pi^2 eB\gamma_0^2 \Delta^2} \left(\frac{12\pi^2 f(\gamma_0) m^* \Delta}{ah^2} \right)^{4/3} . \tag{21.17}$$

However, at $B = 0$ this formula is physically incorrect, and to determine σ_0, we must extend formula (21.17) to the case of weak magnetic fields or their absence. We proceed as follows. We will consider that depending on magnetic field induction the mode of charge carrier scattering on the short-range impurity potential is determined by the larger of the two quantities: equivalent sphere radius k_0 or magnetic pulse $\sqrt{2\pi eB/h}$. Therefore, coefficient C in a weak magnetic field or in its absence takes the form:

$$C = \frac{h^2 Na}{2\pi\gamma_0^2 \Delta^2} \left(\frac{12\pi^2 f(\gamma_0) m^* \Delta}{ah^2} \right)^{2/3} . \tag{21.18}$$

To determine the minimum permissible value of N in these conditions, we estimate the average value $\omega_c\tau$ on the FS, with regard to formula (186). This value is equal to:

$$\langle \omega_c \tau \rangle_{FS} = \frac{2\pi bNa^2}{\gamma_0 \arccos(1-\gamma_0)} \left(\frac{12\pi^2 f(\gamma_0) m^* \Delta}{ah^2} \right)^{2/3} . \tag{21.19}$$

Permissible minimum of N is taken to be such a value whereby $\langle \omega_c \tau \rangle_{FS} = 25$ at $b = 0.04$ (it corresponds to the lower limit of magnetic field range under study). Analysis shows that this condition is reliably fulfilled for all γ_0 both for a real layered crystal, and in the effective mass approximation at $N = 10000$.

Then for σ_0 in the case of a real layered crystal we get the following expression:

$$\sigma_0 = \frac{16\pi^2 e^2 m^* a^3 N\Delta}{h^3} \left(\frac{12\pi^2 f(\gamma_0) m^* \Delta}{ah^2} \right)^{2/3} \left(1 - \frac{\gamma_0}{3} \right) . \tag{21.20}$$

In the effective mass approximation formula (188) acquires the form:

$$\sigma_0 = \frac{16\pi^2 e^2 m^* a^3 N\Delta}{h^3}\left(\frac{12\pi^2 f(\gamma_0) m^* \Delta}{ah^2}\right)^{2/3}. \tag{21.21}$$

So, for instance, with previously specified problem parameters, $N = 10000$ and $\gamma_0 = 0.5;1;1.5;2$, respectively, for σ_0 in the case of a real layered crystal we get the values 5800; 9490; 10970; 10170S/m, respectively, and the effective mass approximation we get the values 6960; 14230; 21930; 30520S/m, respectively.

Let us now establish possible asymptotic laws that can be followed by the longitudinal electric conductivity of a layered crystal in the model $\tau \propto v_z$ in the ultra-quantum limit. Consider this issue in two ways: without regard to effect of magnetic field on charge carrier scattering and with account of this effect. In the former case we must consider that the relaxation time of charge carriers in a strong quantizing field is the same as in the absence of a magnetic field. Then, proceeding similarly to other relaxation time models considered above, in the case of charge carrier scattering on the deformation potential of acoustic phonons, using formula (21.8) for C, we get:

$$\sigma_{zz} = 7.427 \cdot 10^{-3} \frac{e^2 m^* a^2 \rho s^4 \Delta^7 f^4(\gamma_0)}{h\Xi^2 (kT)^4 (\mu^* B)^3}. \tag{21.22}$$

Therefore, with previously specified problem parameters, $B = 60\,\mathrm{T}$, and $\gamma_0 = 0.5;1;1.5;2$, respectively, according to this formula we get $\sigma_{zz} = 2.321 \cdot 10^{-3};0.169;2.262;16.445$ S/m, respectively. And it is 3 to 6 orders of magnitude smaller than in the absence of a magnetic field. Hence, in this case we have the law $\rho_{zz} \propto T^4 B^3$.

Consider now the law which results if formula (21.11) is applied to coefficient C. Then we have the following expression for the longitudinal electric conductivity of a real layered crystal in the ultra-quantum limit:

$$\sigma_{zz} = 2.09 \cdot 10^{-5} \frac{e^2 h m^* \rho s^6 \Delta^7 \arccos^2(1-\gamma_0) f^4(\gamma_0)}{\Xi^2 (\mu^* B)^3 (kT)^6}. \tag{21.23}$$

In the effective mass approximation formula (21.23) acquires the form:

$$\sigma_{zz} = 4.18 \cdot 10^{-5} \frac{e^2 h m^* \rho s^6 \Delta^7 \gamma_{0em} f^4(\gamma_0)}{\Xi^2 (\mu^* B)^3 (kT)^6}. \tag{21.24}$$

Therefore, for a real layered crystal with previously specified problem parameters, $B = 60\,\mathrm{T}$, and $\gamma_0 = 0.5;1;1.5;2$, respectively, according to this formula we get

$\sigma_{zz} = 0.046;7.516;179.023;2.928\cdot10^3$ S/m, respectively. In the effective mass approximation we have $\sigma_{zz} = 0.043;6.336;130.832;1.324\cdot10^3$ S/m, respectively. And it is again 3 to 6 orders of magnitude smaller than in the absence of a magnetic field. Hence, in this case we get the law $\rho_{zz} \propto T^6B^3$.

Now consider scattering on a short-range potential of ionized impurities. In this case, using for C formula (21.18) yields:

$$\sigma_{zz} = 29.748\frac{e^2m^*a^3\Delta^5f^4(\gamma_0)}{h^3\gamma_0^2kT(\mu^*B)^3}\left(\frac{f(\gamma_0)m^*\Delta}{ah^2}\right)^{2/3}N. \quad (21.25)$$

In the effective mass approximation formula (21.26) acquires the form:

$$\sigma_{zz} = 29.748\frac{e^2m^*a^3\Delta^5f^4(\gamma_0)}{h^3\gamma_{0em}^2kT(\mu^*B)^3}\left(\frac{f(\gamma_0)m^*\Delta}{ah^2}\right)^{2/3}N. \quad (21.26)$$

Hence, for a real layered crystal with previously specified problem parameters, $B = 60$ T, and $\gamma_0 = 0.5;1;1.5;2$, respectively, according to this formula we get $\sigma_{zz} = 2.478\cdot10^{-3};0.102;0.937;5.332$ S/m, respectively. In the effective mass approximation we get $\sigma_{zz} = 2.651\cdot10^{-3};0.094;0.820;4.286$ S/m, respectively. And it is 3 to 6 orders of magnitude smaller than in the absence of a magnetic field. In this case we get the law $\rho_{zz} \propto TB^3$.

In conclusion, we consider for all the cases the effect of a magnetic field on scattering. We proceed similarly to constant relaxation time model in two approximations: considering the only Landau subband as partially filled and considering it as completely filled.

In this case, charge carrier scattering on the deformation potential of acoustic phonons shall be considered spontaneous [36], so coefficient C shall be written as:

$$C = \frac{3h^4\rho s^4}{8\pi^3\Xi^2(kT)^3}. \quad (21.27)$$

So, if we consider the only Landau subband as partially filled, then:

$$\sigma_{zz} = \frac{3e^2m^*a^2\rho s^4\Delta^7f^4(\gamma_0)}{16h\Xi^2(kT)^3(\mu^*B)^4}. \quad (21.28)$$

Hence, with previously specified problem parameters and $\gamma_0 = 0.5;1;1.5;2$ we get both for a real layered crystal and in the effective mass approximation

$\sigma_{zz} = 4.359\cdot10^{-5};3.170\cdot10^{-3};0.042;0.309$ S/m, respectively. And it is 4 to 8 orders of magnitude smaller than in the absence of a magnetic field. In this case we get the law $\rho_{zz} \propto T^3 B^4$.

But if we consider the only Landau subband as completely filled, we obtain:

$$\sigma_{zz} = \frac{e^2 m^* a^2 \rho s^4 \Delta^9 f^6(\gamma_0)}{256 h \Xi^2 (kT)^3 (\mu^* B)^6}. \tag{21.29}$$

In this case with previously specified problem parameters and $\gamma_0 = 0.5;1;1.5;2$, we get both for a real layered crystal and in the effective mass approximation $\sigma_{zz} = 8.799\cdot10^{-11};5.458\cdot10^{-8};2.677\cdot10^{-6};5.247\cdot10^{-5}$ S/m, respectively. And it is 9 to 14 orders of magnitude smaller than in the absence of a magnetic field. In this case we have the law $\rho_{zz} \propto T^3 B^6$.

Consider now charge carrier scattering on ionized impurities. In this case coefficient C is equal to double value (21.17). Therefore, considering the only Landau subband as partially filled, for a real layered crystal we obtain:

$$\sigma_{zz} = \frac{\pi e m^* a^3 \Delta^5 f^4(\gamma_0)}{4h^2 B \gamma_0^2 (\mu^* B)^4} \left(\frac{12\pi^2 f(\gamma_0) m^* \Delta}{a h^2} \right)^{4/3} N. \tag{21.30}$$

In the effective mass approximation this formula acquires the form:

$$\sigma_{zz} = \frac{\pi e m^* a^3 \Delta^5 f^4(\gamma_0)}{4h^2 B \gamma_{0em}^2 (\mu^* B)^4} \left(\frac{12\pi^2 f(\gamma_0) m^* \Delta}{a h^2} \right)^{4/3} N. \tag{21.31}$$

According to these formulae, with previously specified problem parameters and $\gamma_0 = 0.5;1;1.5;2$ for a real layered crystal we obtain $\sigma_{zz} = 3.862\cdot10^{-6};2.931\cdot10^{-4};4.132\cdot10^{-3};0.033$ S/m, respectively. In the effective mass approximation we obtain $\sigma_{zz} = 3.726\cdot10^{-6};2.710\cdot10^{-4};3.631\cdot10^{-3};0.026$ S/m, respectively. And it is 5 to 9 orders of magnitude smaller than in the absence of a magnetic field. In this case we have the law $\rho_{zz} \propto B^5$.

But if the only Landau subband is considered as completely filled, then formulae (21.2) and (21.4), respectively, will acquire the form:

$$\sigma_{zz} = \frac{\pi e m^* a^3 \Delta^7 f^6(\gamma_0)}{192 h^2 B \gamma_0^2 (\mu^* B)^6} \left(\frac{12\pi^2 f(\gamma_0) m^* \Delta}{a h^2} \right)^{4/3} N. \tag{21.32}$$

$$\sigma_{zz} = \frac{\pi e m^* a^3 \Delta^7 f^6(\gamma_0)}{192 h^2 B \gamma_{0em}^2 (\mu^* B)^6} \left(\frac{12\pi^2 f(\gamma_0) m^* \Delta}{a h^2} \right)^{4/3} N. \tag{21.33}$$

According to these formulae, with previously specified problem parameters and $\gamma_0 = 0.5;1;1.5;2$ for a real layered crystal we obtain $\sigma_{zz} = 7.796 \cdot 10^{-12};5.046 \cdot 10^{-9};2.612 \cdot 10^{-7};5.580 \cdot 10^{-6}$ S/m, respectively. In the effective mass approximation we obtain $\sigma_{zz} = 7.521 \cdot 10^{-12};4.665 \cdot 10^{-9};2.288 \cdot 10^{-7};4.485 \cdot 10^{-6}$ S/m, respectively. And it is 9 to 13 orders of magnitude smaller than in the absence of a magnetic field. In this case we have the law $\rho_{zz} \propto B^7$.

Chapter 22

EFFECT OF INTERLAYER CHARGE ORDERING ON LONGITUDINAL ELECTRIC CONDUCTIVITY OF LAYERED CRYSTALS UNDER STRONG DEGENERACY CONDITIONS

Effect of interlayer charge ordering on longitudinal electric conductivity of layered crystals in a quantizing magnetic field perpendicular to layers under strong degeneracy conditions will be considered using band spectrum (4.2) and chemical potential and order parameter dependences on the induction of a quantizing magnetic field obtained in paragraph 7. Using general formulae (16.4) and (16.5), we consider longitudinal electric conductivity of layered charge-ordered crystals in the framework of two models: constant relaxation time and relaxation time proportional to longitudinal velocity. Passing from integration over longitudinal quasi-pulse to integration over energy, in the approximation of constant relaxation time we obtain the following expressions for longitudinal electric conductivity components [29, 32]:

$$\sigma_0 = \frac{16\pi^2 e^2 m^* a \tau_0 \Delta^2}{h^4} \int\limits_{-(\gamma-b)}^{\sqrt{w^2\delta^2+1}} \left|y^{-1}\right| \sqrt{\left(w^2\delta^2+1-y^2\right)\left(y^2-w^2\delta^2\right)}\,dy, \tag{22.1}$$

$$\sigma_{os} = \frac{32\pi^2 e^2 m^* a \tau_0 \Delta^2}{h^4} \sum_{l=1}^{\infty} (-1)^l f_l^\sigma \int\limits_{-(\gamma-b)}^{\sqrt{w^2\delta^2+1}} \left|y^{-1}\right| \sqrt{\left(w^2\delta^2+1-y^2\right)\left(y^2-w^2\delta^2\right)} \cos\left[\pi l b^{-1}(\gamma+y)\right] dy \cdot \tag{22.2}$$

In the approximation of relaxation time proportional to longitudinal velocity, i.e. at $\tau(x) \propto v_z(x) = C_0\left|W'(x)\right|$ formulae (22.1) and (22.2) acquire the form [29, 44-48]:

$$\sigma_0 = \frac{16\pi^2 e^2 m^* a C_0 \Delta^3}{h^4} \int\limits_{-(\gamma-b)}^{\sqrt{w^2\delta^2+1}} y^{-2}\left(w^2\delta^2+1-y^2\right)\left(y^2-w^2\delta^2\right) dy, \tag{22.3}$$

$$\sigma_{os} = \frac{32\pi^2 e^2 m^* a C_0 \Delta^3}{h^4} \sum_{l=1}^{\infty} (-1)^l f_l^{\sigma} \int_{-(\gamma-b)}^{\sqrt{w^2\delta^2+1}} y^{-2}\left(w^2\delta^2+1-y^2\right)\left(y^2-w^2\delta^2\right)\cos\left[\pi l b^{-1}(\gamma+y)\right]dy \cdot \quad (22.4)$$

The results of calculation of longitudinal electric conductivity of charge-ordered layered crystal by formulae (22.1) – (22.4) are presented in Figures 71 – 72.

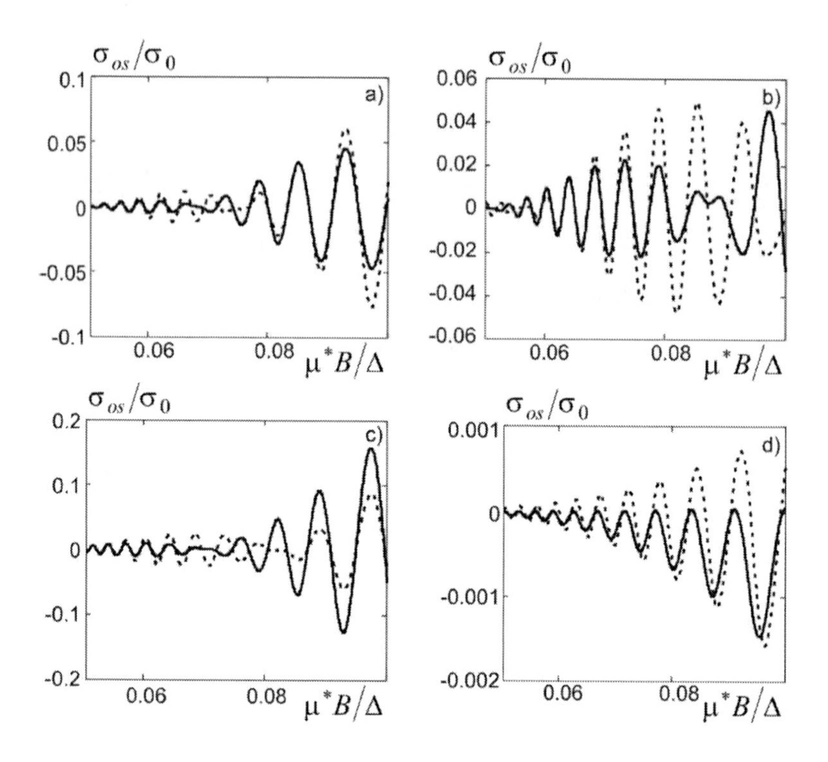

Figure 71. Field dependence of the oscillating part of longitudinal electric conductivity of a layered charger-ordered crystal at $\zeta_{02D}/\Delta = 1, kT/\Delta = 0.03, 0.04 \le \mu^* B/\Delta \le 0.1$ and: a) $W_0/\zeta_{02D} = 1.5$; b) $W_0/\zeta_{02D} = 2$; c) $W_0/\zeta_{02D} = 2.5$; d) $\delta \equiv 0$. Solid curves are for the model $\tau \propto v_z(x)$, dashed curves are for the model $\tau \equiv \tau_0$.

The figures demonstrate that even if the FS of a layered crystal in the disordered state is closed or transient, as in the case under consideration $\zeta_{02D}/\Delta = 1$, in the ordered state ShdH oscillations become biperiodic due to a topological transition from a closed FS to an open one. The frequencies, as in the case of dHvA oscillations, are related to five extreme sections of FS: three maximal by planes $k_z = 0$ and $k_z = \pm\pi/a$ and two minimal by planes $k_z = \pm\pi/2a$. ShdH oscillation frequencies are the same as dHvA oscillation frequencies and are determined by formulae (86) – (89). In so doing, the relative contribution of ShdH oscillations in quasi-classical magnetic field increases with increasing the effective attractive interaction leading to charge ordering. It occurs because as the effective attractive interaction increases, the conduction minibands into which the initial miniband is split when passing into the ordered state are narrowed, and the rate of change in the FS section as a function of longitudinal quasi-pulse is reduced. As a result, the contribution of non-extreme sections becomes comparable to that of extreme sections, which increases the oscillating part of crystal longitudinal electric conductivity.

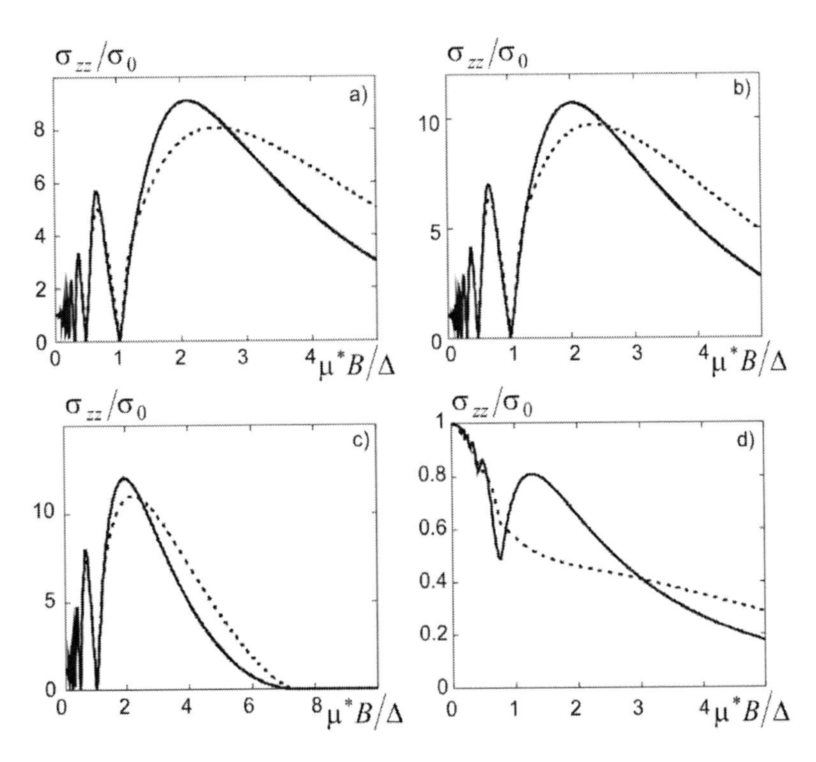

Figure 72. Field dependence of the oscillating part of longitudinal electric conductivity of a layered charge-ordered crystal at $\zeta_{02D}/\Delta = 1, kT/\Delta = 0.03, 0 \le \mu^* B/\Delta \le 5$ and: a) $W_0/\zeta_{02D} = 1.5$; b) $W_0/\zeta_{02D} = 2$; c) $W_0/\zeta_{02D} = 2.5$; d) $\delta \equiv 0$. Solid curves are for the model $\tau \propto v_z(x)$, dashed curves are for the model $\tau \equiv \tau_0$.

Moreover, in the model of constant relaxation time the contribution of oscillations to full electric conductivity is larger than in the model of relaxation time proportional to longitudinal velocity. The reason is that with constant relaxation time in quasi-classic magnetic fields the relative contribution of oscillations is proportional to $(\mu^* B/\Delta^*)^{3/2}$, whereas with relaxation time proportional to longitudinal velocity this contribution is proportional to $(\mu^* B/\Delta^*)^2$. Here, $\Delta^* = 0.5\left(\sqrt{W_0^2\delta^2 + \Delta^2} - W_0\delta\right)$ – half-width of a miniband in the ordered state. Moreover, in the model of constant relaxation time there is phase delay of oscillations as compared to the model of relaxation time proportional to longitudinal velocity. In the considered range of quasi-classic magnetic fields $0.04 \le \mu^* B/\Delta \le 0.1$ the maximum contribution of oscillations to full longitudinal conductivity of a layered crystal in the ordered state with selected problem parameters reaches 5 to 15%. However, in the disordered state ShdH oscillations remain monoperiodic, and their relative contribution does not exceed 0.15%.

With increase in magnetic field, as follows from Figure 71, the field dependence of full conductivity of a layered charge-ordered crystal acquires the form typical of a quasi-one-dimensional crystal with a finite conduction bandwidth. It occurs because with increase in magnetic field induction the populated minibands are separated completely, and dependence of electron gas chemical potential on magnetic field induction becomes essential. Owing to this, longitudinal conductivity goes to zero, whenever with a change in magnetic field the level of electron gas chemical potential passes through the ceiling or bottom of any of these minibands and, on the contrary, attains its maximum, whenever the level of electron gas chemical potential passes through the middle of any of these minibands. The last zero of full longitudinal electric conductivity in the ordered state in both models of relaxation time for all problem parameters is attained at point of inverse topological transition from an open FS to a closed one due to magnetic field effect. Afterwards, the last maximum of full longitudinal electric conductivity is attained. At this maximum, full longitudinal electric conductivity of a layered crystal with selected problem parameters exceeds its value in the absence of a magnetic field by a factor of 9 to 12. After the maximum, longitudinal electric conductivity of a layered crystal in both models of relaxation time decays, tending to zero according to certain asymptotic laws which will be discussed below. The intersection of plots is attributable to the fact that in a model of relaxation time proportional to longitudinal velocity, the maximum of full longitudinal electric conductivity (with respect to its value in the absence of a magnetic field) is larger than in a model of constant relaxation time, as well as to the fact that a decay in full longitudinal electric conductivity after the maximum with increase in magnetic field in the model of relaxation time proportional to electron longitudinal velocity occurs faster than in a model of constant relaxation time.

The same situation takes place in the disordered state, but in this state the electric conductivity at point of local minimum does not go to zero, which is attributable to transient character of the FS in this state.

Let us now estimate the electric conductivity of a layered charge-ordered crystal in the absence of a magnetic field for both models of relaxation time. We will start from the model of constant relaxation time which is suitable for dominant electron scattering on ionized impurities. For this purpose, the radius of equivalent sphere which substitutes the real FS of a charge-ordered layered crystal so that charge carrier scattering could be considered to be isotropic, will be defined as follows:

$$k_0 = \sqrt[3]{\frac{6\pi^2 m^*}{ah^2}\int_0^{\pi}\left(\zeta + \sqrt{W_0^2\delta^2 + \Delta^2\cos^2 x}\right)dx}. \tag{22.5}$$

Then, by analogy with the case of charge ordering absence, the relaxation time will be defined as follows:

$$\tau_0 = \frac{Nahk_0}{4\pi\left(\zeta + \sqrt{W_0^2\delta^2 + \Delta^2}\right)}. \tag{22.6}$$

Analysis shows that in the case under consideration the Dingle factor can be considered to be close to unity, if $N \geq 50000$. Assuming $N = 50000$ with previously stipulated problem parameters and $W_0/\zeta_0 = 1.5;2;2.5$, respectively, we obtain $\sigma_0 = 7.414 \cdot 10^3;4.329 \cdot 10^3;2.746 \cdot 10^3$ S/m, respectively. Whereas in the disordered state $\sigma_0 = 1.856 \cdot 10^5$ S/m. Hence, in the model $\tau(x) \equiv \tau_0$ on transition to the charge-ordered state in the model $\tau(x) \equiv \tau_0$ with selected problem parameters the electric conductivity of a layered crystal is reduced by a factor of 25 to 68, which is not only in qualitative, but also in quantitative agreement with the experimental data on phase transitions in the intercalated graphite compounds.

We will now consider the model $\tau(x) = C_0 |W'(x)|$ which is more adequate with charge carrier scattering on acoustic phonons. Assuming, for instance,

$$C_0 = \frac{4h^5 \rho s_0^6}{31\Gamma(5)\zeta(5)\pi^3 a(kT)^5 \Xi^2}, \tag{22.7}$$

with previously specified problem parameters and $W_0/\zeta_0 = 1.5;2;2.5$, respectively, we obtain $\sigma_0 = 5.061 \cdot 10^4;2.087 \cdot 10^4;1.067 \cdot 10^4$ S/m, respectively. Whereas in the disordered state $\sigma_0 = 2.668 \cdot 10^6$ S/m. Hence, in the model $\tau(x) \propto v_z(x)$ on passing to the charge-ordered state with selected problem parameters the electric conductivity of a layered crystal is reduced by a factor of 50 to 250.

Let us now establish the asymptotic laws, according to which the longitudinal electric conductivity of a charge-ordered layered crystal tends to zero in the ultra-quantum limit on the assumption that in a strong magnetic field charge ordering is not destroyed. Proceeding as before and taking into account the relation (7.2), in the model of constant relaxation time with charge carrier scattering on ionized impurities we obtain:

$$\sigma_{zz} = \frac{1.015 e^2 m^* a^2 k_0 \Delta^4 \zeta_{02D}^{-3} N}{h^3 \left(\zeta + \sqrt{W_0^2 \delta_1^2 + \Delta^2}\right)\left(W_0^2 \delta_2^2 + \Delta^2\right) kT \left(\mu^* B\right)^2}. \tag{22.8}$$

In this formula, the values k_0, ζ and δ_1 for each value W_0/ζ_0 should be taken in the absence of a magnetic field, whereas the value δ_1 should be taken in a strong magnetic field. Therefore, with previously stipulated problem parameters, $B = 60\,\mathrm{T}$ and $W_0/\zeta_0 = 1.5;2;2.5$, respectively, we will obtain $\sigma_{zz} = 107.315;47.457;21.151$ S/m, respectively. This would imply that in a strong quantizing magnetic field the electric conductivity of a layered charge-ordered crystal in the model $\tau(x) \equiv \tau_0$ is reduced by a factor of 69 to 130 as compared to its value in the absence of a magnetic field. In the disordered state we have $\sigma_{zz} = 212$ S/m. This would imply that in a strong quantizing magnetic field the electric conductivity of a layered crystal in the absence of ordering is reduced by a factor of 871. Thus, we get the law $\rho_{zz} \propto TB^2$.

In a similar manner one can consider the asymptotic laws of change in longitudinal electric conductivity as a function of magnetic field in the framework of approach whereby electric conductivity in the ultra-quantum limit with charge carrier scattering on ionized impurities is not a function of temperature [36]. In this case, considering the only Landau subband with the number $n = 0$ as completely filled, we obtain:

$$\sigma_{zz} = \frac{1.015 e^2 m^* a^2 k_0 \Delta^4 \zeta_{02D}^3 N}{h^3 \left(\zeta + \sqrt{W_0^2 \delta_1^2 + \Delta^2}\right)\left(W_0^2 \delta_2^2 + \Delta^2\right) kT \left(\mu^* B\right)^2}. \tag{22.9}$$

Therefore, with previously stipulated problem parameters, $B = 60\,\text{T}$ and $W_0/\zeta_0 = 1.5;2;2.5$, respectively, we obtain $\sigma_{zz} = 0.370;0.164;0.087\,\text{S/m}$, respectively. In the disordered state we have $\sigma_{zz} = 353.270\,\text{S/m}$. This would imply that in a strong quantizing magnetic field the electric conductivity of a layered crystal both with and without ordering is reduced by 4 orders of magnitude as compared to its value in the absence of a magnetic field. Therefore, we get the law $\rho_{zz} \propto B^3$.

If, however, we consider the only Landau subband with the number $n = 0$ as the only filled with the effective half-width equal to

$$\Delta_{ef} = \frac{\pi^2 \zeta_{02D}^2 \Delta^2}{64\left(\mu^* B\right)^2 \sqrt{W_0^2 \delta_2^2 + \Delta^2}}, \tag{22.10}$$

it yields the following expression for longitudinal electric conductivity of crystal in the ultra-quantum limit:

$$\sigma_{zz} = \frac{\pi^6 e^2 m^* a^2 k_0 \Delta^4 \zeta_{02D}^4 N}{512 h^3 \left(\zeta + \sqrt{W_0^2 \delta_1^2 + \Delta^2}\right)\left(W_0^2 \delta_2^2 + \Delta^2\right)\left(\mu^* B\right)^4}. \tag{22.11}$$

Therefore, with previously stipulated problem parameters, $B = 60\,\text{T}$ and $W_0/\zeta_0 = 1.5;2;2.5$, respectively, we obtain $\sigma_{zz} = 4.922 \cdot 10^{-3};2.177 \cdot 10^{-3};1.153 \cdot 10^{-3}\,\text{S/m}$, respectively. In the disordered state we have $\sigma_{zz} = 0.152\,\text{S/m}$. This would imply that in a strong quantizing magnetic field the electric conductivity of a layered crystal both with and without ordering is reduced by 6 orders of magnitude as compared to its value in the absence of a magnetic field. Thus, we get the law $\rho_{zz} \propto B^4$.

Consider now the case of charge carrier scattering on deformation potential of acoustic phonons. Using for C_0 the expression (208), we find:

$$\sigma_{zz} = \frac{1.217 \cdot 10^{-3} e^2 h m^* \rho s_0^6 \Delta^6 \zeta_{02D}^4}{\Xi^2 \left(W_0^2 \delta^2 + \Delta^2\right)^{3/2} (kT)^6 \left(\mu^* B\right)^3}. \tag{22.12}$$

Therefore, with previously stipulated problem parameters, $B = 60\,\text{T}$ and $W_0/\zeta_0 = 1.5;2;2.5$, respectively, we obtain $\sigma_{zz} = 20.523;8.803;4.541\,\text{S/m}$, respectively. In the disordered state we have $\sigma_{zz} = 73.018\,\text{S/m}$. This would imply that in a strong quantizing magnetic field the electric conductivity of a layered crystal in the presence of charge ordering is reduced by 3 orders of magnitude and in the absence of charge ordering – by 4 orders of magnitude as compared to its value in the absence of a magnetic field. Thus, we obtain the law of the type $\rho_{zz} \propto T^6 B^3$.

If, however, we take into account the effect of magnetic field on charge carrier scattering on acoustic phonons, this scattering should be considered spontaneous and coefficient C_0 taken as follows:

$$C_0 = \frac{3h^3 a\rho s_0^4}{4\pi^2 \Xi^2 (kT)^3}. \tag{22.13}$$

Therefore, with previously stipulated problem parameters, $B = 60\,\text{T}$ and $W_0/\zeta_0 = 1.5;2;2.5$, respectively, we get $\sigma_{zz} = 20.523;8.803;4.541\,\text{S/m}$, respectively. In the disordered state we have $\sigma_{zz} = 73.018\,\text{S/m}$. This would imply that in a strong quantizing magnetic field the electric conductivity of a layered crystal in the presence of charge ordering is reduced by 3 orders of magnitude and in the absence of charge ordering – by 4 orders of magnitude as compared to its value in the absence of a magnetic field. Hence, we get:

$$\sigma_{zz} = \frac{0.571 e^2 m^* a^2 \rho s_0^4 \Delta^6 \zeta_0^4}{h\left(\Delta^2 + W_0^2 \delta^2\right)^{3/2} \Xi^2 (kT)^4 \left(\mu^* B\right)^3}. \tag{22.14}$$

Therefore, with previously stipulated problem parameters, $B = 60\,\text{T}$ and $W_0/\zeta_0 = 1.5;2;2.5$, respectively, we get $\sigma_{zz} = 2.019;0.866;0.447\,\text{S/m}$, respectively. In the disordered state we have $\sigma_{zz} = 7.183\,\text{S/m}$. This would imply that in a strong quantizing magnetic field the electric conductivity of a layered crystal in the presence of charge ordering is reduced by 4 orders of magnitude and in the absence of charge ordering – by 5 orders of magnitude as compared to its value in the absence of a magnetic field. In this case we get the law of the type $\rho_{zz} \propto T^4 B^3$.

Chapter 23

SIMPLE FORMULAE FOR LONGITUDINAL ELECTRIC CONDUCTIVITY OF LAYERED CHARGE-ORDERED CRYSTALS

In the model of relaxation time proportional to longitudinal velocity the longitudinal electric conductivity of layered charge-ordered crystals allows for explicit representation comprising no integrals over quasi-pulse, since integrals over y in formulae (204) and (205) can be calculated analytically. Let us make this calculation for the simplest case $W_0/\zeta_{02D}=2$, when in the absence of a magnetic field the level of electron gas chemical potential in the charge-ordered state lies in the middle of pseudo-gap between minibands into which the initial conduction miniband is split when passing into the charge-ordered state. This yields the following formulae for longitudinal conductivity components of a layered crystal [49]:

$$\sigma_0=\frac{32\pi^2e^2m^*aC_0}{3h^4}\left(\Delta_\delta-W_0\delta\right)^3, \tag{23.1}$$

$$\begin{aligned}\sigma_{os}=&\frac{32\pi^2e^2m^*aC_0}{h^4}\sum_{l=1}^{\infty}(-1)^l f_l^{\sigma}\left\{\cos\left(\frac{\pi l\Delta_\delta}{\mu^*B}\right)\left[W_0^2\delta^2\Delta_\delta-2\left(\frac{\mu^*B}{\pi l}\right)^2\Delta_\delta\right]+\cos\left(\frac{\pi lW_0\delta}{\mu^*B}\right)\times\right.\\&\times\left[2W_0\delta\left(\frac{\mu^*B}{\pi l}\right)^2-W_0\delta\Delta_\delta^2\right]+\sin\left(\frac{\pi l\Delta_\delta}{\mu^*B}\right)\left[\frac{\mu^*B}{\pi l}W_0^2\delta^2+2\left(\frac{\mu^*B}{\pi l}\right)^3\right]-\sin\left(\frac{\pi lW_0\delta}{\mu^*B}\right)\times\\&\left.\times\left[\frac{\mu^*B}{\pi l}\Delta_\delta^2+2\left(\frac{\mu^*B}{\pi l}\right)^3\right]+\frac{\pi lW_0^2\delta^2\Delta_\delta^2}{\mu^*B}\left[\mathrm{Si}\left(\frac{\pi l\Delta_\delta}{\mu^*B}\right)-\mathrm{Si}\left(\frac{\pi lW_0\delta}{\mu^*B}\right)\right]\right\}.\end{aligned} \tag{23.2}$$

In formula (23.2), $\mathrm{Si}(x)$ – integral sine.

In terms of miniband effective width Δ^* caused by charge ordering, formula (215) can be re-written as follows:

$$\sigma_0 = \frac{256\pi^2 e^2 m^* a C_0 \Delta^{*3}}{3h^4}. \tag{23.3}$$

Further we will show that the oscillating part of electric conductivity can be also with a reasonable degree of accuracy calculated in terms of Δ^* and formula (216) will acquire a much simpler form. With this in mind, we primarily note that function $W(x)=\sqrt{\Delta^2\cos^2 x + W_0^2\delta^2}$ is expandable into Fourier series as follows:

$$W(x) = \Delta_0 + \sum_{m=1}^{\infty} \Delta_m \cos 2mx, \tag{23.4}$$

in so doing, for instance, at $W_0\delta/\Delta = 2$, i.e. for the case of a transient FS of a charge-ordered layered crystal, Δ_0 to an accuracy of 0.1% coincides with $W_0\delta + \Delta^*$, and Δ_1 to the same accuracy coincides with Δ^*. In its value, the second harmonic of series (23.4) is 1.4% of the first, the third – $3\cdot10^{-2}$% of the first, the fourth – $1.4\cdot10^{-3}$% of the first. Therefore, using the results of [49], one can assert that at $\zeta \equiv 0$ the frequencies of ShdH oscillations are determined by the values $\Delta_0 - \Delta_1 = W_0\delta$ and $\Delta_0 + \Delta_1 = \Delta_\delta$. These frequencies are related, accordingly, to minimum sections of the FS by planes $k_z = \pm\pi/2a$ and maximum sections of the FS by planes $k_z = 0$ and $k_z = \pm\pi/a$. Restricting ourselves in formula (23.2) to the first nonvanishing approximations with respect to parameter $\Delta^*/W_0\delta$, for the oscillating part of longitudinal conductivity we obtain the following expression:

$$\frac{\sigma_{os}}{\sigma_0} = 3\sum_{l=1}^{\infty}(-1)^l f_l^{\sigma}\left\{\left(\frac{\mu^* B}{\pi l\Delta^*}\right)^3\left[\sin\left(\frac{\pi l\Delta_\delta}{\mu^* B}\right) - \sin\left(\frac{\pi l W_0\delta}{\mu^* B}\right)\right] - \right.$$
$$\left. -\left(\frac{\mu^* B}{\pi l\Delta^*}\right)^2\left[\cos\left(\frac{\pi l\Delta_\delta}{\mu^* B}\right) + \cos\left(\frac{\pi l W_0\delta}{\mu^* B}\right)\right]\right\}. \tag{23.5}$$

Numerical analysis shows that this formula coincides with (23.2) with a high degree of precision.

Chapter 24

General Formula for the Seebeck Coefficient of Layered Crystals in a Quantizing Magnetic Field

We now turn to deriving the general formula for the Seebeck coefficient of layered crystals in a quantizing magnetic field under electron gas strong degeneracy conditions. Note beforehand that thermoelectric properties of many materials are actively studied today both theoretically and experimentally. The investigation is concerned with metals, alloys, semiconductors [50, 51], fullerenes [52], composite materials, including biomorphous [53, 54], graphene [55], etc. As regards graphene, it should be noted that it is of particular interest as a material for high-efficiency solar elements. The reason for this interest is special photothermoelectric effect inherent in them. Earlier this effect was considered to be purely photovoltaic.

At first sight it may seem that study of thermoelectric material properties at low, for instance, helium temperatures is of academic interest only. Indeed, on the one hand, from the general thermodynamic considerations it follows that, at least in the absence of a magnetic field, the Seebeck coefficient with decrease in temperature must tend to zero. On the other hand, thermoelectric generators, heaters and thermal recuperators operate by no means at helium, but, on the contrary, at high temperatures. The idea of creating some "thermoelectric supercooler" for particularly deep cooling of something cannot inspire anybody, for even at relatively high figure of merit of thermoelectric material because of necessarily small temperature difference on thermoelements the efficiency of such hypothetical device is negligibly low. However, it should be remembered that thermoelectric phenomena are a powerful additional metrological tool for studying, for instance, the band structure of materials and charge carriers scattering mechanisms in them. Such use of thermoelectricity is possible and quite reasonable, because the Seebeck coefficient oscillations in a quantizing magnetic field, as will be shown further, can be observed directly. At the same time, observation of dHvA and SdH oscillations requires application of special radio engineering methods and signal filtering tools for a reliable separation of the oscillating parts of diamagnetic susceptibility and conductivity from their monotonic parts. Moreover, study of SdH effects is involved with a series of additional problems. The first problem is that the oscillating part of electric conductivity in the quasi-classical area of magnetic fields for the majority of materials, including layered ones, is low as compared to the monotonic part. The second problem lies in the necessity of a reliable distinction between the oscillating currents due to SdH effect proper and the oscillating eddy induction currents caused by oscillations of magnetic field induction in material due to dHvA effect. Study of dHvA oscillations requires a complicated measuring device such as torsion balance. As to the Seebeck coefficient oscillations, their study is not involved with these problems.

In deriving the general formula for the longitudinal Seebeck coefficient α_{zz} of layered crystals, we will proceed from the kinetic Boltzmann equation. We will consider the case when the quantizing magnetic field and temperature gradient are parallel to each other and normal to the layers. If the Seebeck coefficient is determined as the ratio of electron gas electrochemical potential gradient to temperature gradient under no electric current in the sample, then in the relaxation time approximation this approach yields the following general expression for α_{zz} :

$$\alpha_{zz} \equiv \alpha = -\left[\frac{\partial}{\partial T}\sum_{\gamma}\tau_{\gamma}g_{\gamma}v_{z\gamma}^{2}f^{0}(\varepsilon_{\gamma})\right]\left[e\frac{\partial}{\partial \zeta}\sum_{\gamma}\tau_{\gamma}g_{\gamma}v_{z\gamma}^{2}f^{0}(\varepsilon_{\gamma})\right]^{-1}. \quad (24.1)$$

Taking into account in Eq.(32) only the first two terms, from (220) we obtain the following final expression for the longitudinal Seebeck coefficient of layered crystal in a quantizing magnetic field under electron gas strong degeneracy conditions [56]:

$$\alpha_{zz} = \pi\alpha_0 \frac{A^{(\alpha)}}{B^{(\alpha)} + C^{(\alpha)}}. \quad (24.2)$$

In formula (24.2) $\alpha_0 = k/e$, and coefficients A, B, C have the following values:

$$A = \sum_{l=1}^{\infty}(-1)^{l-1} f_l^{th} \int\limits_{W(x)\le\zeta} \tau(x)W'(x)^2 \sin\left[\pi l \frac{\zeta - W(x)}{\mu^* B}\right]dx, \quad (24.3)$$

$$B = 0.5 \int\limits_{W(x)\le\zeta} \tau(x)W'(x)^2\, dx, \quad (24.4)$$

$$C = \sum_{l=1}^{\infty}(-1)^{l} f_l^{\sigma} \int\limits_{W(x)\le\zeta} \tau(x)W'(x)^2 \cos\left[\pi l \frac{\zeta - W(x)}{\mu^* B}\right]dx. \quad (24.5)$$

In so doing

$$f_l^{th} = \left[\sinh\left(\pi^2 lkT/\mu^* B\right)\right]^{-1}\left[1 - \left(\pi^2 lkT/\mu^* B\right)\coth\left(\pi^2 lkT/\mu^* B\right)\right]. \quad (24.6)$$

The rest of designations in formulae (24.2) – (24.5) are generally accepted or explained above.

Later on we will apply formula (24.2) with account of (24.3) – (24.6) for the analysis of the field dependences of the Seebeck coefficient with different charge carrier scattering mechanisms and for different models of band spectrum $W(x)$.

Chapter 25

FIELD DEPENDENCE OF THE SEEBECK COEFFICIENT IN LAYERED CRYSTALS IN THE APPROXIMATION OF CONSTANT RELAXATION TIME

In the approximation of constant relaxation time the dimensionless coefficients A, B, C for a real layered crystal have the following values:

$$A^{(\alpha)} = \sum_{l=1}^{\infty}(-1)^{l-1} f_l^{th}\left\{\sin\left(\pi l\frac{\gamma-1}{b}\right)\left[\left(C_0^\sigma - C_2^\sigma\right)J_0\left(\pi l b^{-1}\right) + \sum_{r=1}^{\infty}(-1)^r\left(2C_{2r}^\sigma - C_{2r+2}^\sigma - C_{2r-2}^\sigma\right)J_{2r}\left(\pi l b^{-1}\right)\right] + \right.$$
$$\left. + \cos\left(\pi l\frac{\gamma-1}{b}\right)\sum_{r=0}^{\infty}(-1)^r\left(2C_{2r+1}^\sigma - C_{2r+3}^\sigma - C_{|2r-1|}^\sigma\right)J_{2r+1}\left(\pi l b^{-1}\right)\right\}, \qquad (25.1)$$

$$B^{(\alpha)} = 0.5\left(C_0^\sigma - C_2^\sigma\right), \qquad (25.2)$$

$$C^{(\alpha)} = \sum_{l=1}^{\infty}(-1)^{l} f_l^{\sigma}\left\{\cos\left(\pi l\frac{\gamma-1}{b}\right)\left[\left(C_0^\sigma - C_2^\sigma\right)J_0\left(\pi l b^{-1}\right) + \sum_{r=1}^{\infty}(-1)^r\left(2C_{2r}^\sigma - C_{2r+2}^\sigma - C_{2r-2}^\sigma\right)J_{2r}\left(\pi l b^{-1}\right)\right] - \right.$$
$$\left. - \sin\left(\pi l\frac{\gamma-1}{b}\right)\sum_{r=0}^{\infty}(-1)^r\left(2C_{2r+1}^\sigma - C_{2r+3}^\sigma - C_{|2r-1|}^\sigma\right)J_{2r+1}\left(\pi l b^{-1}\right)\right\}, \qquad (25.3)$$

In the effective mass approximation the values of these coefficients are as follows:

$$A^{(\alpha)} = \sum_{l=1}^{\infty}(-1)^{l-1} f_l^{th}\left\{\frac{b}{\pi l}\sqrt{2(\gamma-b)} - \frac{1}{\pi}\left(\frac{b}{l}\right)^{3/2}\left[\cos\left(\frac{\pi l\gamma}{b}\right)C\left(\sqrt{2l\left(\frac{\gamma}{b}-1\right)}\right) + \right.\right.$$
$$\left.\left. + \sin\left(\frac{\pi l\gamma}{b}\right)S\left(\sqrt{2l\left(\frac{\gamma}{b}-1\right)}\right)\right]\right\}. \qquad (25.4)$$

$$B^{(\alpha)} = \frac{1}{6}\left[2(\gamma-b)\right]^{3/2}. \qquad (25.5)$$

$$C^{(\alpha)} = \sum_{l=1}^{\infty} (-1)^l f_l^{\sigma} \frac{1}{\pi}\left(\frac{\mu^* B}{l\Delta}\right)^{3/2} \left[\sin\left(\frac{\pi l \zeta}{\mu^* B}\right) C\left(\sqrt{2l\left(\frac{\zeta}{\mu^* B} - 1\right)}\right) - \right.$$
$$\left. - \cos\left(\frac{\pi l \zeta}{\mu^* B}\right) S\left(\sqrt{2l\left(\frac{\zeta}{\mu^* B} - 1\right)}\right)\right] . \qquad (25.6)$$

The results of the Seebeck coefficient calculations for a real layered crystal and in the effective mass approximation are depicted in Figures 73-74.

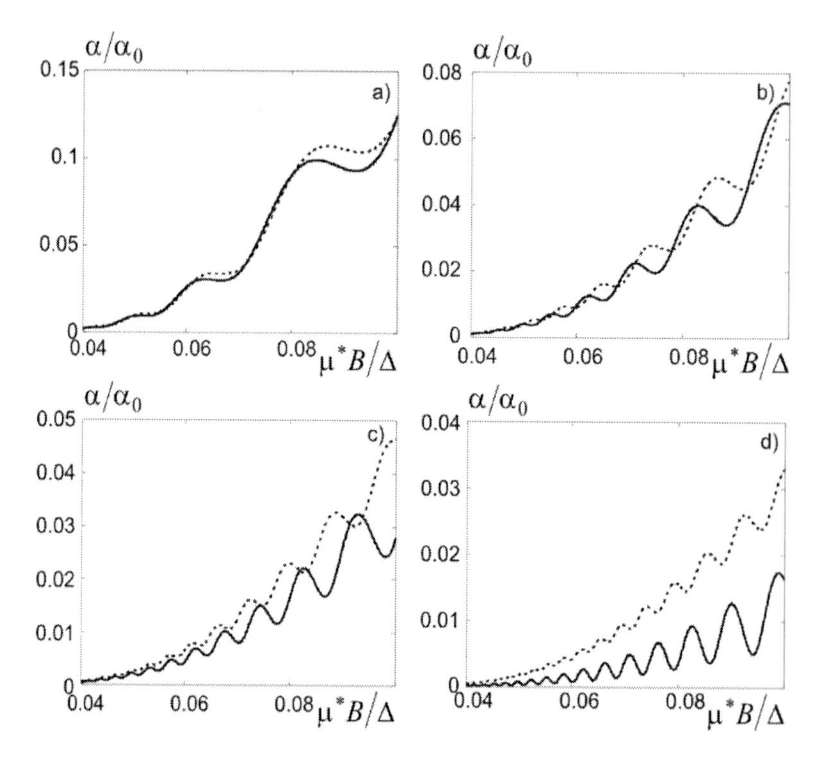

Figure 73. Field dependences of the Seebeck coefficient at $\tau(x) \equiv \tau_0$ for a real layered crystal and in the effective mass approximation with $kT/\Delta = 0.03; 0.04 \le \mu^* B/\Delta \le 0.1$.

The figures demonstrate that the Seebeck coefficient oscillations are directly observable, and the decisive reason for this is vanishing of the Seebeck coefficient at $T = 0$. With increasing electron concentration, hence the degree of miniband filling, the phase lag of solid curves from dashed curves is increased. Also increased is the amplitude of oscillations on solid curves with respect to dashed curves. Here lies the influence of crystal layered structure effects on the Seebeck coefficient oscillations in the quasi-classical area of magnetic fields. The greatest difference between the solid and dashed curves takes place at $\zeta/\Delta = 2$, i.e. for the transient FS. As in the cases with electron gas diamagnetic susceptibility and longitudinal electric conductivity, it is attributable to the fact that in a real layered crystal there are no planes that are tangent to such FS and simultaneously normal to magnetic field direction under study. At the same time, in the effective mass approximation such planes are always present. In the quasi-classical range of magnetic fields, with increasing degree of miniband filling ζ/Δ from 0.5 to 2, the maximum value of the Seebeck coefficient in a real layered crystal drops from 10.8 to 1.55μV/K, and in the effective mass approximation – from 10.8 to 2.85μV/K. Besides, as could be expected, with increasing degree of miniband filling, the oscillation frequencies of the Seebeck coefficient increase.

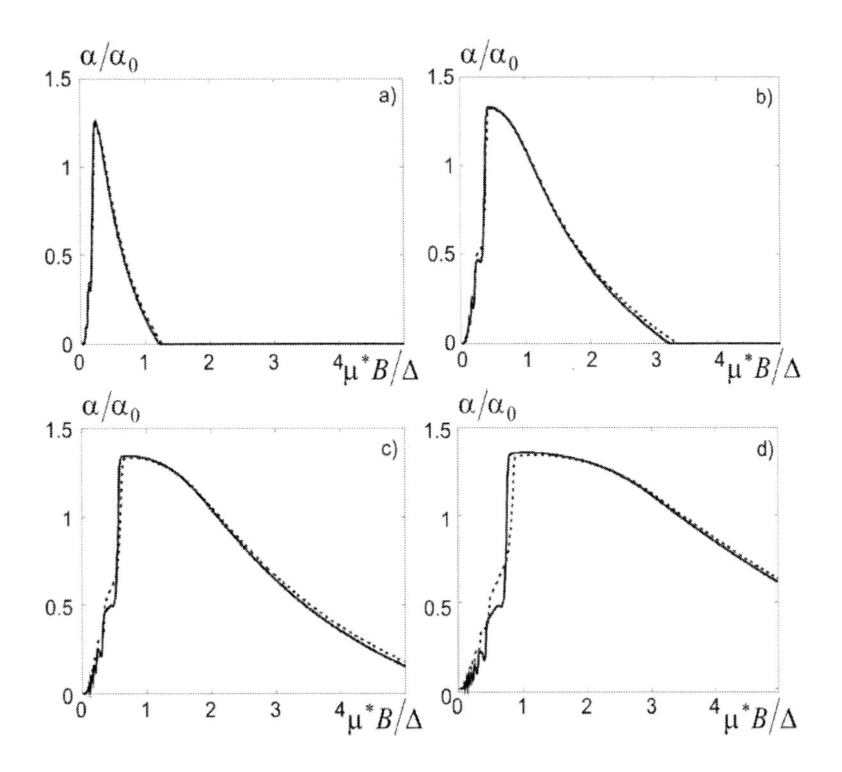

Figure 74. Field dependences of the Seebeck coefficient at $\tau(x) \equiv \tau_0$ for a real layered crystal and in the effective mass approximation with $kT/\Delta = 0.03; 0 \leq \mu^* B/\Delta \leq 5$.

In stronger magnetic fields the Seebeck coefficient first increases to maximum value, and then drops to zero by a certain asymptotic law. The existence of the Seebeck coefficient maximum is due to the following physical considerations. On the one hand, at low temperatures and weak magnetic fields the Seebeck coefficient should tend to zero in conformity with general thermodynamic considerations. On the other hand, it should also tend to zero in the ultraquantum limit by virtue of free charge carrier condensation on the bottom of the lowest Landau subband with the number $n = 0$ and the FS compression toward the magnetic field. However, the Seebeck coefficient cannot be identically zero in the system of free charge carriers. Therefore, there must be such magnetic field value whereby the Seebeck coefficient modulus reaches a maximum. At the same time, as is evident from Figure 74, there is an optimal range of magnetic field inductions wherein the layered structure effects are most pronounced. These effects consist in the presence in a real layered crystal of a series of local maxima and minima of the Seebeck coefficient that are absent in the effective mass approximation. Moreover, these effects consist in a slight shift of the Seebeck coefficient maximum in a layered crystal toward weaker magnetic fields as compared to the effective mass approximation. This shift is attributable to the same physical reason as the phase delay of oscillations, but maximum shift in a real layered crystal is lesser due to the fact that in strong magnetic fields the differences between the tight-binding approximation and the effective mass approximation used for the description of interlayer motion of charge carriers are smoothed over. As the degree of band filling ζ/Δ increases from 0.5 to 2, the Seebeck coefficient maximum slowly grows from 108 to 121μV/K.

We now establish the asymptotic law of the Seebeck coefficient reduction in the ultraquantum limit. For this purpose at $\mu^* B/kT >> 1$ we represent f_l^{th} as below:

$$f_l^{th} = \frac{\left(\pi^2 lkT/\mu^* B\right)^2}{2\mathrm{sh}\left(\pi^2 lkT/\mu^* B\right)}, \tag{25.7}$$

and we consider in addition that the numerical analysis shows the validity of the following relationship which is correct at low δ:

$$\sum_{l=1}^{\infty} \frac{\delta l^3}{\mathrm{sh}(\delta l)} = \frac{12.176}{\delta^3}. \tag{25.8}$$

Proceeding in the same way as in deriving the asymptotic expression for electric conductivity, we obtain the following asymptotic law of the Seebeck coefficient reduction in the ultraquantum limit:

$$\alpha = 0.123\alpha_0 f^2(\gamma_0)\frac{\Delta^3}{\left(\mu^* B\right)^2 kT}. \tag{25.9}$$

At first sight, this law is hard to understand, since the Seebeck coefficient does not tend to zero at $T = 0$, as it should be from thermodynamic considerations. However, at real low temperatures this law does not yield incorrect physical results. Indeed, for instance, at $B = 60$ T, $T = 3$ K and previously stipulated crystal parameters we obtain that the Seebeck in the ultraquantum limit drops by 2 – 3 orders as compared to its maximum value.

Chapter 26

FIELD DEPENDENCE OF THE SEEBECK COEFFICIENT IN A LAYERED CRYSTAL IN THE MODEL OF CONSTANT MEAN FREE PATH

In the model of constant mean free path of carriers the coefficients $A^{(\alpha)}$, $B^{(\alpha)}, C^{(\alpha)}$ for a real layered crystal have the following values:

$$A^{(\alpha)} = \sum_{l=1}^{\infty} (-1)^{l-1} f_l^{th} \int_0^{\gamma-b} \sqrt{\frac{2y - y^2}{A(\gamma - y) + 2y - y^2}} \sin\left(\pi l \frac{\gamma - y}{b}\right) dy, \tag{26.1}$$

$$B^{(\alpha)} = 0.5 \int_0^{\gamma-b} \sqrt{\frac{2y - y^2}{A(\gamma - y) + 2y - y^2}} dy, \tag{26.2}$$

$$C^{(\alpha)} = \sum_{l=1}^{\infty} (-1)^{l} f_l^{\sigma} \int_0^{\gamma-b} \sqrt{\frac{2y - y^2}{A(\gamma - y) + 2y - y^2}} \cos\left(\pi l \frac{\gamma - y}{b}\right) dy. \tag{26.3}$$

In the effective mass approximation these coefficients are as follows:

$$A^{(\alpha)} = \sum_{l=1}^{\infty} (-1)^{l-1} f_l^{th} \int_0^{\gamma-b} \sqrt{\frac{2y}{A(\gamma - y) + 2y}} \sin\left(\pi l \frac{\gamma - y}{b}\right) dy, \tag{26.4}$$

$$B^{(\alpha)} = 0.5 \int_0^{\gamma-b} \sqrt{\frac{2y}{A(\gamma - y) + 2y}} dy, \tag{26.5}$$

$$C^{(\alpha)} = \sum_{l=1}^{\infty} (-1)^{l} f_l^{\sigma} \int_0^{\gamma-b} \sqrt{\frac{2y}{A(\gamma - y) + 2y}} \cos\left(\pi l \frac{\gamma - y}{b}\right) dy. \tag{26.6}$$

In this model, as before, the magnetic field effect on scattering is taken into account through the magnetic field dependence of γ. If, however, the magnetic field effect on scattering is disregarded, then under the square root signs in formulae (235) – (240) one should take the value γ calculated in the absence of a magnetic field for a real layered crystal and in the effective mass approximation, respectively. The results of the Seebeck coefficient calculation according to formulae (26.1) – (26.6) are given in Figures 75 to 85.

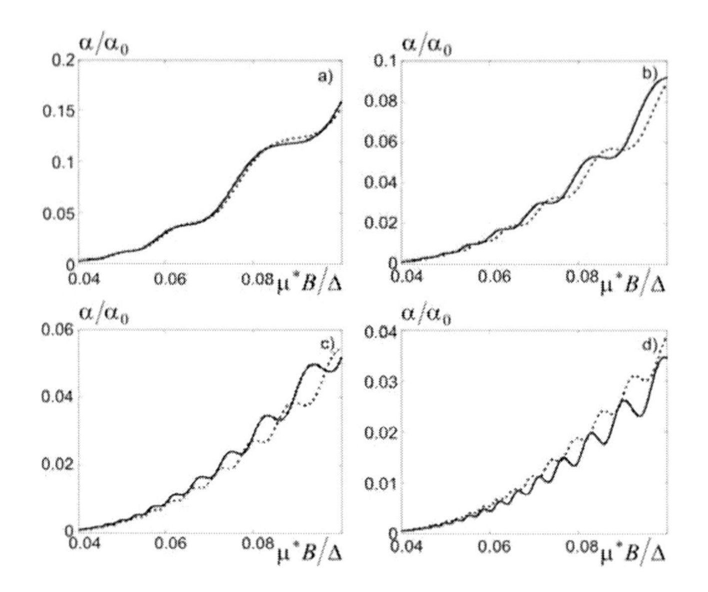

Figure 75. Field dependences of the Seebeck coefficient in the model $l = const$ at $A = 5; kT/\Delta = 0.03; 0.04 \leq \mu^* B/\Delta \leq 0.1$.

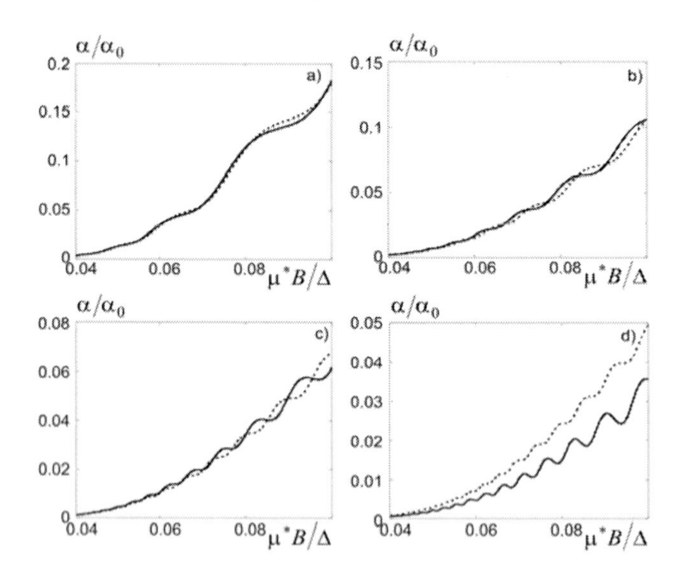

Figure 76. Field dependences of the Seebeck coefficient in the model $l = const$ at $A = 15; kT/\Delta = 0.03; 0.04 \leq \mu^* B/\Delta \leq 0.1$.

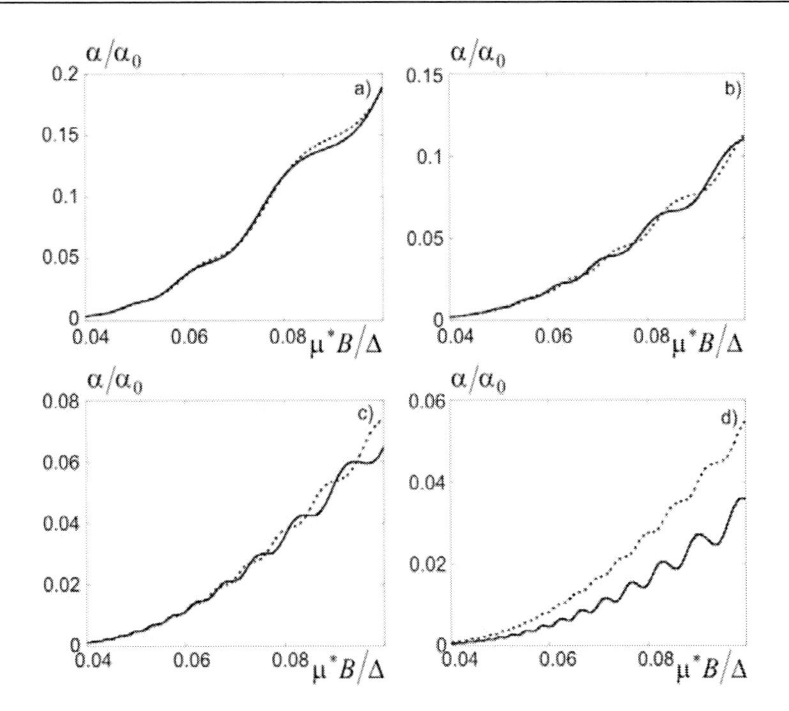

Figure 77. Field dependences of the Seebeck coefficient in the model $l = const$ at $A = 25; kT/\Delta = 0.03; 0.04 \le \mu^* B/\Delta \le 0.1$.

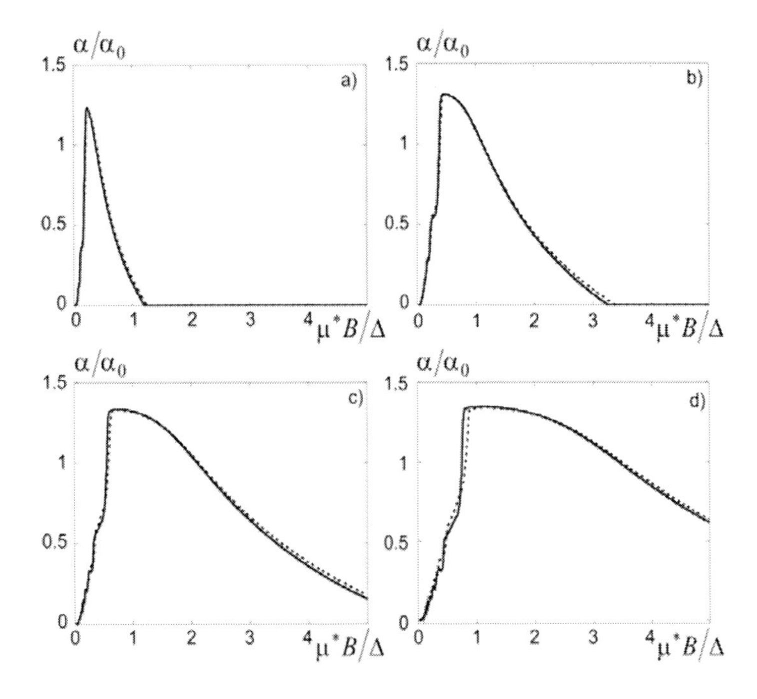

Figure 78. Field dependences of the Seebeck coefficient in the model $l = const$ without regard to magnetic field effect on scattering at $A = 5; kT/\Delta = 0.03; 0 \le \mu^* B/\Delta \le 5$.

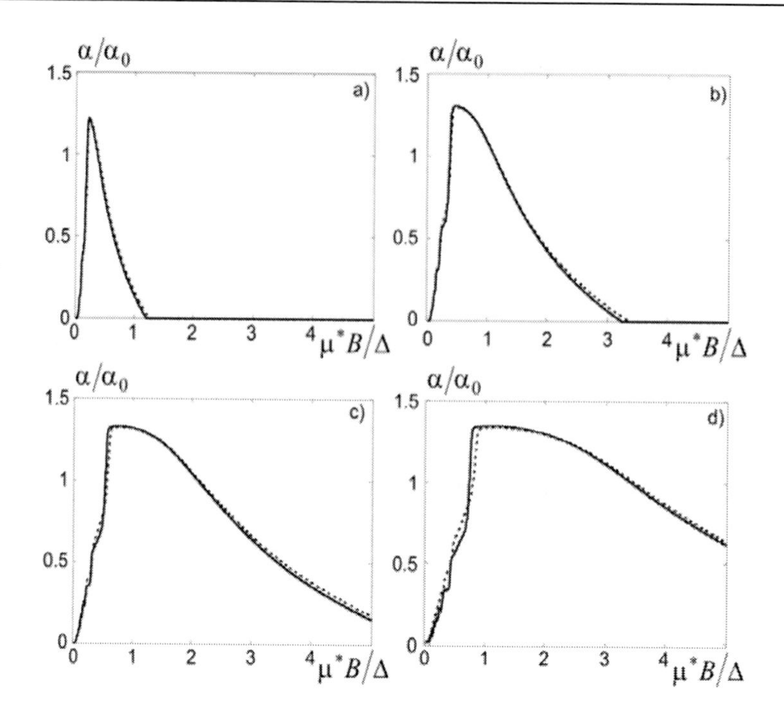

Figure 79. Field dependences of the Seebeck coefficient in the model $l = const$ without regard to magnetic field effect on scattering at $A = 15; kT/\Delta = 0.03; 0 \le \mu^* B/\Delta \le 5$.

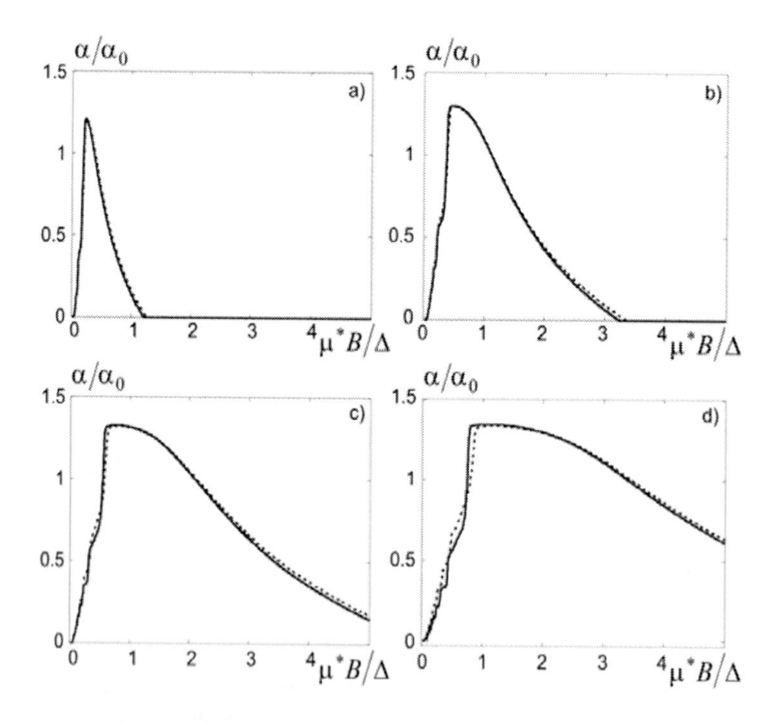

Figure 80. Field dependences of the Seebeck coefficient in the model $l = const$ without regard to magnetic field effect on scattering at $A = 25; kT/\Delta = 0.03; 0 \le \mu^* B/\Delta \le 5$.

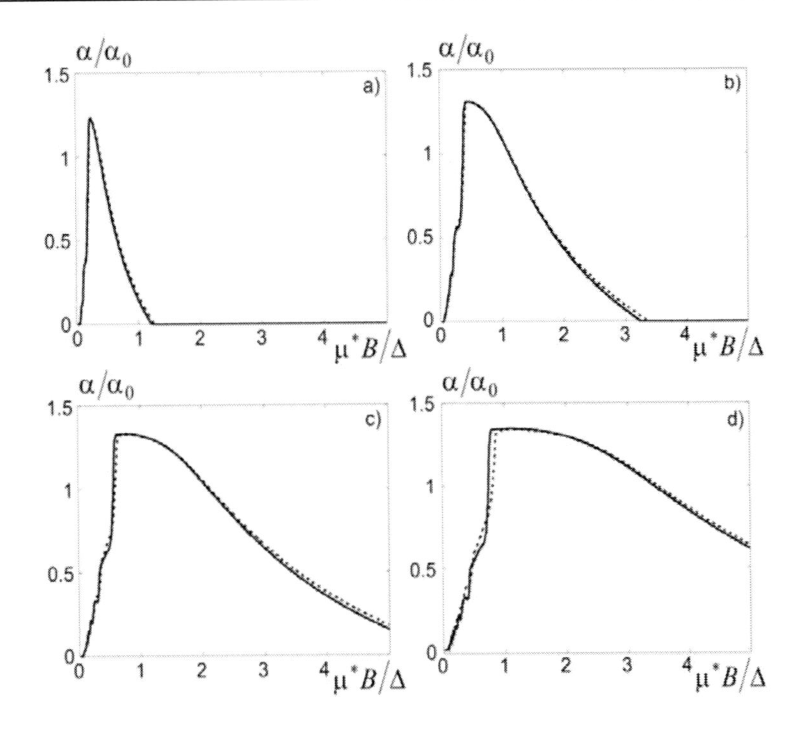

Figure 81. Field dependences of the Seebeck coefficient in the model $l = const$ with regard to magnetic field effect on scattering at $A = 5; kT/\Delta = 0.03; 0 \leq \mu^* B/\Delta \leq 5$.

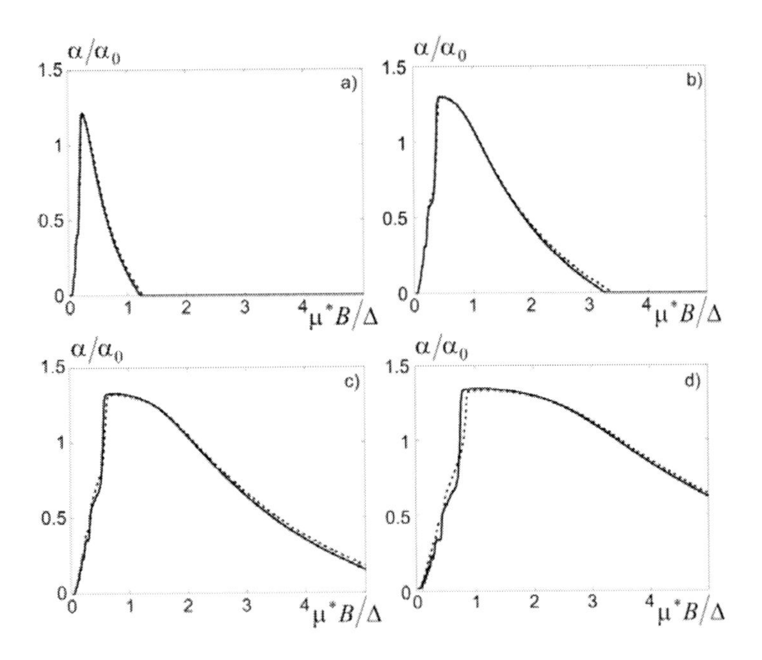

Figure 82. Field dependences of the Seebeck coefficient in the model $l = const$ with regard to magnetic field effect on scattering at $A = 15; kT/\Delta = 0.03; 0 \leq \mu^* B/\Delta \leq 5$.

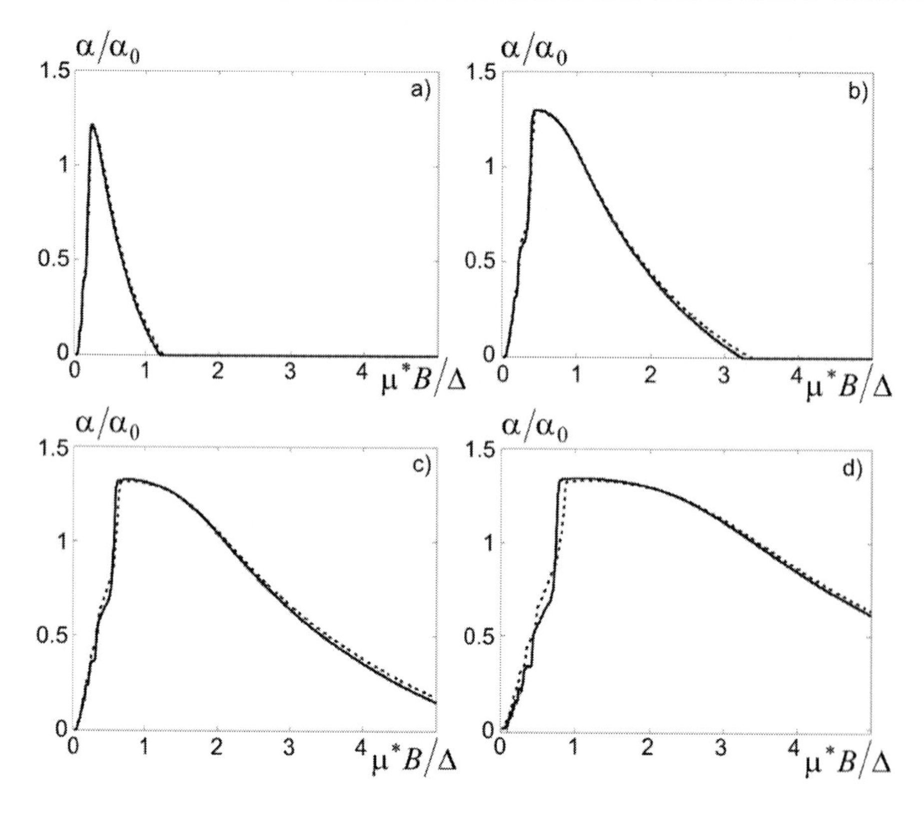

Figure 83. Field dependences of the Seebeck coefficient in the model $l = const$ with regard to magnetic field effect on scattering at $A = 25; kT/\Delta = 0.03; 0 \le \mu^* B/\Delta \le 5$.

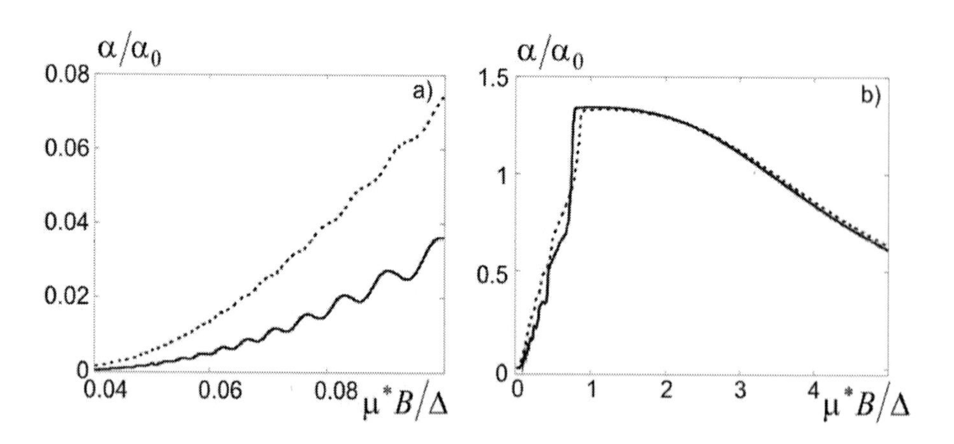

Figure 84. Field dependences of the Seebeck coefficient in the model $l = const$ with regard to magnetic field effect on scattering at $\zeta_0/\Delta = 2;\ A = 10^5; kT/\Delta = 0.03$ and: a) $0.04 \le \mu^* B/\Delta \le 0.1$; b) $0 \le \mu^* B/\Delta \le 5$.

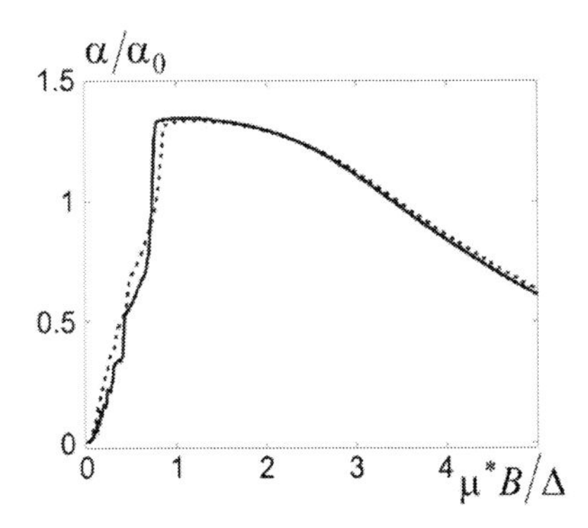

Figure 85. Field dependence of the Seebeck coefficient in the model $l = const$ without regard to magnetic field effect on scattering at $\zeta_0/\Delta = 2;\ A = 10^5; kT/\Delta = 0.03$ and $0 \le \mu^* B/\Delta \le 5$.

The figures show that in the model $l = const$ in the quasi-classical range of magnetic fields, just as in the case of constant relaxation time, the layered structure effects become apparent chiefly in the phase lag of the Seebeck coefficient oscillations and their amplitude increase as compared to the effective mass approximation. However, the model $l = const$ is significantly different from the model $\tau = const$. This difference lies in the fact that with a relatively low degree of miniband filling, for instance, at $\zeta_0/\Delta \le 1$, in a real layered crystal the oscillating function of a magnetic field is not the Seebeck coefficient itself, but its first derivative with respect to a magnetic field. And only with increase in the degree of miniband filling, for instance, at $\zeta_0/\Delta \ge 1.5$, the oscillating function of a magnetic field in a real layered crystal becomes the Seebeck coefficient itself. In the effective mass approximation, however, the Seebeck coefficient remains a monotonically increasing function of a magnetic field for all degrees of miniband filling up to conversion of a closed FS to a transient one, and the oscillating function is its first derivative with respect to a magnetic field.

A change in the anisotropy parameter within $A = 5 \div 10^5$ leads to a slight increase in the Seebeck coefficient value in the quasi-classical range of magnetic fields. In so doing, as in the model $\tau = const$, the Seebeck coefficient of a real layered crystal in the quasi-classical range of magnetic fields drops with increase in the degree of miniband filling in the range of $0.5 \le \zeta_0/\Delta \le 2$ from 13 to 3μV/K, and in the effective mass approximation – from 13 to 3÷4.7μV/K depending on the value of A in the range of 5÷25. With the value of $A = 10^5$ and $\zeta_0/\Delta = 2$ the Seebeck coefficient maximum in the considered quasi-classical range of magnetic fields is 3.2μV/K for a real layered crystal and 6.5μV/K in effective mass approximation.

With further increase in magnetic field induction, the Seebeck coefficient increases to maximum value which is little affected by the value of anisotropy parameter, as well as by the magnetic field dependence of relaxation time, but is mostly affected by the degree of miniband filling. The weak effect of anisotropy parameter and the magnetic field dependence of relaxation time on the maximum value of the Seebeck coefficient is attributable to the fact that condensation of free charge carriers begins close to this maximum on the bottom of the only filled lowest Landau subband with the number $n = 0$. Therefore, in the range of

$0.5 \le \zeta_0/\Delta \le 2$ the Seebeck coefficient maximum in the model $l = const$ varies as in the model $\tau = const$, i.e. increases from 108 to 121μV/K. Moreover, as in the model $\tau = const$, there exists such optimal range of magnetic field inductions wherein the layered structure effects are most pronounced. However, unlike the model $\tau = const$, in the model $l = const$ in this range of magnetic fields the layered structure effects manifest themselves not in the presence of local maxima and minima, but in more sharp kinks of the field dependence of the Seebeck coefficient as compared to the effective mass approximation.

Chapter 27

FIELD DEPENDENCE OF THE SEEBECK COEFFICIENT IN A LAYERED CRYSTAL IN THE MODEL OF RELAXATION TIME PROPORTIONAL TO FULL ELECTRON VELOCITY

Within this model the coefficients $A^{(\alpha)}, B^{(\alpha)}, C^{(\alpha)}$ for a real layered crystal have the following values:

$$A^{(\alpha)} = \sum_{l=1}^{\infty} (-1)^{l-1} f_l^{th} \int_0^{\gamma-b} \sqrt{(2y-y^2)[A(\gamma-y)+2y-y^2]} \sin\left(\pi l \frac{\gamma-y}{b}\right) dy, \quad (27.1)$$

$$B^{(\alpha)} = 0.5 \int_0^{\gamma-b} \sqrt{(2y-y^2)[A(\gamma-y)+2y-y^2]} dy, \quad (27.2)$$

$$C^{(\alpha)} = \sum_{l=1}^{\infty} (-1)^{l} f_l^{\sigma} \int_0^{\gamma-b} \sqrt{(2y-y^2)[A(\gamma-y)+2y-y^2]} \cos\left(\pi l \frac{\gamma-y}{b}\right) dy, \quad (27.3)$$

In the effective mass approximation these formulae acquire the form as follows:

$$A^{(\alpha)} = \sum_{l=1}^{\infty} (-1)^{l-1} f_l^{th} \int_0^{\gamma-b} \sqrt{2y[A(\gamma-y)+2y]} \sin\left(\pi l \frac{\gamma-y}{b}\right) dy, \quad (27.4)$$

$$B^{(\alpha)} = 0.5 \int_0^{\gamma-b} \sqrt{2y[A(\gamma-y)+2y]} dy, \quad (27.5)$$

$$C^{(\alpha)} = \sum_{l=1}^{\infty} (-1)^{l-1} f_l^{\sigma} \int_0^{\gamma-b} \sqrt{2y[A(\gamma-y)+2y]} \cos\left(\pi l \frac{\gamma-y}{b} \right) dy. \quad (27.6)$$

The results of calculations of the field dependences of the Seebeck coefficient according to these formulae are given in Figures 86 – 96.

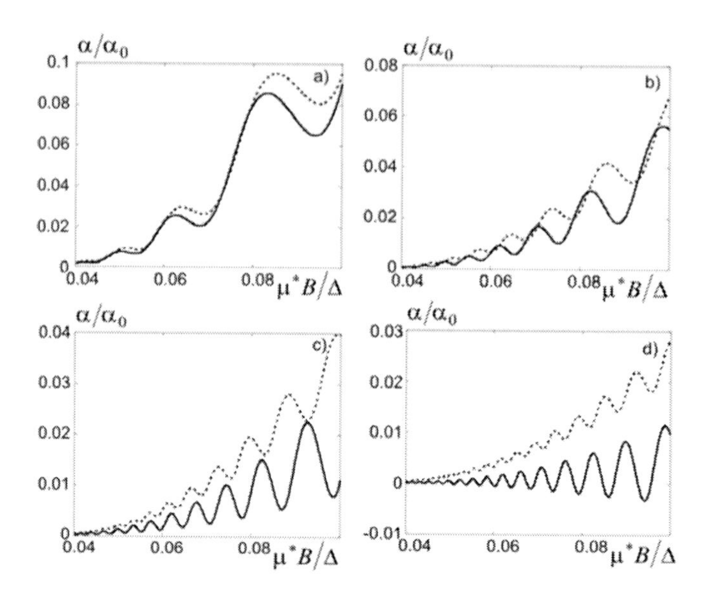

Figure 86. Field dependences of the Seebeck coefficient in the model $\tau \propto v_f$ at $A = 5; kT/\Delta = 0.03; 0.04 \le \mu^* B/\Delta \le 0.1$.

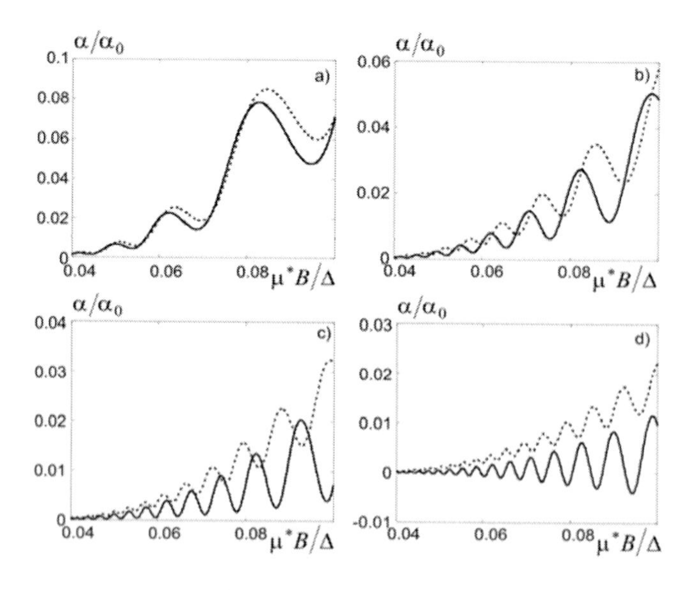

Figure 87. Field dependences of the Seebeck coefficient in the model $\tau \propto v_f$ at $A = 15; kT/\Delta = 0.03; 0.04 \le \mu^* B/\Delta \le 0.1$.

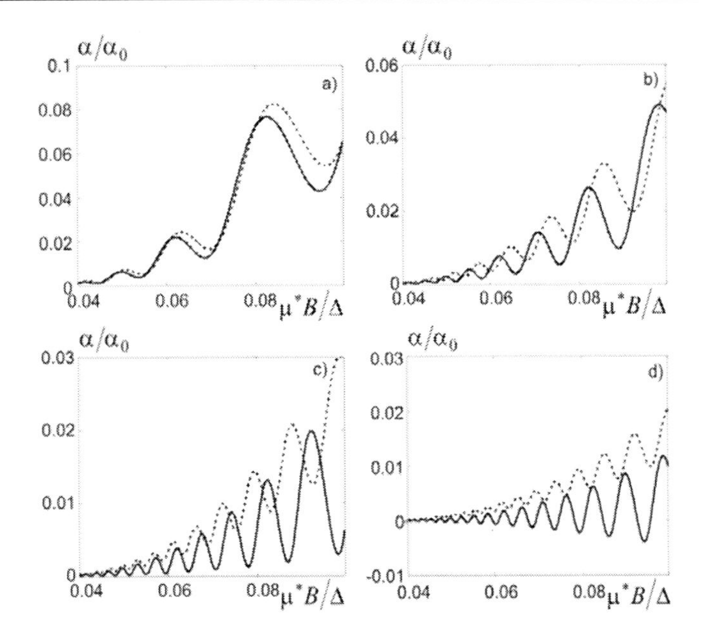

Figure 88. Field dependences of the Seebeck coefficient in the model $\tau \propto v_f$ at $A = 25; kT/\Delta = 0.03; 0.04 \le \mu^* B/\Delta \le 0.1$.

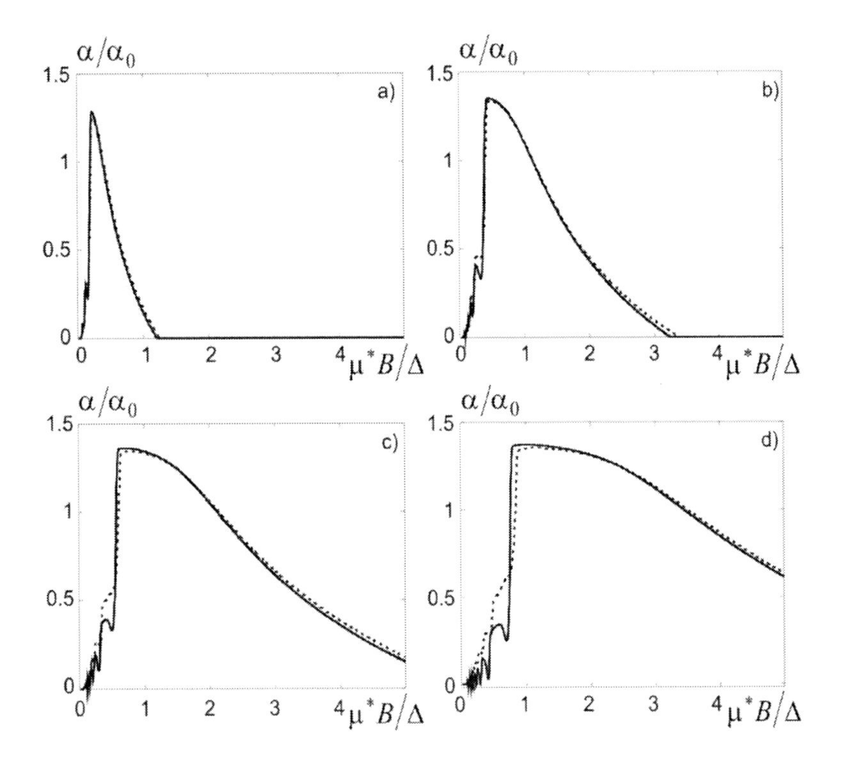

Figure 89. Field dependences of the Seebeck coefficient in the model $\tau \propto v_f$ at $A = 5; kT/\Delta = 0.03; 0 \le \mu^* B/\Delta \le 5$ without regard to magnetic field dependence of τ.

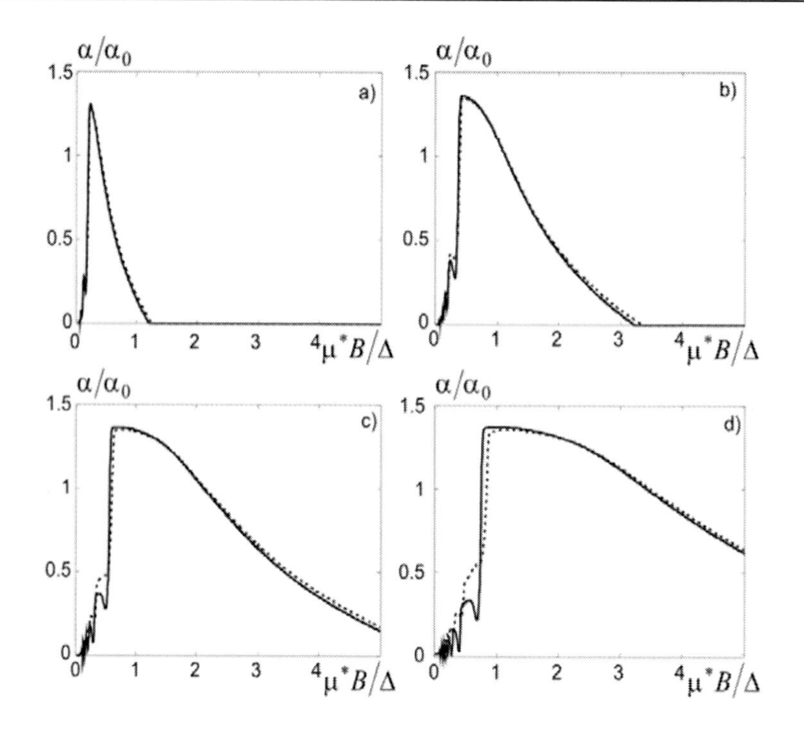

Figure 90. Field dependences of the Seebeck coefficient in the model $\tau \propto v_f$ at $A = 15; kT/\Delta = 0.03; 0 \le \mu^* B/\Delta \le 5$ without regard to magnetic field dependence of τ.

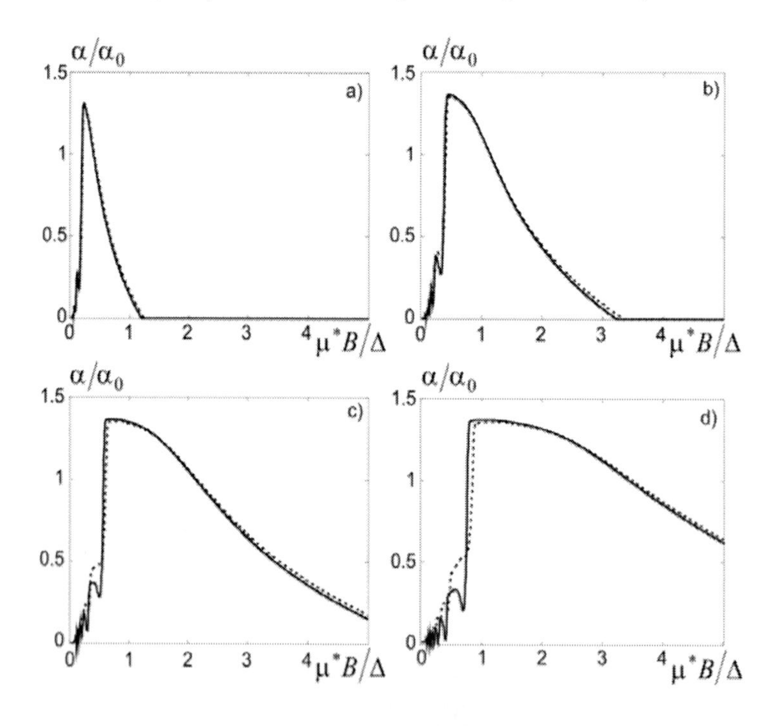

Figure 91. Field dependences of the Seebeck coefficient in the model $\tau \propto v_f$ at $A = 25; kT/\Delta = 0.03; 0 \le \mu^* B/\Delta \le 5$ without regard to magnetic field dependence of τ.

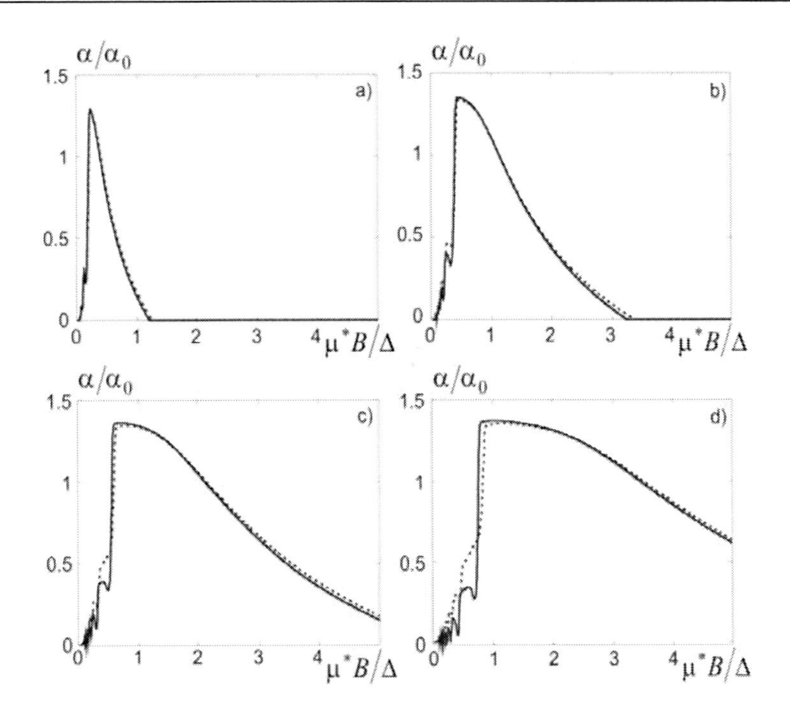

Figure 92. Field dependences of the Seebeck coefficient in the model $\tau \propto v_f$ at $A = 5; kT/\Delta = 0.03; 0 \leq \mu^* B/\Delta \leq 5$ with regard to magnetic field dependence of τ.

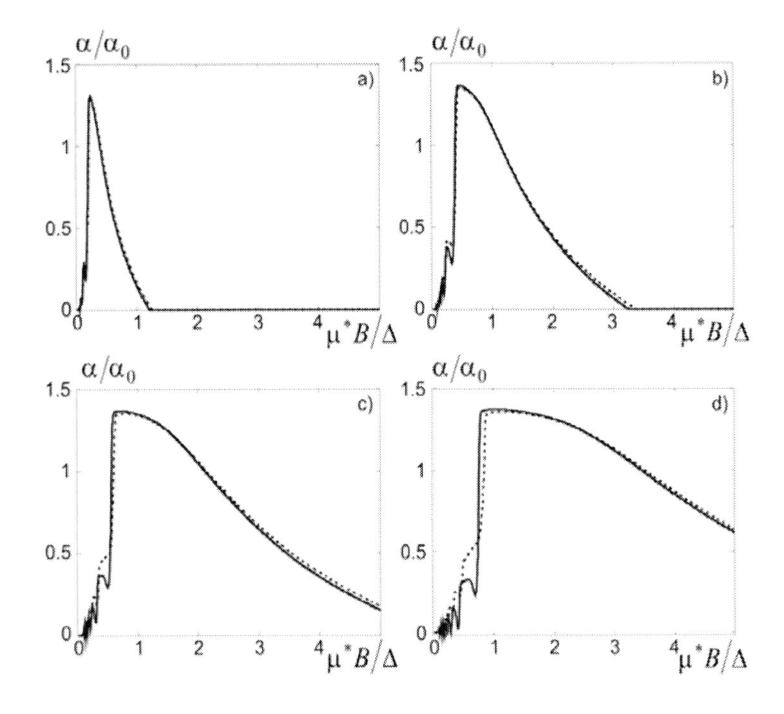

Figure 93. Field dependences of the Seebeck coefficient within the model $\tau \propto v_f$ at $A = 15; kT/\Delta = 0.03; 0 \leq \mu^* B/\Delta \leq 5$ with regard to magnetic field dependence of τ.

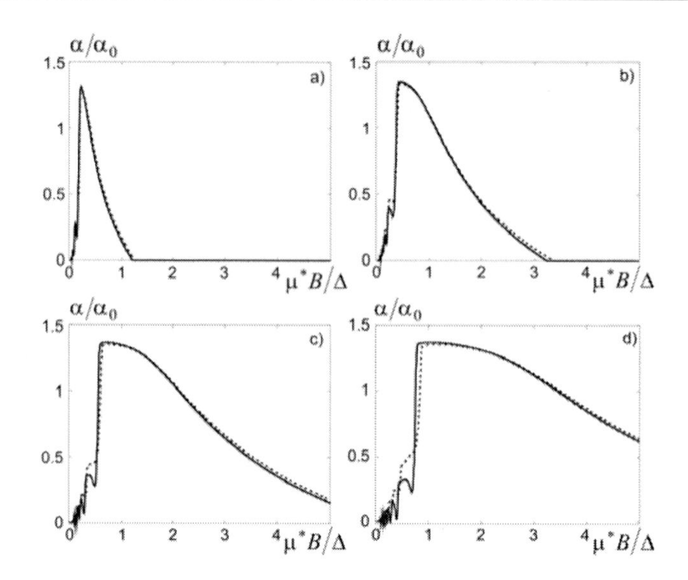

Figure 94. Field dependences of the Seebeck coefficient in the model $\tau \propto v_f$ at $A = 25; kT/\Delta = 0.03; 0 \le \mu^* B/\Delta \le 5$ with regard to magnetic field dependence of τ.

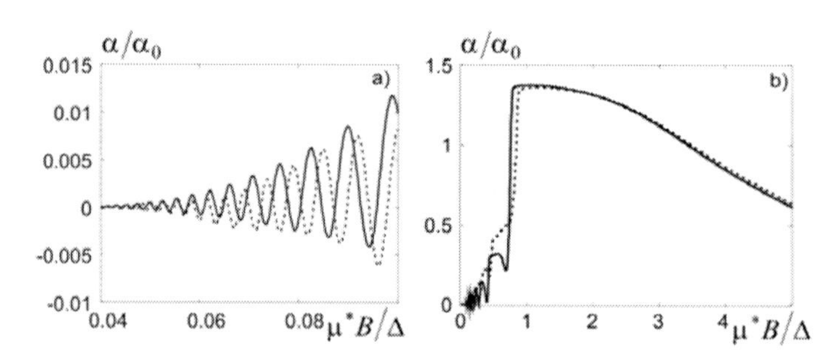

Figure 95. Field dependences of the Seebeck coefficient in the model $\tau \propto v_f$ with regard to magnetic field effect on scattering at $\zeta_0/\Delta = 2;\ A = 10^5; kT/\Delta = 0.03$ and: a) $0.04 \le \mu^* B/\Delta \le 0.1$; b) $0 \le \mu^* B/\Delta \le 5$.

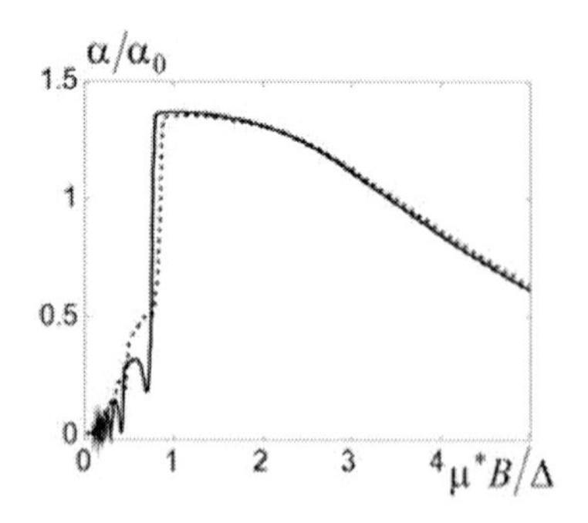

Figure 96. Field dependence of the Seebeck coefficient in the model $\tau \propto v_f$ without regard to magnetic field effect on scattering at $\zeta_0/\Delta = 2;\ A = 10^5; kT/\Delta = 0.03$ and $0 \le \mu^* B/\Delta \le 5$.

First and foremost, the figures show that in the model $\tau \propto v_f$, as in the model $\tau = const$, in the quasi-classical range of magnetic fields the oscillating function of the magnetic field is the Seebeck coefficient itself. However, in the model $\tau \propto v_f$ the absolute value of the Seebeck coefficient is lower than in the model $\tau = const$. With increase in anisotropy parameter A in this range of magnetic fields, the Seebeck coefficient modulus slightly drops. In the range of anisotropy parameters $5 \leq A \leq 25$ in the model $\tau \propto v_f$, as in the models $\tau = const$ and $l = const$, the layered structure effects become apparent primarily in the phase lag of the Seebeck coefficient oscillations and their amplitude increase as compared to the effective mass approximation. In so doing, the Seebeck coefficient modulus in the model $\tau \propto v_f$, as in the other considered models, in the effective mass approximation at $5 \leq A \leq 25$ is somewhat higher than for a real layered crystal. However, at $A = 10^5$ and $\zeta_0/\Delta = 2$, i.e. at rather high anisotropy parameters for a transient FS the situation is changed: the Seebeck coefficient modulus in a layered crystal becomes larger than in the effective mass approximation. At $5 \leq A \leq 25$ and $0.5 \leq \zeta_0/\Delta \leq 2$ in the quasi-classical range of magnetic fields the maximum Seebeck coefficient modulus in a real layered crystal changes from 7.8 to 0.86μV/K, and in the effective mass approximation – from 8.2 to 1.73μV/K, dropping with increase in the degree of miniband filling.

In stronger magnetic fields the Seebeck coefficient reaches maximum, which in a real layered crystal is somewhat larger than in the effective mass approximation. According to the analysis, this maximum has only a weak dependence on the anisotropy parameter and the ratio ζ_0/Δ, and is close to 121μV/K. The degree of miniband filling, as in the other considered models of charge carrier scattering, affects mainly the position of the Seebeck coefficient peak on the scale of magnetic field inductions, as well as the form and width of this peak, i.e. the rate of increase to maximum and the rate of decrease after maximum with increase in magnetic field induction. Thus, it turns out that in a quantizing magnetic field at low temperatures the Seebeck coefficient maximum is largely determined by the thermodynamic features of electron gas, rather than by subtle details of scattering mechanisms or charge carrier band spectrum. It is mainly due to the fact that at low temperatures the thermodiffusion flux and electric current are affected by charge carrier scattering close to the FS.

Besides, the figures show that in the model $\tau \propto v_f$, as in the other considered models, there is an optimal range of magnetic field inductions wherein the layered structure effects are most pronounced. In this model, as in the model $\tau = const$, they consist in the presence of local maxima and minima of the Seebeck coefficient which in the model $\tau \propto v_f$, unlike the model $\tau = const$, are present not only in a real layered crystal, but also in the effective mass approximation. But in a real layered crystal these maxima and minima are much more pronounced than in the effective mass approximation. Moreover, the layered structure effects, as in the other considered models, are evident as a slight shift of the Seebeck coefficient maximum toward weaker magnetic fields as compared to the effective mass approximation and in a faster decay of the Seebeck coefficient in a real layered crystal as compared to the effective mass approximation after the maximum. Note also that in the model $\tau \propto v_f$, as in the model $l = const$, account of the magnetic field effect on scattering through the magnetic field dependence of electron gas chemical potential exerts a weak influence on the character of the magnetic field dependence of the Seebeck coefficient, because the magnetic field dependence of electron gas chemical potential is most essential near the area of current carrier condensation to the bottom of the lowest filled Landau subband, where the relaxation time is weakly dependent on the longitudinal quasi-pulse, hence, it can be considered practically constant. By virtue of this, in the models $l = const$ and $\tau \propto v_f$ the asymptotic law of the Seebeck coefficient tending to zero in the ultra-quantum limit is the same as in the model $\tau = const$, i.e. it is determined by formula (25.9).

Chapter 28

FIELD DEPENDENCE OF THE SEEBECK COEFFICIENT IN THE MODEL OF RELAXATION TIME PROPORTIONAL TO LONGITUDINAL VELOCITY

Within this model, coefficients $A^{(\alpha)}, B^{(\alpha)}, C^{(\alpha)}$ for a real layered crystal have the following values:

$$A^{(\alpha)} = \sum_{l=1}^{\infty} f_l^{th}\left[\frac{(\mu^* B)^3}{\pi l\Delta^3} + \frac{\mu^* B\zeta^2}{\pi l\Delta^3} - 2\left(\frac{\mu^* B}{\pi l\Delta}\right)^3 - \frac{2(\zeta-\Delta)(\mu^* B)^2}{\pi l\Delta^3} - \frac{2\mu^* B\zeta}{\pi l\Delta^2}\right] + \\ + 2\sum_{l=1}^{\infty} f_l^{th}\left[\left(\frac{\mu^* B}{\pi l\Delta}\right)^3 \cos\left(\frac{\pi l\zeta}{\mu^* B}\right) + \left(\frac{\mu^* B}{\pi l\Delta}\right)^2 \sin\left(\frac{\pi l\zeta}{\mu^* B}\right)\right], \tag{28.1}$$

$$B^{(\alpha)} = \frac{(\zeta-\mu^* B)^2(3\Delta-\zeta+\mu^* B)}{6\Delta^3}, \tag{28.2}$$

$$C^{(\alpha)} = 2\sum_{l=1}^{\infty} f_l^{\sigma}\left[\frac{1}{\pi^2 l^2}\left(\frac{\mu^* B}{\Delta}\right)^3 + \frac{(\mu^* B)^2(\Delta-\zeta)}{\pi^2 l^2\Delta^3}\right] + \\ + 2\sum_{l=1}^{\infty}(-1)^l f_l^{\sigma}\left[\left(\frac{\mu^* B}{\pi l\Delta}\right)^3 \sin\left(\frac{\pi l\zeta}{\mu^* B}\right) - \left(\frac{\mu^* B}{\pi l\Delta}\right)^2 \cos\left(\frac{\pi l\zeta}{\mu^* B}\right)\right]. \tag{28.3}$$

In the effective mass approximation these coefficients are as follows:

$$A^{(\alpha)} = 2\sum_{l=1}^{\infty}(-1)^l f_l^{th}\left(\frac{\mu^* B}{\pi l\Delta}\right)^2 \sin\left(\frac{\pi l\zeta}{\mu^* B}\right) - 2\sum_{l=1}^{\infty}\frac{\mu^* B(\zeta-\mu^* B)f_l^{th}}{\pi l\Delta^2}, \tag{28.4}$$

$$B^{(\alpha)} = \frac{(\zeta-\mu^* B)^2}{2\Delta^2}, \tag{28.5}$$

$$C^{(\alpha)} = 2\sum_{l=1}^{\infty}\left(\frac{\mu^* B}{\pi l \Delta}\right)^2 f_l^{\sigma} - 2\sum_{l=1}^{\infty}(-1)^l f_l^{\sigma}\left(\frac{\mu^* B}{\pi l \Delta}\right)^2 \cos\left(\frac{\pi l \zeta}{\mu^* B}\right). \tag{28.6}$$

The results of calculations of the field dependences of the Seebeck coefficient by formulae (28.1) – (28.6) in the framework of this model are given in Figures 97 – 98.

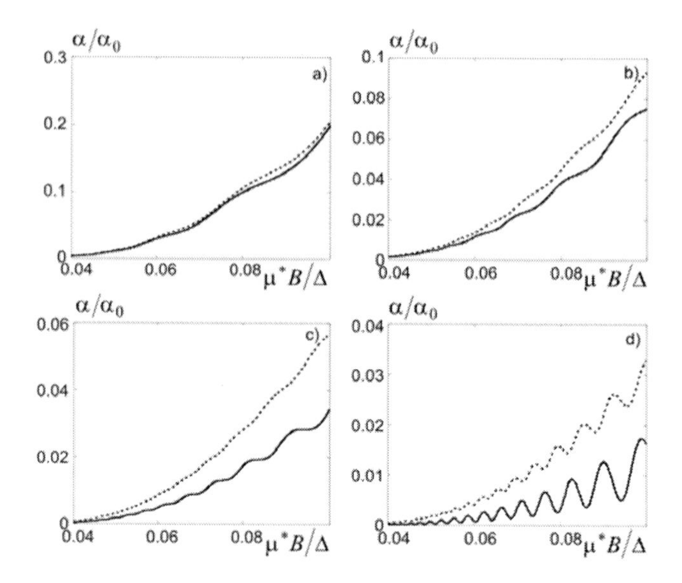

Figure 97. Field dependences of the Seebeck coefficient in the model $\tau \propto v_z$ at $kT/\Delta = 0.03; 0.04 \le \mu^* B/\Delta \le 0.1$.

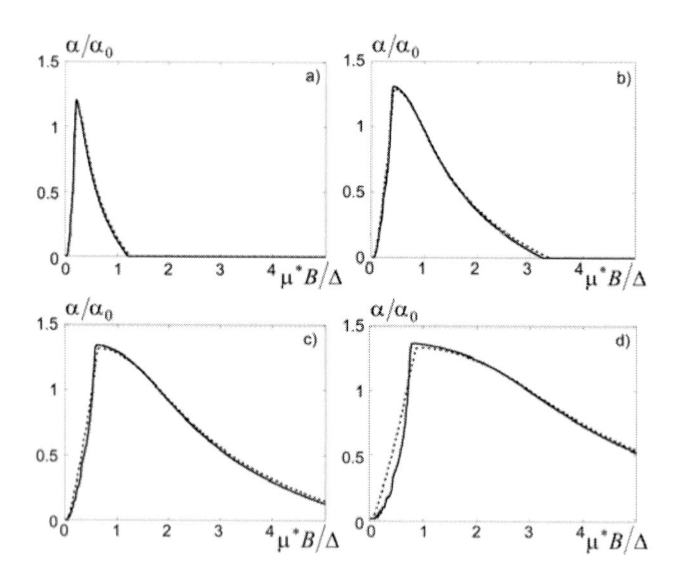

Figure 98. Field dependences of the Seebeck coefficient in the model $\tau \propto v_z$ at $kT/\Delta = 0.03; 0.04 \le \mu^* B/\Delta \le 0.1$.

The figures demonstrate that in the model $\tau \propto v_z$, in the quasi-classical range of magnetic fields, at $0.5 \le \zeta_0/\Delta \le 1.5$, the oscillating function of a magnetic field is not the Seebeck coefficient itself, but its first derivative with respect to a magnetic field. But in the case of a transient FS, i.e. at $\zeta_0/\Delta = 2$, the oscillating function of a magnetic field becomes the Seebeck coefficient itself. It should be noted that in the model $\tau \propto v_z$ the Seebeck coefficient oscillations at $\zeta_0/\Delta = 2$ occur both for a real layered crystal and in the effective mass approximation. The layered structure effects in the model $\tau \propto v_z$ in the quasi-classical range of magnetic fields become apparent in the phase lag of oscillations of the first derivative of the Seebeck coefficient with respect to a magnetic field or the Seebeck coefficient proper and in the amplitude increase of these oscillations in a real layered crystal as compared to the effective mass approximation. Moreover, the layered structure effects in the quasi-classical range of magnetic fields in this model of charge carrier scattering, as in the other considered models, become apparent in the reduction of the Seebeck coefficient modulus. As in the other considered models of charge carrier scattering, with increase in the degree of miniband filling in the range of $0.5 \le \zeta_0/\Delta \le 2$, the oscillation frequency of the first derivative of the Seebeck coefficient with respect to a magnetic field or the Seebeck coefficient itself, is increased, and its modulus is decreased. The maximum Seebeck coefficient value in the quasi-classical range of magnetic fields with increase in the degree of miniband filling in the range of $0.5 \le \zeta_0/\Delta \le 2$ is reduced from 17.3 to 1.6μV/K in a real layered crystal and from 17.3 to 2.8μV/K in the effective mass approximation.

With further increase in magnetic field induction, the Seebeck coefficient is at first drastically increased, reaching maximum, and then decays, tending to zero by a certain asymptotic law. The layered structure effects and the degree of miniband filling have a relatively weak influence on the Seebeck coefficient maximum value, and with increase in the degree of miniband filling, in the range of $0.5 \le \zeta_0/\Delta \le 2$, this maximum increases from 108 to 121μV/K. Besides, the layered structure effects lead to slight increase of this maximum and its shift toward weaker magnetic fields with respect to the effective mass approximation. As before, there is an optimal range of magnetic field inductions wherein the layered structure effects are most pronounced and manifest themselves in the emergence in a real layered crystal of kinks on the field dependence of the Seebeck coefficient that are absent in the effective mass approximation.

Chapter 29

EFFECT OF CHARGE ORDERING ON THE SEEBECK COEFFICIENT OF LAYERED CRYSTALS

Effect of charge ordering on the Seebeck coefficient of layered crystals will be considered within the model of constant relaxation time. In this case the coefficients $A^{(\alpha)}, B^{(\alpha)}, C^{(\alpha)}$ have the following values [57]:

$$A^{(\alpha)} = \sum_{l=1}^{\infty} (-1)^l f_l^{th} \int_{-(\gamma-b)}^{\sqrt{w^2\delta^2+1}} |y^{-1}| \sqrt{(1+w^2\delta^2-y^2)(y^2-w^2\delta^2)} \sin[\pi l b^{-1}(\gamma-y)] dy, \quad (29.1)$$

$$B^{(\alpha)} = 0.5 \int_{-(\gamma-b)}^{\sqrt{w^2\delta^2+1}} |y^{-1}| \sqrt{(1+w^2\delta^2-y^2)(y^2-w^2\delta^2)} dy, \quad (29.2)$$

$$C^{(\alpha)} = \sum_{l=1}^{\infty} (-1)^l f_l^{\sigma} \int_{-(\gamma-b)}^{\sqrt{w^2\delta^2+1}} |y^{-1}| \sqrt{(1+w^2\delta^2-y^2)(y^2-w^2\delta^2)} \cos[\pi l b^{-1}(\gamma-y)] dy. \quad (29.3)$$

The results of calculations of the Seebeck coefficient of charge-ordered layered crystals by formulae (29.1) – (29.3) are given in Figures 99 and 100.

The figures show that in the ordered state in the quasi-classical range of magnetic fields the Seebeck coefficient oscillations take double periodic form. The reason for this, as in the case of dHvA and SdH effects, is that in the charge-ordered state there is a topological transition from a transient FS to an open one. Moreover, in the charge-ordered state the Seebeck coefficient oscillations have variable polarity, whereas in the disordered state the polarity of these oscillations is constant, and, for instance, in the case of electrons, negative. With increase in the value of effective interaction, the amplitude of the Seebeck oscillations and their high "carrier" frequency are increased, and the beat frequency of the Seebeck coefficient is reduced. The reason for this, as in the case of dHvA and SdH effects, is that in the charge-ordered state due to narrowing of conduction minibands, the areas of extreme sections of an open FS become close to each other. In the charge-ordered state at considered values of effective interaction the maximum values of the Seebeck coefficient in the quasi-classical range of magnetic fields are 12.9, 8.63 and 19.8μV/K, respectively. However, in the disordered state in the quasi-classical range of magnetic fields the Seebeck coefficient does not exceed 1.47μV/K, i.e. it is a factor of 5.9÷13.5 lower than in the ordered state. Thus, transition into the charge-ordered state increases the Seebeck coefficient considerably.

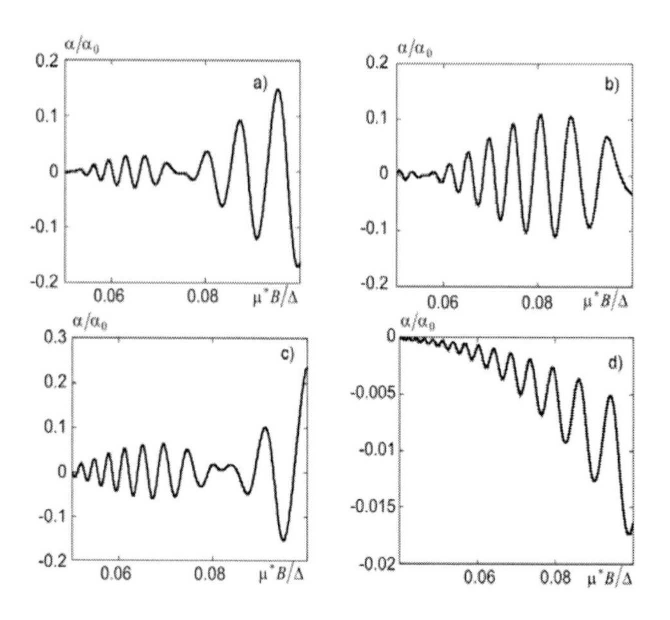

Figure 99. Field dependences of the Seebeck coefficient at $0.04 \le \mu^* B/\Delta \le 0.1$, $\zeta_{02D}/\Delta = 1, kT/\Delta = 0.03$ and: a) $W_0/\zeta_{02D} = 1.5$; b) $W_0/\zeta_{02D} = 2$; c) $W_0/\zeta_{02D} = 2.5$; d) in the disordered state.

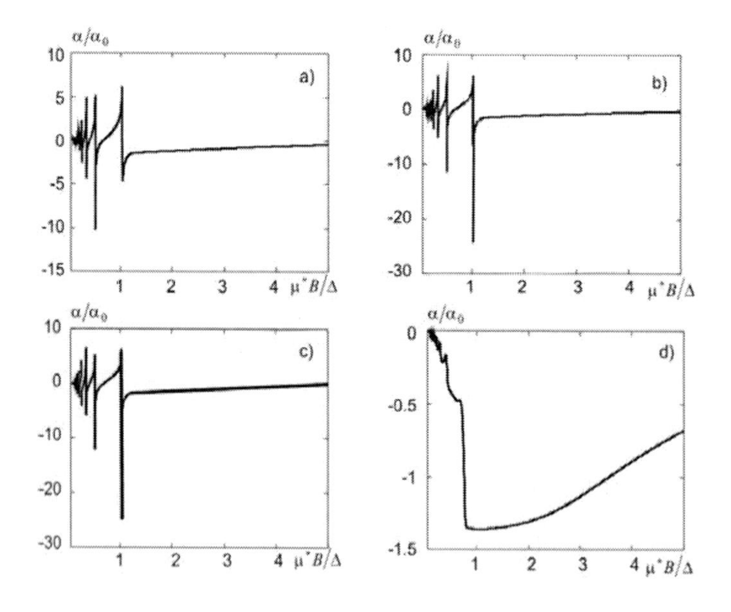

Figure 100. Field dependences of the Seebeck coefficient at $0 \le \mu^* B/\Delta \le 5$, $\zeta_{02D}/\Delta = 1, kT/\Delta = 0.03$ and: a) $W_0/\zeta_{02D} = 1.5$; b) $W_0/\zeta_{02D} = 2$; c) $W_0/\zeta_{02D} = 2.5$; d) in the disordered state.

These results are confirmed by the results of studying the field dependences of the Seebeck coefficient in stronger magnetic fields. In this case in the charge-ordered state multiple inversion of the Seebeck coefficient polarity takes place with a change in magnetic field. With increase in effective interaction value, the inversion frequency and the amplitude of the Seebeck coefficient increase. Polarity inversions of the Seebeck coefficient are synchronized with changes in electron gas chemical potential as a function of a magnetic field. Namely, to drastic jumps of chemical potential as a function of a magnetic field correspond drastic polarity inversions, whereas to portions of linear change in chemical potential correspond curvilinear portions of a smooth change in the Seebeck coefficient polarity. At a point of inverse topological transition from an open FS to a closed one for each value of effective interaction there is the last polarity inversion of the Seebeck coefficient with the largest amplitude, following which the Sebeck coefficient drastically decays, tending to zero by the inverse square law, i.e.:

$$|\alpha| = 0.305\alpha_0 \frac{\zeta_{02D}^2 \Delta^2}{kT(\mu^* B)^2 \sqrt{W_0^2 \delta^2 + \Delta^2}}. \tag{29.4}$$

At the same time, in the disordered state the Seebeck coefficient for a closed or transient FS in the model of constant relaxation time in the entire range of magnetic fields retains its polarity, remaining, for instance, for electrons, negative. The Seebeck coefficient maximum with considered values of effective interaction is 890, 2093 and 2147μV/K, respectively, whereas in the disordered state the Seebeck coefficient maximum with selected problem parameters does not exceed 117μV/K, which is a factor of 7.6 to 18.4 lower than in the charge-ordered state. At the same time, according to formula (256) at $B = 60$ T and previously specified values of effective interaction and other problem parameters we get 0.475, 0.358 and 0.287μV/K, which is three orders of magnitude lower than the respective maxima. In the disordered state at $B = 60$ T the Seebeck coefficient is 0.725μV/K, which is a factor of 162 lower than the respective maximum. Therefore, at real low temperatures the asymptotic law (29.4) does not lead to incorrect physical results.

Chapter 30

POWER FACTOR OF A LAYERED CRYSTAL IN THE APPROXIMATION OF CONSTANT RELAXATION TIME

Power factor of thermoelectric material, including layered one, is an important integral characteristic of a subsystem of free charge carriers in it. This factor is a product of the Seebeck coefficient square by the electric conductivity. In this section we will consider the influence of layered structure effects on the field dependence of power factor of a layered crystal in the approximation of constant relaxation time. In the most general form power factor of a layered crystal can be written as follows [56]:

$$P = \widetilde{P}_0 \frac{A^{(\alpha)2}}{B^{(\alpha)} + C^{(\alpha)}}, \tag{30.1}$$

where the value $\widetilde{P}_0$ and the coefficients $A^{(\alpha)}, B^{(\alpha)}, C^{(\alpha)}$ are determined by specific dispersion law $W(x)$ describing the interlayer motion of charge carriers and specific model of relaxation time corresponding to some or other mechanism of charge carrier scattering. In particular, in the approximation of constant relaxation time the coefficients $A^{(\alpha)}, B^{(\alpha)}, C^{(\alpha)}$ for a real layered crystal are determined by formulae (25.1) – (25.3), and the value P_0 is equal to:

$$\widetilde{P}_0 = \frac{2\pi^3 k^2 m^* a^2 \Delta}{h^3 \gamma_0} \sqrt[3]{\frac{12\pi^2 m^* \Delta f(\gamma_0)}{ah^2}} N. \tag{30.2}$$

In the effective mass approximation the coefficients $A^{(\alpha)}, B^{(\alpha)}, C^{(\alpha)}$ are determined by formulae (25.4) – (25.6), and the value P_0 is equal to:

$$\widetilde{P}_0 = \frac{2\pi^3 k^2 m^* a^2 \Delta}{h^3 \gamma_{0em}} \sqrt[3]{\frac{12\pi^2 m^* \Delta f(\gamma_0)}{ah^2}} N. \tag{30.3}$$

The field dependences of power factor in the approximation of constant relaxation time are shown in Figures 101 and 102. In the construction of plots it is assumed that $P_0 = 2\tilde{P}_0/\pi^2$

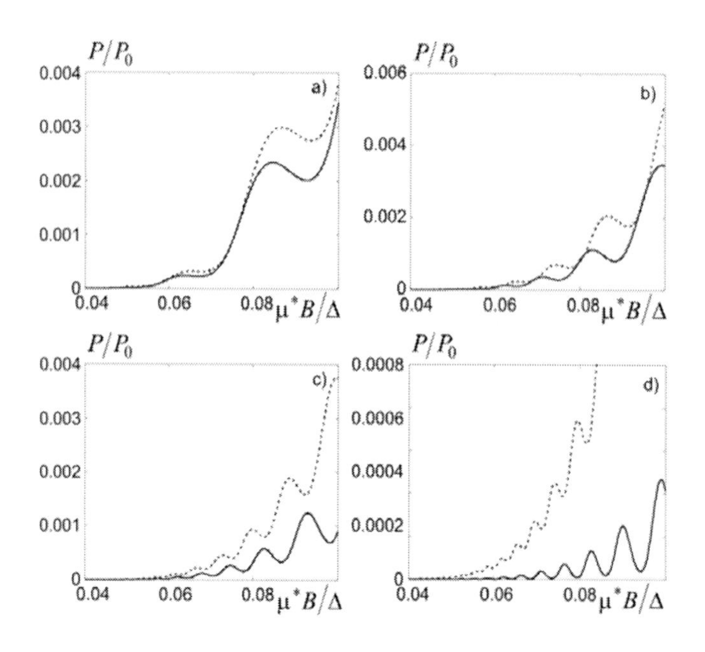

Figure 101. Field dependences of power factor in the model $\tau = const$ at $kT/\Delta = 0.03$ $0.04 \le \mu^* B/\Delta \le 0.1$.

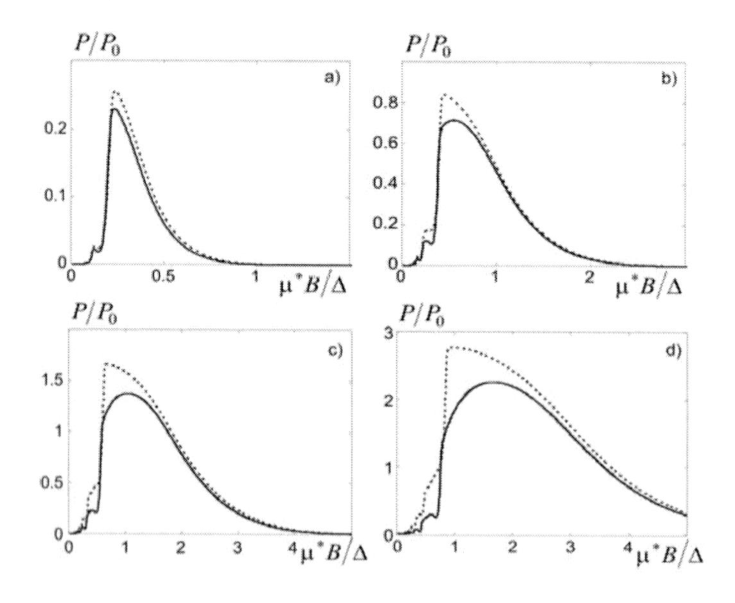

Figure 102. Field dependences of power factor in the model $\tau = const$ at $kT/\Delta = 0.03$ $0 \le \mu^* B/\Delta \le 5$.

The figures demonstrate that in the approximation of constant relaxation time the layered structure effects manifest themselves in the lag of power factor oscillations of a real layered crystal as compared to the effective mass approximation and in the reduction of this factor value as compared to the effective mass approximation.

In the quasi-classical approximation, crystal power factor is determined as follows:

$$P = P_0 \frac{3b^3}{(2\gamma)^{3/2}} \left[(-1)^l l^{-3/2} f_l^{th} \cos\left(\pi l \gamma b^{-1} - \frac{\pi}{4} \right) \right]^2 . \qquad (30.4)$$

From this formula it is again obvious that with a constant Fermi energy it is impossible to identify the influence of layered structure effects on power factor. Nevertheless, phase lag of power factor oscillations in a layered crystal is explained by a larger value of electron density of states in a layered crystal, and power factor reduction - by a decrease in longitudinal electric conductivity of crystal due to restriction of the longitudinal motion of charger carriers. In so doing, the difference in solid and dashed curves drastically increases with increase in charge carrier concentration, i.e. the ratio ζ_0/Δ.

The figures also show that with increase in magnetic field induction, power factor first increases and reaches maximum, and then starts to decay, tending to zero according to a certain asymptotic law that will be discussed below. For each degree of miniband filling there is an optimal range of magnetic field inductions where the influence of layered structure effects on power factor is most pronounced. Before achievement of maximum these effects show themselves in the presence in a real layered crystal of local maxima and minima of power factor, whereas in the effective mass approximation in the model $\tau = const$ on the field dependence of power factor there are only kinks. Moreover, for a real layered crystal the peak of power factor is rounded, whereas in the effective mass approximation it has almost a triangular form. The existence of power factor maximum is attributable to the same physical reasons that the existence of the Seebeck coefficient maximum, though positions of these maxima on the scale of magnetic field inductions, generally speaking, do not coincide. The relative and absolute values of power factor maximum in a real layered crystal with the same problem parameters are lower than in the effective mass approximation by virtue of the fact that restriction of the longitudinal motion of current carriers in a real layered crystal as compared to the effective mass approximation has a weak effect on the Seebeck coefficient, but reduces crystal electric conductivity. According to numerical calculations, with previously specified parameters of a layered crystal, $N = 10^4$ and $\zeta_0/\Delta = 0.5;1;1.5;2$, respectively, maximum power factor for a real layered crystal is $5.81 \cdot 10^{-5};1.32 \cdot 10^{-4};2.19 \cdot 10^{-4};3.04 \cdot 10^{-4}$ W/(m·K^2), respectively, and in the effective mass approximation – $6.48 \cdot 10^{-5};1.49 \cdot 10^{-4};2.48 \cdot 10^{-4};3.47 \cdot 10^{-4}$ W/(m·K^2), respectively. Thus, for closed FS maximum power factor in the effective mass approximation is 12 to 14% larger than for a real layered crystal. With the specified problem parameters these maxima are achieved in the range of magnetic fields 0.4÷2T.

Let us now establish the asymptotic law of change in power factor as a function of a magnetic field in the ultra-quantum limit. Proceeding in the same way as when deriving the asymptotic formulae for the longitudinal electric conductivity and the Seebeck coefficient, in the case of a real layered crystal we will get:

$$P = 1.929 \cdot 10^{-2} k^2 f^7(\gamma_0) \frac{m^* a^2 \Delta^{10}}{h^3 (\mu^* B)^6 (kT)^3 \gamma_0} \sqrt[3]{\frac{m^* \Delta f(\gamma_0)}{ah^2}} N. \tag{30.5}$$

In the effective mass approximation we will obtain:

$$P = 1.929 \cdot 10^{-2} k^2 f^7(\gamma_0) \frac{m^* a^2 \Delta^{10}}{h^3 (\mu^* B)^6 (kT)^3 \gamma_{0em}} \sqrt[3]{\frac{m^* \Delta f(\gamma_0)}{ah^2}} N. \tag{30.6}$$

With previously specified problem parameters $kT/\Delta = 0.03$ and $B = 60\,\mathrm{T}$ we get a reduction of power factor by 6 orders of magnitude as compared to its maximum values. Therefore, though the resulting asymptotic law $P \propto B^{-6} T^{-3}$ is somewhat hard to understand by virtue of the fact that power factor does not tend to zero at $T = 0$, at real low temperatures this law does not lead to incorrect physical results.

Chapter 31

POWER FACTOR OF A LAYERED CRYSTAL IN THE APPROXIMATION OF CONSTANT MEAN FREE PATH OF CHARGE CARRIERS

In the approximation of constant mean free path of charge carriers the coefficients $A^{(\alpha)}, B^{(\alpha)}, C^{(\alpha)}$ are determined by formulae (26.1) – (26.6). The value $\widetilde{P}_0$ is determined as:

$$\widetilde{P}_0 = \frac{8\pi k^2 N}{aAh}. \tag{31.1}$$

The results of calculations of power factor of a layered crystal in the approximation of constant mean free path are given in Figures 103 – 113. As before, when constructing the plots, it is assumed that $P_0 = 2P_0'/\pi^2$.

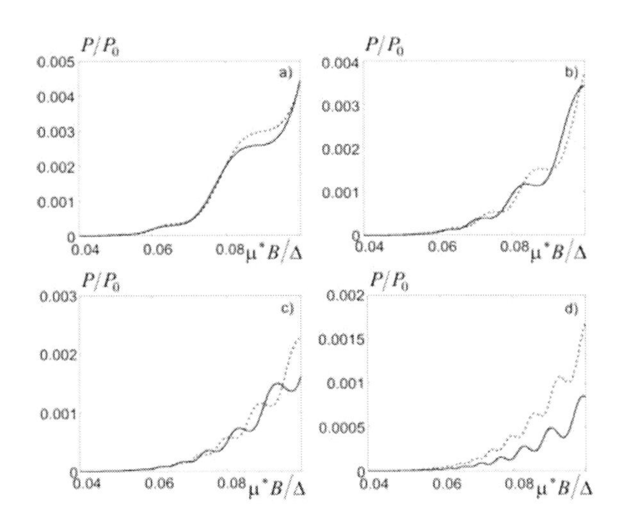

Figure 103. Field dependence of power factor of a layered crystal in the model $l = const$ at $A = 5; kT/\Delta = 0.03; 0.04 \le \mu^* B/\Delta \le 0.1$.

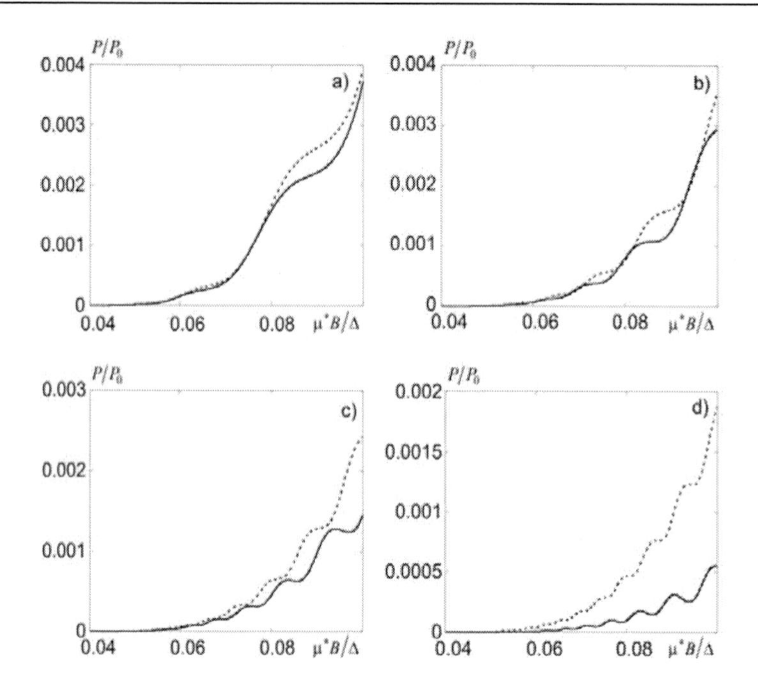

Figure 104. Field dependence of power factor of a layered crystal in the model $l = const$ at $A = 15; kT/\Delta = 0.03; 0.04 \le \mu^* B/\Delta \le 0.1$.

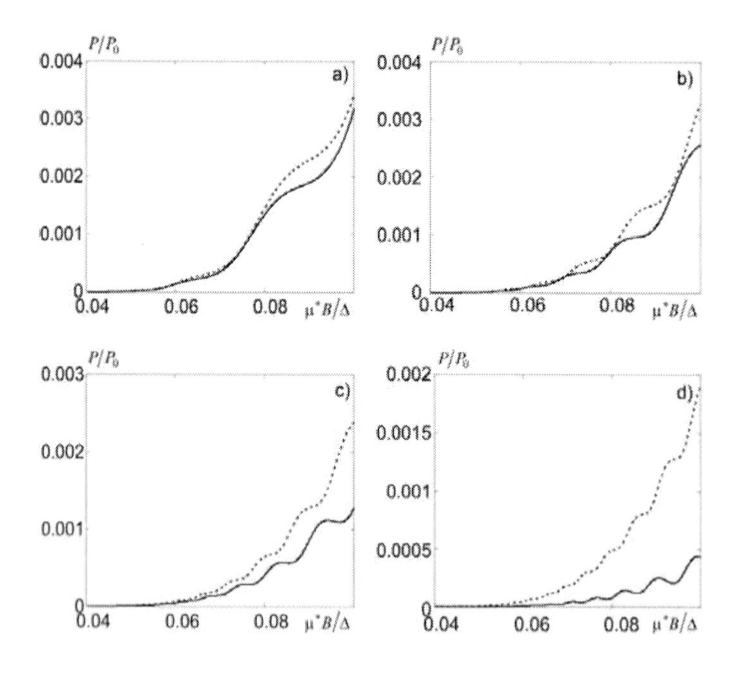

Figure 105. Field dependence of power factor of a layered crystal in the model $l = const$ at $A = 25; kT/\Delta = 0.03; 0.04 \le \mu^* B/\Delta \le 0.1$.

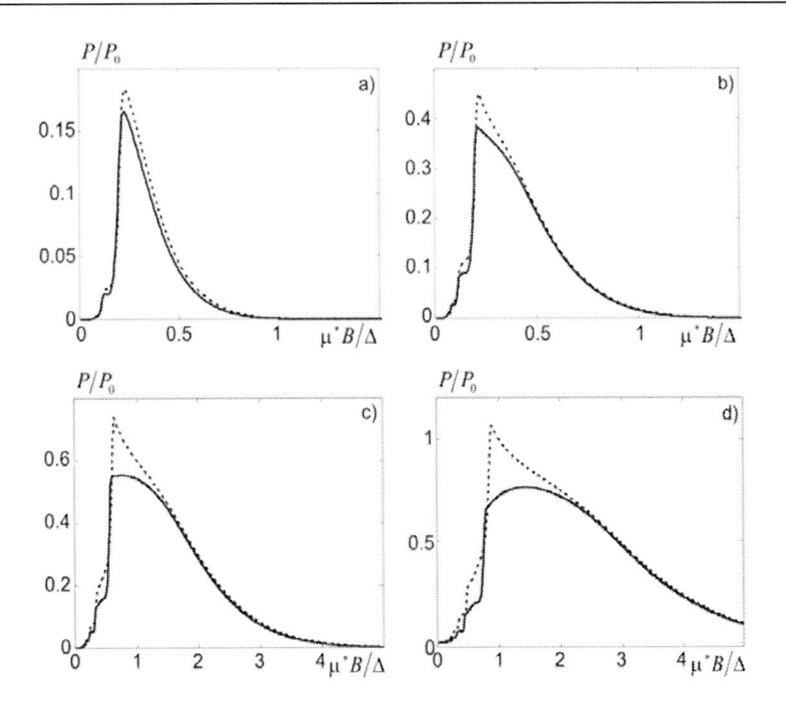

Figure 106. Field dependence of power factor of a layered crystal in the model $l = const$ at $A = 5; kT/\Delta = 0.03; 0 \le \mu^* B/\Delta \le 5$ without regard to magnetic field dependence of τ.

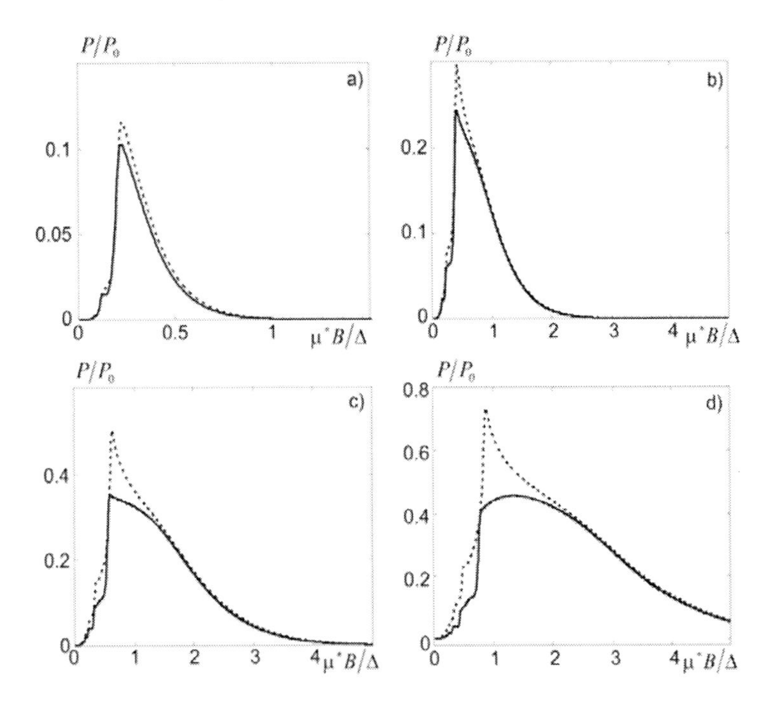

Figure 107. Field dependence of power factor of a layered crystal in the model $l = const$ at $A = 15; kT/\Delta = 0.03; 0 \le \mu^* B/\Delta \le 5$ without regard to magnetic field dependence of τ.

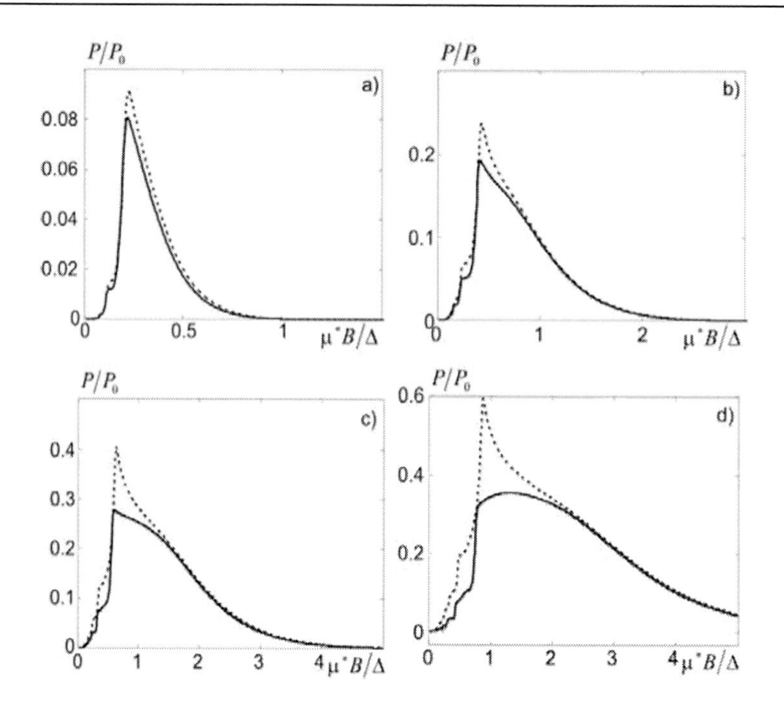

Figure 108. Field dependence of power factor of a layered crystal in the model $l = const$ at $A = 25; kT/\Delta = 0.03; 0 \le \mu^* B/\Delta \le 5$ without regard to magnetic field dependence of τ.

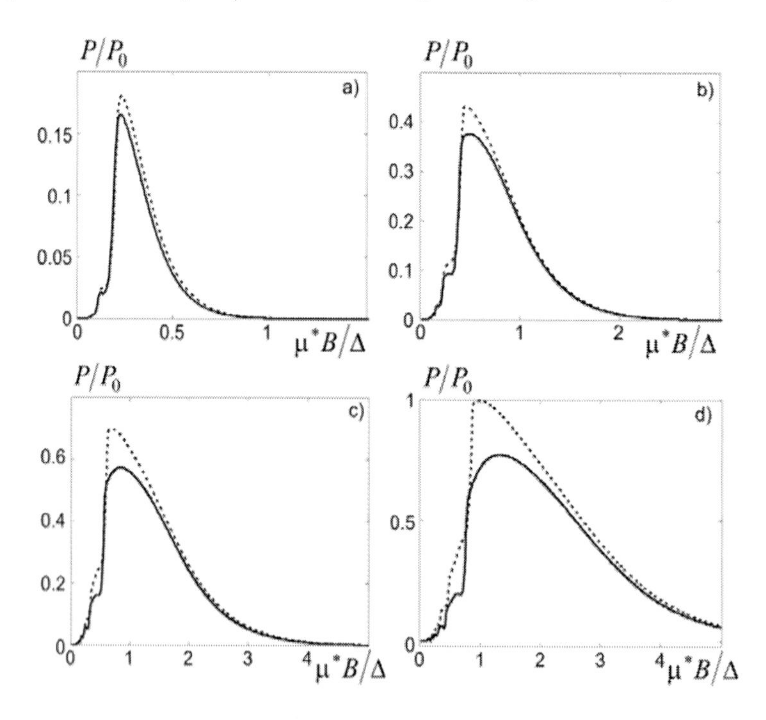

Figure 109. Field dependence of power factor of a layered crystal in the model $l = const$ at $A = 5; kT/\Delta = 0.03; 0 \le \mu^* B/\Delta \le 5$ with regard to magnetic field dependence of τ.

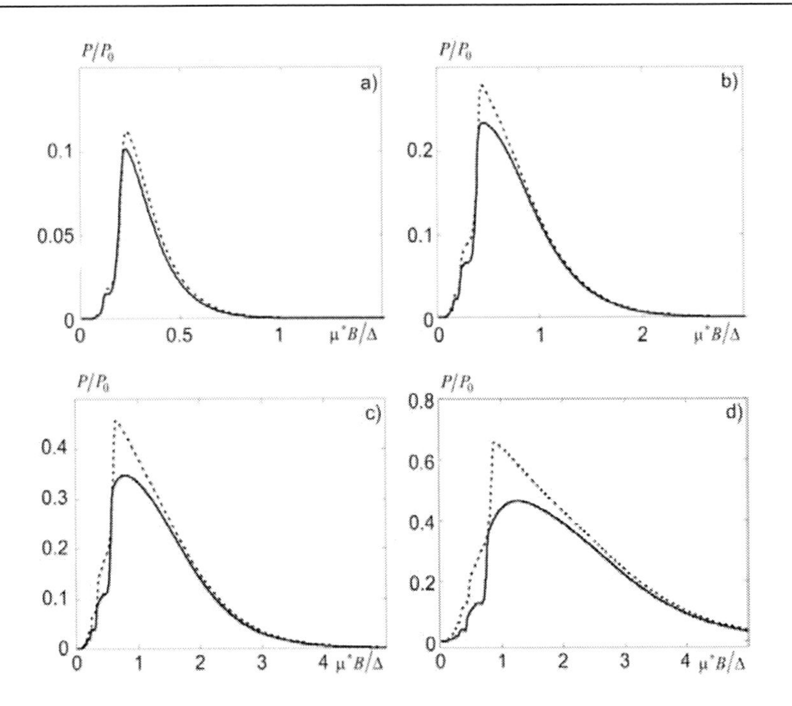

Figure 110. Field dependence of power factor of a layered crystal in the model $l = const$ at $A = 15; kT/\Delta = 0.03; 0 \le \mu^* B/\Delta \le 5$ with regard to magnetic field dependence of τ.

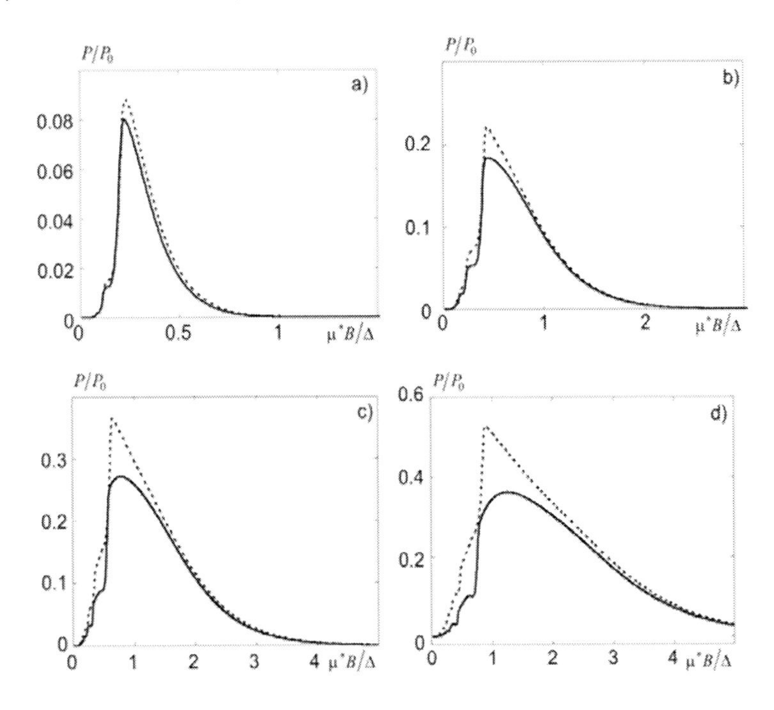

Figure 111. Field dependence of power factor of a layered crystal in the model $l = const$ at $A = 25; kT/\Delta = 0.03; 0 \le \mu^* B/\Delta \le 5$ with regard to magnetic field dependence of τ.

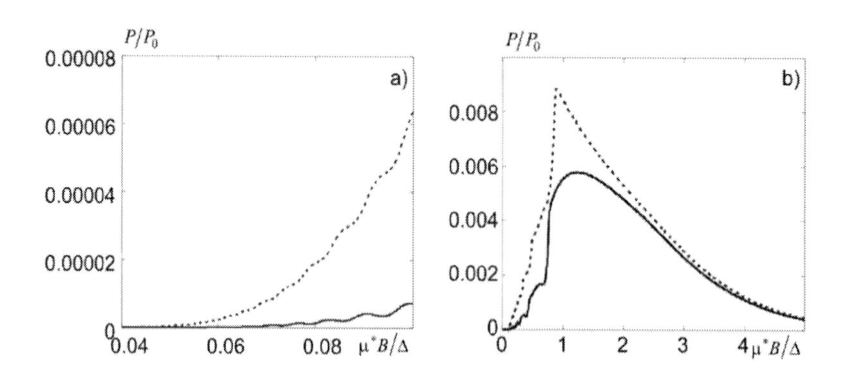

Figure 112. Field dependence of power factor of a layered crystal in the model $l = const$ for $A = 10^5; kT/\Delta = 0.03$ at: a) $0.04 \le \mu^* B/\Delta \le 0.1$; b) $0 \le \mu^* B/\Delta \le 5$ with regard to magnetic field dependence of τ.

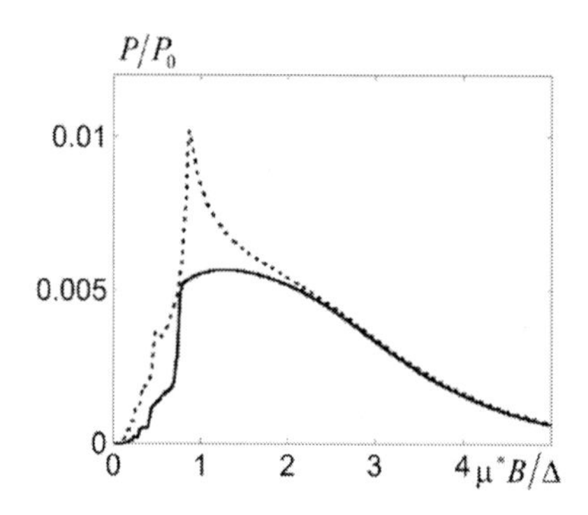

Figure 113. Field dependence of power factor of a layered crystal in the model $l = const$ for $A = 10^5; kT/\Delta = 0.03$ at $0 \le \mu^* B/\Delta \le 5$ without regard to magnetic field dependence of τ.

The figures demonstrate that in the approximation of constant mean free path, in the quasi-classical range of magnetic fields, with relatively low degrees of miniband filling the oscillating function of a magnetic field is not crystal power factor itself, but its first derivative with respect to a magnetic field. However, with increase in the degree of miniband filling, power factor itself becomes the oscillating function of a magnetic field. With increase in the anisotropy parameter A, the degree of miniband filling, beginning with which power factor itself becomes the oscillating function of a magnetic field, is increased. The layered structure effects in the quasi-classical range of magnetic fields are evident as the phase lag in the oscillations of power factor or its first derivative with respect to magnetic field as compared to the effective mass approximation. These effects are also evident as the reduction of power factor value by virtue of restriction of interlayer motion of electrons and, hence, the longitudinal conductivity. Moreover, layered structure effects manifest themselves in different dependences of power factor oscillations character on the anisotropy parameter A. Thus, in a real layered crystal at $A = 5$ and $\zeta_0/\Delta = 0.5$ the oscillating function of a magnetic field is the first derivative of power factor with respect to a magnetic field, and at

$A=5$ and $\zeta_0/\Delta=1;1.5;2$ – the power factor itself. At $A=15;25$ and $\zeta_0/\Delta=0.5;1$ the oscillating function of a magnetic field is the first derivative of power factor with respect to a magnetic field, and at $A=15;25$ and $\zeta_0/\Delta=1.5;2$ – the power factor itself. At $A=10^5$ and $\zeta_0/\Delta=2$ the oscillations of power factor are retained. In the effective mass approximation the oscillations of power factor at $A=5$ take place only when $\zeta_0/\Delta=1;1.5;2$. At $A=15;25;10^5$ in the effective mass approximation the oscillating function of a magnetic field is the first derivative of power factor in a magnetic field. With increase in the degree of miniband filling, the value of power factor drops, all other parameters being equal, and the oscillation frequency of power factor or its first derivative with respect to a magnetic field is increased. In the quasi-classical range of magnetic fields at $N=625\sqrt{A\gamma_0}$ (see section 18) and $A=5;15;25$ power factor of a layered crystal does not exceed $1.301\cdot10^{-3}$, $6.176\cdot10^{-4}$ and $4.267\cdot10^{-4}$ W/(m·K^2), respectively. At first sight, such power factor values seem to be high, but they are due to large average mean free path of charge carriers which is needed to make valid equating of the Dingle factor to unity in the entire studied range of quantizing magnetic fields. At $A=10^5, \zeta_0/\Delta=2$ and the same restrictions for mean free path of charge carriers, the power factor of crystal in the quasi-classical range of magnetic fields does not exceed $2.617\cdot10^{-7}$ W/(m·K^2).

With a rise in quantizing magnetic field induction, power factor, as in the model $\tau=const$, starts to increase, and on achievement of maximum, it drops, tending to zero. Moreover, this model also has an optimal range of magnetic fields, wherein the layered structure effects are most pronounced. However, the figures show that the shape of peaks in the model $l=const$ differs rather considerably from the shape of peaks in the model $\tau=const$. In strong magnetic fields in this model, as in the model $\tau=const$, peaks in the effective mass approximation are sharp, and in the case of a real layered crystal these peaks are rounded. Moreover, without regard to magnetic field dependence of relaxation time, peaks in the effective mass approximation are sharper, and for a real layered crystal – more rounded than with regard to magnetic field dependence of relaxation time. Till the achievement of maximum power factors, the layered structure effects are expressed in a larger number of local maxima and minima of power factor than in the effective mass approximation.

Without regard to magnetic field dependence of relaxation time at $A=5$ and $\zeta_0/\Delta=0.5;1;1.5;2$ power factor maxima for a real layered crystal are $0.046;0.155;0.275;0.434$ W/(m·K^2), respectively, and in the effective mass approximation $0.052;0.184;0.341;0.636$ W/(m·K^2), respectively. At $A=15$ and $\zeta_0/\Delta=0.5;1;1.5;2$ power factor maxima for a real layered crystal are $0.017;0.059;0.107;0.150$ W/(m·K^2), respectively, and in the effective mass approximation $0.020;0.071;0.145;0.250$ W/(m·K^2), respectively. At $A=25$ and $\zeta_0/\Delta=0.5;1;1.5;2$ power factor maxima for a real layered crystal are $0.01;0.035;0.063;0.091$ W/(m·K^2), respectively, and in the effective mass approximation $0.012;0.044;0.09;0.155$ W/(m·K^2), respectively. At $A=10^5$ and $\zeta_0/\Delta=2$ power factor maximum for a real layered crystal is $2.249\cdot10^{-5}$ W/(m·K^2), and in the effective mass approximation – $4.089\cdot10^{-5}$ W/(m·K^2).

With regard to magnetic field dependence of relaxation time at $A=5$ and $\zeta_0/\Delta=0.5;1;1.5;2$ power factor maxima for a real layered crystal are 0.046;0.155;0.290;0.434 W/(m·K^2), respectively, and in the effective mass approximation 0.052;0.176;0.361;0.578 W/(m·K^2), respectively. At $A=15$ and $\zeta_0/\Delta=0.5;1;1.5;2$ power factor maxima for a real layered crystal are 0.017;0.054;0.095;0.150 W/(m·K^2), respectively, and in the effective mass approximation 0.018;0.066;0.133;0.217 W/(m·K^2), respectively. At $A=25$ and $\zeta_0/\Delta=0.5;1;1.5;2$ power factor maxima for a real layered crystal are 0.01;0.033;0.063;0.096 W/(m·K^2), respectively, and in the effective mass approximation 0.012;0.052;0.085;0.142 W/(m·K^2), respectively. At $A=10^5$ and $\zeta_0/\Delta=2$ power factor maximum for a real layered crystal is $2.331\cdot10^{-5}$ W/(m·K^2), and in the effective mass approximation – $3.680\cdot10^{-5}$ W/(m·K^2). Thus, account of magnetic field dependence of relaxation time slightly changes downward the value of power factor maximum. The asymptotic law of power factor variation as a function of a magnetic field in the ultra-quantum limit remains the same as in the model $\tau = const$, since relaxation time is converted into constant with respect to quantum numbers.

Chapter 32

POWER FACTOR OF A LAYERED CRYSTAL WITH RELAXATION TIME PROPORTIONAL TO FULL ELECTRON VELOCITY ON THE FERMI SURFACE

In the model $\tau \propto v_f$ the coefficients $A^{(\alpha)}, B^{(\alpha)}, C^{(\alpha)}$ are determined by formulae (27.1) – (27.6), and the value $\widetilde{P}_0$ with the use of most common approximation (20.8) for the relaxation time of electron longitudinal momentum at acoustic phonon deformation potential scattering at low temperatures in the case of band spectrum sharp anisotropy acquires the form:

$$\widetilde{P}_0 = \frac{4\rho s^4 h^5}{7\pi^6 A^3 k \Xi^2 a^4 m^{*2} T^3 \Gamma(3)\zeta(3)}. \tag{32.1}$$

The field dependences of power factor of a layered crystal in the model $\tau \propto v_f$ are represented in Figures 114 – 124.

The figures demonstrate that in the model $\tau \propto v_f$ in the quasi-classical range of magnetic fields for all degrees of miniband filling the oscillating function of magnetic field is crystal power factor itself. The layered structure effects in the model $\tau \propto v_f$ in the quasi-classical range of magnetic fields manifest themselves in the phase lag of oscillations of power factor or its first derivative with respect to magnetic field as compared to the effective mass approximation. Also, these effects become apparent in power factor decrease by virtue of restriction of interlayer motion of electrons and, hence, the longitudinal conductivity. Moreover, in the case $\zeta_0/\Delta = 2$, i.e. in the case of a transient FS, a peculiar "square-law detection" effect appears on the field dependence of power factor, due to the fact that in the model $\tau \propto v_f$ in the case of a transient FS the Seebeck coefficient oscillations have variable polarity, and at a number of points power factor vanishes together with the Seebeck coefficient. In the effective mass approximation in the model $\tau \propto v_f$ the Seebeck coefficient in quasi-classical range of magnetic fields does not change its polarity, and, hence, no vanishing of power factor occurs.

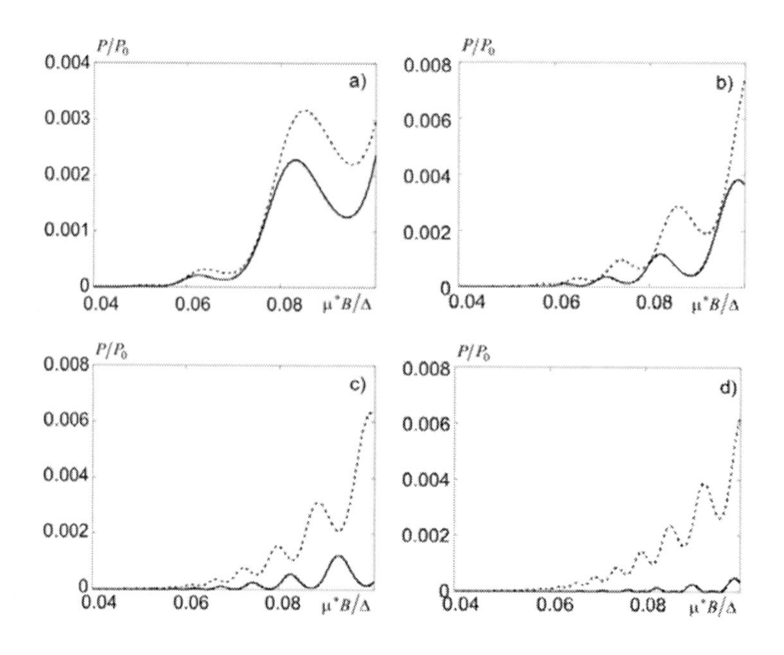

Figure 114. Field dependences of power factor of a layered crystal in the model $\tau \propto v_f$ at $kT/\Delta = 0.03; A = 5; 0.04 \le \mu^* B/\Delta \le 0.1$.

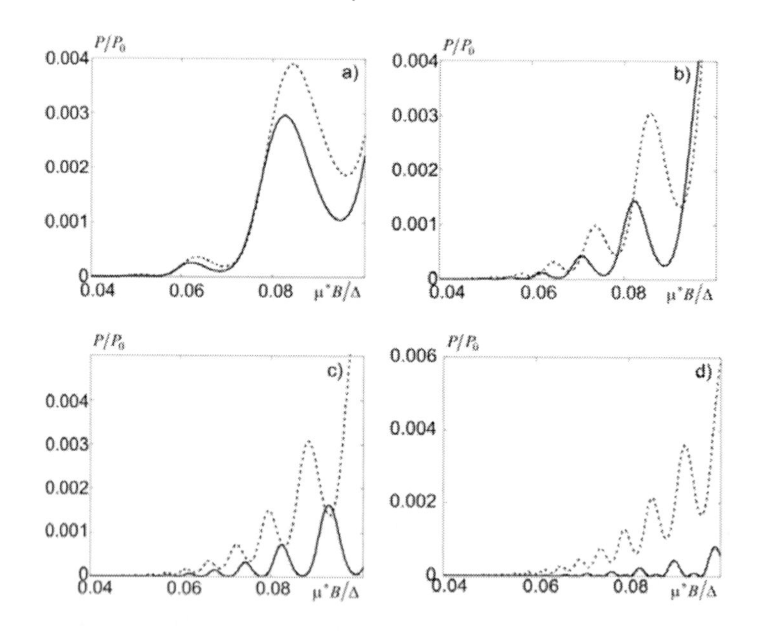

Figure 115. Field dependences of power factor of a layered crystal in the model $\tau \propto v_f$ at $kT/\Delta = 0.03; A = 15; 0.04 \le \mu^* B/\Delta \le 0.1$.

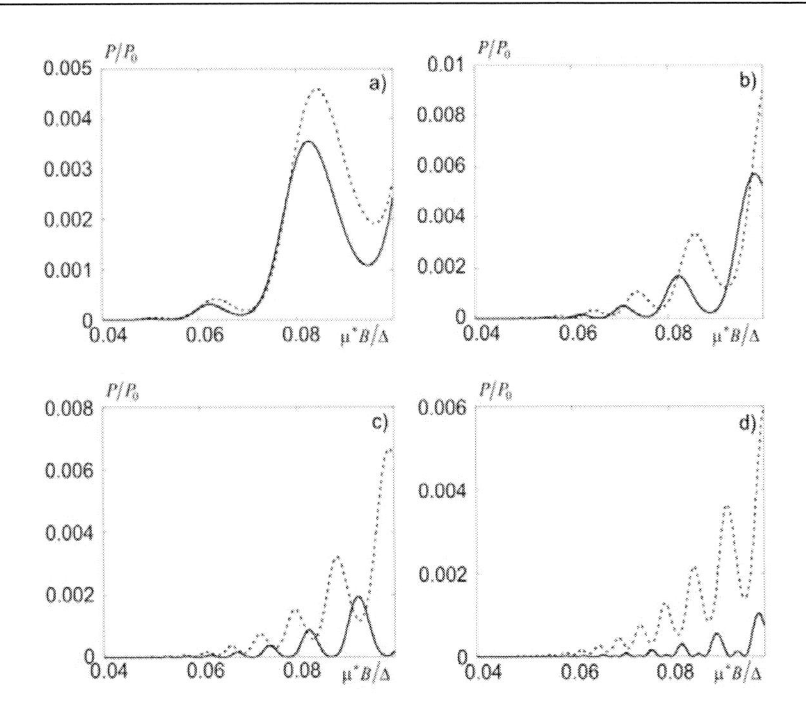

Figure 116. Field dependences of power factor of a layered crystal in the model $\tau \propto v_f$ at $kT/\Delta = 0.03; A = 25; 0.04 \le \mu^* B/\Delta \le 0.1$.

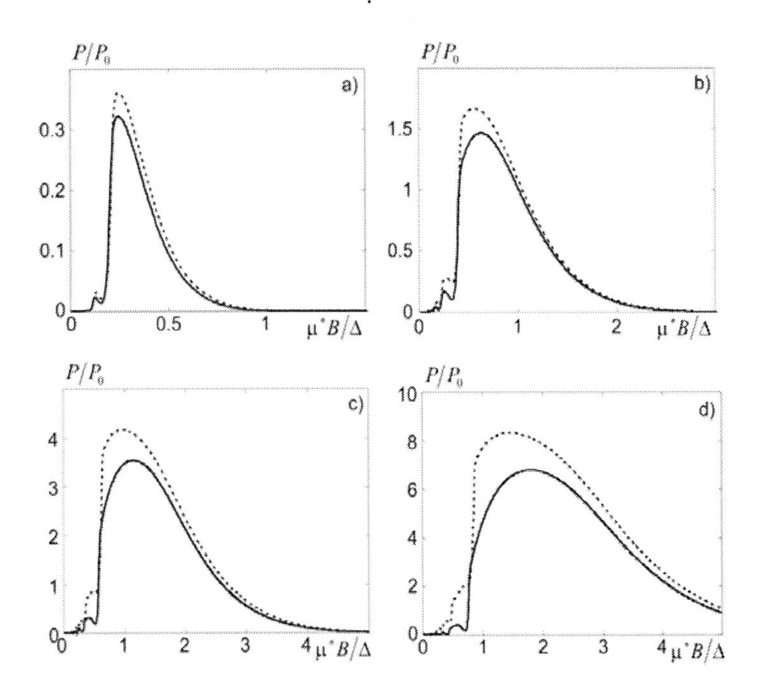

Figure 117. Field dependences of power factor of a layered crystal in the model $\tau \propto v_f$ at $kT/\Delta = 0.03; A = 5; 0 \le \mu^* B/\Delta \le 5$ without regard to magnetic field dependence of τ.

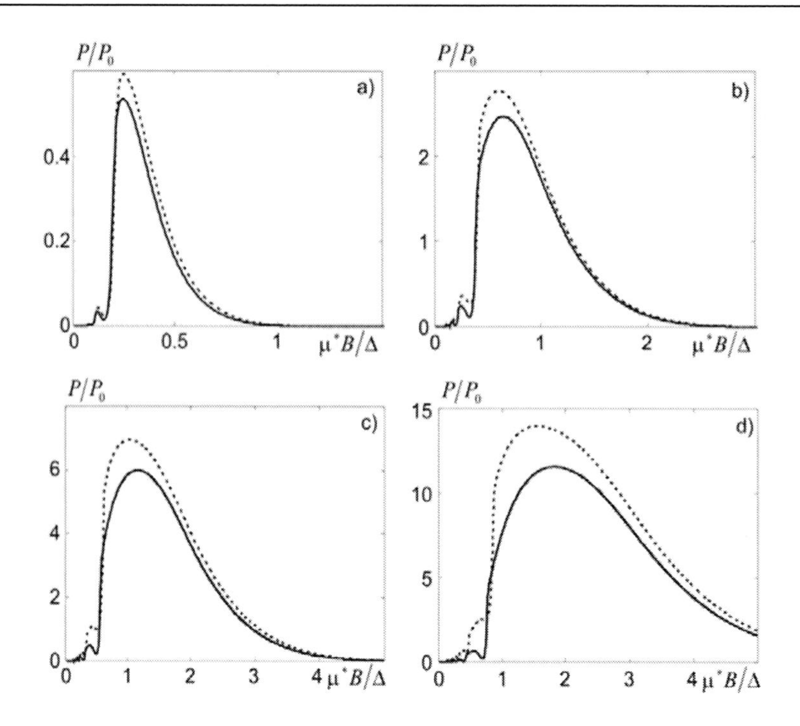

Figure 118. Field dependences of power factor of a layered crystal in the model $\tau \propto v_f$ at $kT/\Delta = 0.03; A = 15; 0 \leq \mu^* B/\Delta \leq 5$ without regard to magnetic field dependence of τ.

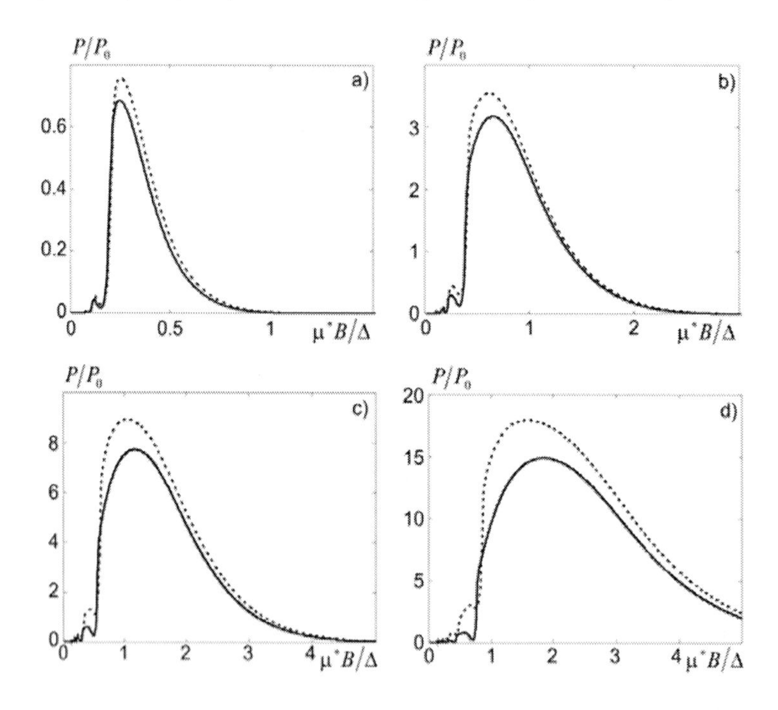

Figure 119. Field dependences of power factor of a layered crystal in the model $\tau \propto v_f$ at $kT/\Delta = 0.03; A = 25; 0 \leq \mu^* B/\Delta \leq 5$ without regard to magnetic field dependence of τ.

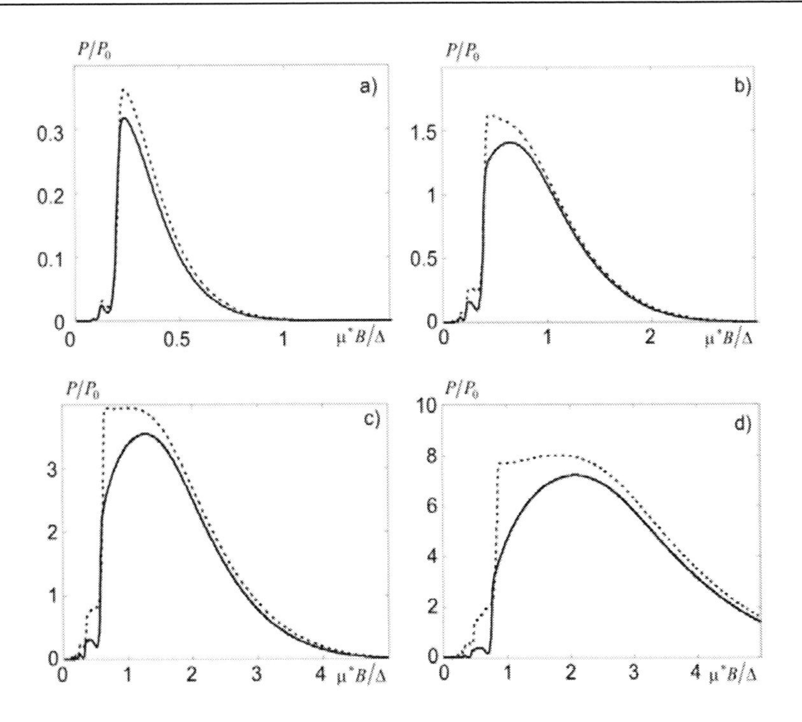

Figure 120. Field dependences of power factor of a layered crystal in the model $\tau \propto v_f$ at $kT/\Delta = 0.03; A = 5; 0 \le \mu^* B/\Delta \le 5$ with regard to magnetic field dependence of τ.

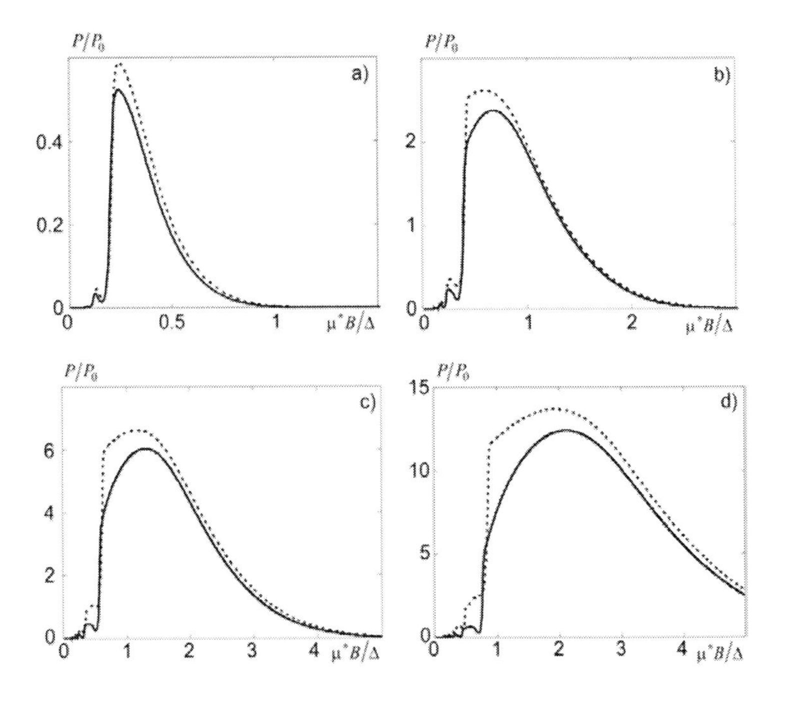

Figure 121. Field dependences of power factor of a layered crystal in the model $\tau \propto v_f$ at $kT/\Delta = 0.03; A = 15; 0 \le \mu^* B/\Delta \le 5$ with regard to magnetic field dependence of τ.

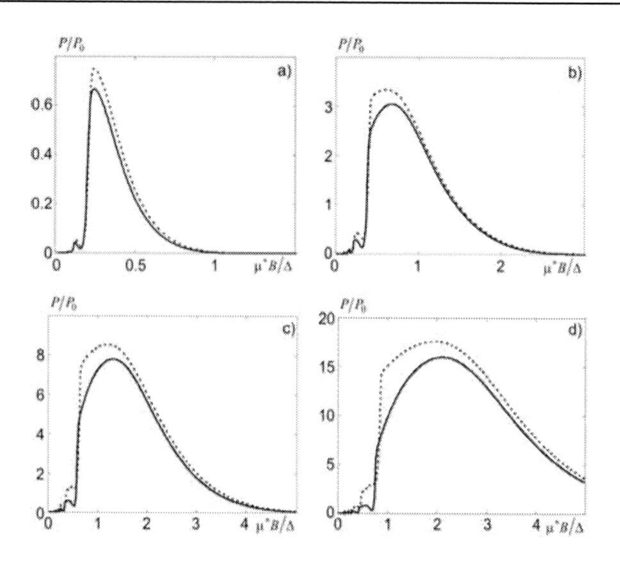

Figure 122. Field dependences of power factor of a layered crystal in the model $\tau \propto v_f$ at $kT/\Delta = 0.03; A = 25; 0 \le \mu^* B/\Delta \le 5$ with regard to magnetic field dependence of τ.

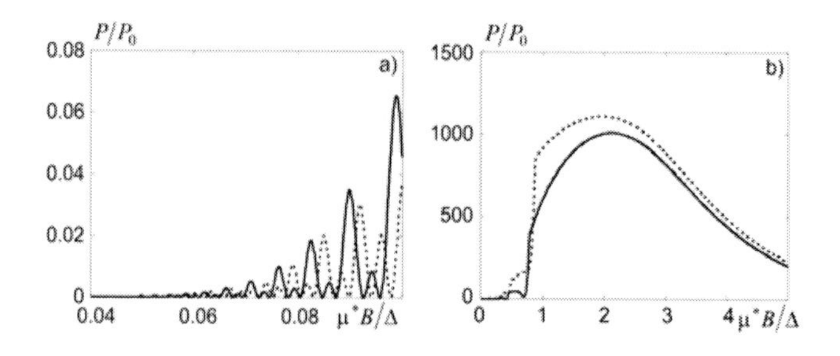

Figure 123. Field dependences of power factor of a layered crystal in the model $\tau \propto v_f$ at $\zeta_0/\Delta = 2; kT/\Delta = 0.03; A = 10^5$ with regard to magnetic field dependence of τ at: a) $0.04 \le \mu^* B/\Delta \le 0.1$; b) $0 \le \mu^* B/\Delta \le 5$.

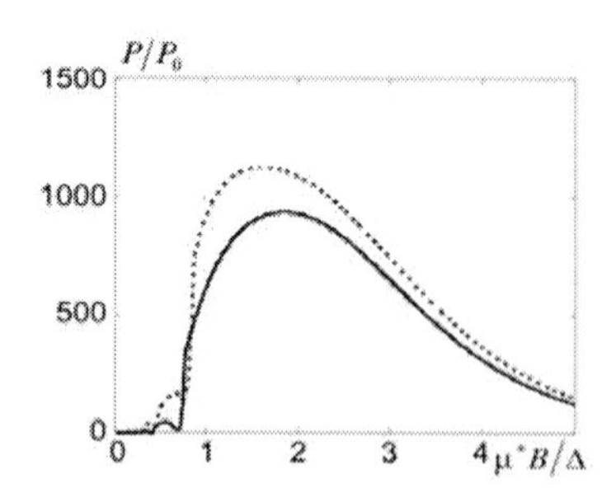

Figure 124. Field dependence of power factor of a layered crystal in the model $\tau \propto v_f$ at $\zeta_0/\Delta = 2; kT/\Delta = 0.03; A = 10^5; 0 \le \mu^* B/\Delta \le 5$ without regard to magnetic field dependence of τ.

With a rise in quantizing magnetic field induction, power factor, as in the other considered models, starts to increase, and on achievement of maximum, it drops, tending to zero. Moreover, this model also has an optimal range of magnetic fields, wherein the layered structure effects are most pronounced. However, the figures show that the shape of peaks in the model $\tau \propto v_f$ has its specific features. In this model, with low degrees of miniband filling, namely at $\zeta_0/\Delta \le 0.5$, the peaks of power factor, both for a real layered crystal, and in the effective mass approximation, are nearly triangular. At the same time, with increase in the degree of miniband filling, in the case of a real layered crystal the peaks of power factor become rounded. In so doing, if the influence of magnetic field on charge carrier scattering, i.e. on relaxation time, is disregarded, the peaks of power factor are rounded, both in the case of a real layered crystal and in the effective mass approximation. If, however, the magnetic field dependence of relaxation time is taken into account, then with all parameters of anisotropy A the peaks of power factor in the effective mass approximation acquire clear-cut kinks. Till the achievement of maximum power factors, the layered structure effects are expressed in a larger number of local maxima and minima of power factor and are more strongly pronounced than in the effective mass approximation.

Without regard to magnetic field dependence of relaxation time at $A = 5$ and $\zeta_0/\Delta = 0.5;1;1.5;2$ power factor maxima for a real layered crystal are $1.457 \cdot 10^{-5};6.376 \cdot 10^{-5};1.594 \cdot 10^{-4};2.960 \cdot 10^{-4}$ W/(m·K^2), respectively, and in the effective mass approximation $1.685 \cdot 10^{-5};7.286 \cdot 10^{-5};1.913 \cdot 10^{-4};3.871 \cdot 10^{-4}$ W/(m·K^2), respectively. At $A = 15$ and $\zeta_0/\Delta = 0.5;1;1.5;2$ power factor maxima for a real layered crystal are $2.505 \cdot 10^{-5};1.139 \cdot 10^{-4};2.732 \cdot 10^{-4};5.465 \cdot 10^{-4}$ W/(m·K^2), respectively, and in the effective mass approximation $2.732 \cdot 10^{-5};1.252 \cdot 10^{-5};3.188 \cdot 10^{-4};6.376 \cdot 10^{-4}$ W/(m·K^2), respectively. At $A = 25$ and $\zeta_0/\Delta = 0.5;1;1.5;2$ power factor maxima for a real layered crystal are $2.915 \cdot 10^{-5};1.412 \cdot 10^{-4};3.552 \cdot 10^{-4};6.376 \cdot 10^{-4}$ W/(m·K^2), respectively, and in the effective mass approximation $3.097 \cdot 10^{-5};1.639 \cdot 10^{-4};4.099 \cdot 10^{-4};8.197 \cdot 10^{-4}$ W/(m·K^2), respectively. At $A = 10^5$ and $\zeta_0/\Delta = 2$ power factor maximum for a real layered crystal is 0.041 W/(m·K^2), and in the effective mass approximation – 0.05W/(m·K^2).

With regard to magnetic field dependence of relaxation time, at $A = 5$ and $\zeta_0/\Delta = 0.5;1;1.5;2$ power factor maxima for a real layered crystal are $1.457 \cdot 10^{-5};6.376 \cdot 10^{-5};1.594 \cdot 10^{-4};3.188 \cdot 10^{-4}$ W/(m·K^2), respectively, and in the effective mass approximation $1.685 \cdot 10^{-5};7.286 \cdot 10^{-5};1.799 \cdot 10^{-4};3.643 \cdot 10^{-4}$ W/(m·K^2), respectively. At $A = 15$ and $\zeta_0/\Delta = 0.5;1;1.5;2$ power factor maxima for a real layered crystal are $2.505 \cdot 10^{-5};1.047 \cdot 10^{-4};2.372 \cdot 10^{-4};5.465 \cdot 10^{-4}$ W/(m·K^2), respectively, and in the effective mass approximation $2.687 \cdot 10^{-5};1.184 \cdot 10^{-4};2.960 \cdot 10^{-4};6.376 \cdot 10^{-4}$ W/(m·K^2), respectively. At $A = 25$ and $\zeta_0/\Delta = 0.5;1;1.5;2$ power factor maxima for a real layered crystal are $2.960 \cdot 10^{-5}; 1.366 \cdot 10^{-4}; 3.552 \cdot 10^{-4}; 6.831 \cdot 10^{-4}$ W/(m·K^2), respectively, and in the effective mass approximation $3.416 \cdot 10^{-5}; 1.548 \cdot 10^{-4}; 3.871 \cdot 10^{-4}; 8.197 \cdot 10^{-4}$ W/(m·K^2), respectively. At $A = 10^5$ and $\zeta_0 / \Delta = 2$ power factor maximum for a real layered crystal is 0.046 W/(m·K^2), and in the effective mass approximation – 0.05W/(m·K^2). Thus, account of magnetic field dependence of relaxation time slightly changes the value of power factor maximum. Increase in power factor value with increase in anisotropy parameter A at first sight may seem incomprehensible. Really, from formula (32.1) it follows that power factor, on the contrary, must drastically decrease with increase in this parameter, as long as to increase in A corresponds miniband narrowing, hence, a reduction of longitudinal conductivity. However, calculations in model $\tau \propto v_f$ were made not at some constant temperature, but on condition of constant Δ / kT *ratio*, i.e. actually at $AT = const$.The asymptotic law of power factor variation as a function of a magnetic field in the ultra-quantum limit remains the same as in the model $\tau = const$, since relaxation time is converted into constant with respect to quantum numbers.

Chapter 33

POWER FACTOR OF A LAYERED CRYSTAL AT RELAXATION TIME PROPORTIONAL TO ELECTRON LONGITUDINAL VELOCITY

In this model, the coefficients $A^{(\alpha)}, B^{(\alpha)}, C^{(\alpha)}$ are determined by formulae (28.1) – (28.6), and the value $\widetilde{P}_0$ in case of acoustic phonon deformation scattering of electrons is:

$$\widetilde{P}_0 = \frac{32k^2 m^* a^2 \Delta^3 \rho s^4}{7h\Xi^2 (kT)^3 \Gamma(3)\zeta(3)}. \tag{33.1}$$

The results of calculations of power factor of a layered crystal in the model $\tau \propto v_z$ are represented in Figures 125, 126. As before, when constructing the plots, it is assumed that $P_0 = 2\widetilde{P}_0/\pi^2$.

The figures demonstrate that in the quasi-classical range of magnetic fields, in the model $\tau \propto v_z$ at $0.5 \le \zeta_0/\Delta \le 1.5$ the oscillating function of a magnetic field is not power factor of a layered crystal itself, but its first derivative with respect to a magnetic field. However, at $\zeta_0/\Delta = 2$, i.e. in the case of a transient FS, the oscillating function of a magnetic field is power factor itself. In the effective mass approximation, for all degrees of miniband filling up to a transient FS, the oscillating function of a magnetic field is the first derivative of power factor with respect to a magnetic field. In so doing, as could be expected, with increase in the degree of miniband filling, the difference between solid and dashed curves is increased, and oscillation frequency of the first derivative or power factor itself is increased. The exception, however, is the case of a transient FS, when in the effective mass approximation the oscillations of the first derivative of power factor with respect to a magnetic field become apparent only for such magnetic fields whereby power factor in the effective mass approximation is considerably higher than power factor of a real layered crystal. Therefore, in Figure 125d the field dependence of power factor in the effective mass approximation looks not only monotonous, but also smooth.

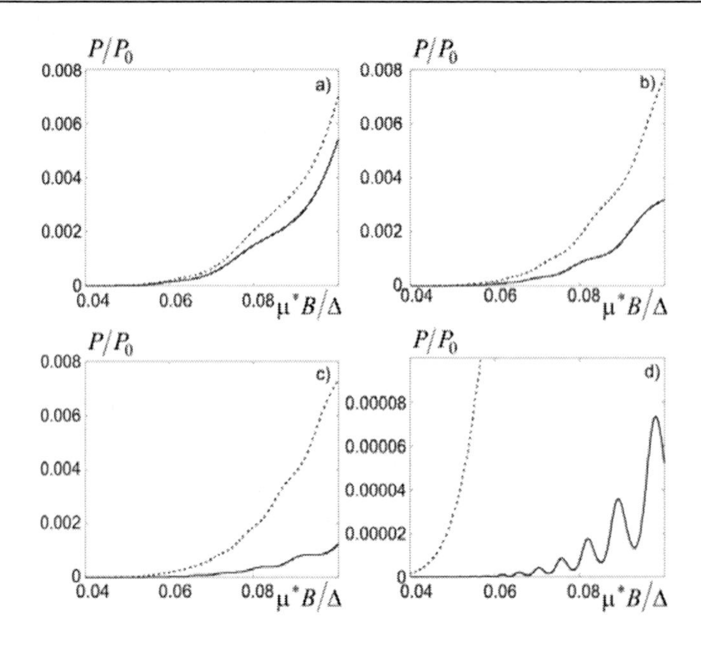

Figure 125. Field dependences of power factor of a layered crystal in the model $\tau \propto v_z$ at $kT/\Delta = 0.03; 0.04 \le \mu^* B/\Delta \le 0.1$.

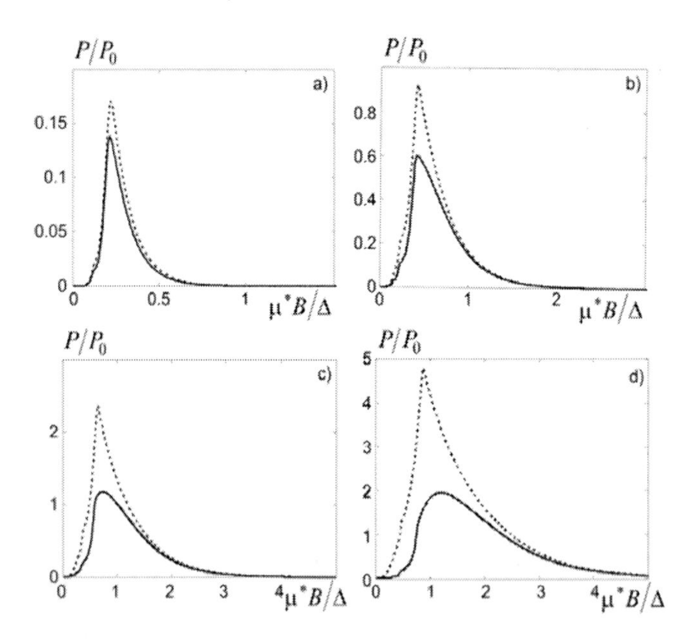

Figure 126. Field dependences of power factor of a layered crystal in the model $\tau \propto v_z$ at $kT/\Delta = 0.03; 0 \le \mu^* B/\Delta \le 5$.

As in the other considered models, with a rise in magnetic field, power factor at first increases rather sharply, and on achievement of maximum, it drops more smoothly, tending to zero. In the case of a real layered crystal, maxima at $\zeta_0/\Delta = 0.5;1$ are sharp, and at

$\zeta_0/\Delta = 1.5;2$ – rounded. In the effective mass approximation, however, these maxima are sharp for all degrees of miniband filling up to the case of a transient FS, i.e. $\zeta_0/\Delta = 2$. In so doing, there is an optimal range of magnetic fields, wherein the layered structure effects are most pronounced. In the model $\tau \propto v_z$ with previously specified problem parameters, power factor maxima for a real layered crystal at $\zeta_0/\Delta = 0.5;1;1.5;2$ are $6.375 \cdot 10^{-6}, 2.732 \cdot 10^{-5}, 5.465 \cdot 10^{-5}, 8.652 \cdot 10^{-5}$ W/(m·K^2), respectively, and in the effective mass approximation – $8.197 \cdot 10^{-6}, 4.326 \cdot 10^{-5}, 1.138 \cdot 10^{-4}, 2.186 \cdot 10^{-4}$ W/(m·K^2), respectively.

Chapter 34

POWER FACTOR OF A CHARGE-ORDERED LAYERED CRYSTAL

Now we will consider the field dependence of power factor of a charge-ordered layered crystal in the approximation of constant relaxation time for the case of charge carrier scattering on ionized impurities and in the approximation of relaxation time proportional to longitudinal electron velocity in case of charge carrier scattering on acoustic phonons. In the approximation of constant relaxation time the coefficients $A^{(\alpha)}, B^{(\alpha)}, C^{(\alpha)}$ are determined by formulae (29.1) – (29.3), and in the approximation of relaxation time proportional to longitudinal velocity by the formulae:

$$A^{(\alpha)} = \sum_{l=1}^{\infty} (-1)^l f_l^{th} \int_{-(\gamma-b)}^{\sqrt{w^2\delta^2+1}} y^{-2}\left(1 + w^2\delta^2 - y^2\right)\left(y^2 - w^2\delta^2\right)\sin\left[\pi l b^{-1}(\gamma - y)\right]dy\,, \quad (34.1)$$

$$B^{(\alpha)} = 0.5 \int_{-(\gamma-b)}^{\sqrt{w^2\delta^2+1}} y^{-2}\left(1 + w^2\delta^2 - y^2\right)\left(y^2 - w^2\delta^2\right)dy\,, \quad (34.2)$$

$$C^{(\alpha)} = \sum_{l=1}^{\infty} (-1)^l f_l^{\sigma} \int_{-(\gamma-b)}^{\sqrt{w^2\delta^2+1}} y^{-2}\left(1 + w^2\delta^2 - y^2\right)\left(y^2 - w^2\delta^2\right)\cos\left[\pi l b^{-1}(\gamma - y)\right]dy\,. \quad (34.3)$$

The value $\widetilde{P}_0$ for a charge-ordered layered crystal in the case of charge carrier scattering on ionized impurities is determined by the formula:

$$\widetilde{P}_0 = \frac{8\pi^3 k^2 m^* a^2 \Delta N}{h^3\left(\gamma + \sqrt{w^2\delta^2 + 1}\right)} \sqrt[3]{\frac{6\pi^2 m^* \Delta}{ah^2} \int_0^{\pi} \left(\gamma + \sqrt{w^2\delta^2 + \cos^2 x}\right)dx}\,, \quad (34.4)$$

and in the case of strongly anisotropic acoustic phonon deformation potential scattering of current carriers by formula (263). The results of calculations of the field dependences of power factor for both relaxation time models are represented in Figures 127 and 128. As before, when constructing the plots, it is assumed that $P_0 = 2\widetilde{P}_0/\pi^2$

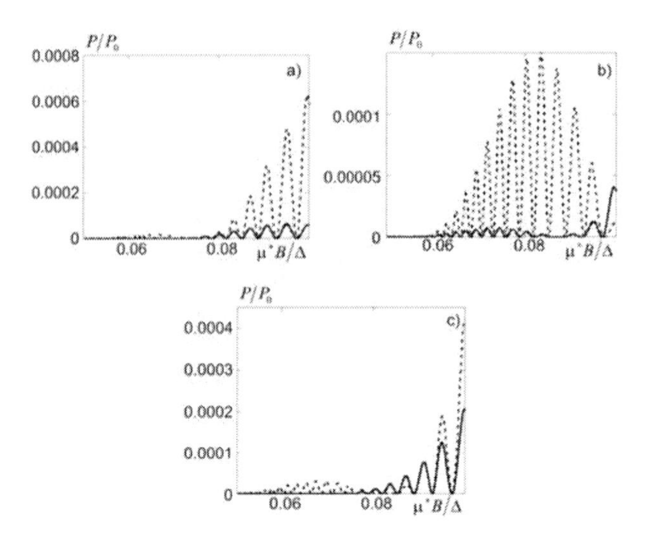

Figure 127. Field dependences of power factor of a charge-ordered crystal at $\zeta_{02D}/\Delta = 1; kT/\Delta = 0.03; 0.04 \le \mu^* B/\Delta \le 0.1$ and: a) $W_0/\zeta_{02D} = 1.5$; b) $W_0/\zeta_{02D} = 2$; c) $W_0/\zeta_{02D} = 2.5$. The solid curve is for the model $\tau \propto v_z$, the dashed curve – for the model $\tau = const$.

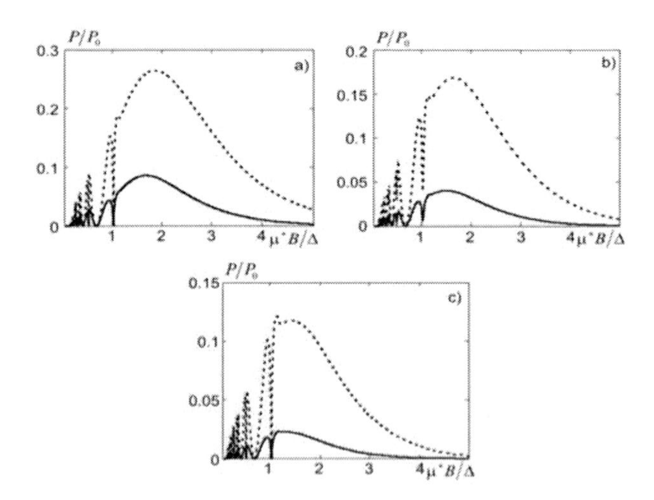

Figure 128. Field dependences of power factor of a charge-ordered crystal at $\zeta_{02D}/\Delta = 1; kT/\Delta = 0.03; 0 \le \mu^* B/\Delta \le 5$ and: a) $W_0/\zeta_{02D} = 1.5$; b) $W_0/\zeta_{02D} = 2$; c) $W_0/\zeta_{02D} = 2.5$, The solid curve is for the model $\tau \propto v_z$, the dashed curve – for the model $\tau = const$.

The figures demonstrate that in the quasi-classical range of magnetic fields the oscillations of power factor of a charge–ordered crystal are biperiodic and their amplitude (in relative units) is increased. In so doing, in the model $\tau = const$ this increase is more intensive than in the model $\tau \propto v_z$. With increase in the value of effective attractive interaction between electrons, the oscillation frequency of power factor of a layered crystal in the quasi-classical range of magnetic fields is increased.

With further magnetic field growth, the field dependence of power factor becomes typical of one-dimensional crystal with finite conduction bandwidth. As long as the longitudinal conductivity of a charge-ordered layered crystal vanishes whenever with a change in magnetic field the level of chemical potential passes through the bottom or ceiling of any filled Landau subband, and achieves maximum when chemical potential level passes through its middle, and the Seebeck coefficient polarity is switched abruptly or smoothly, then at some points crystal power factor becomes zero. Depending on the character of switching the Seebeck coefficient polarity, a drop of power factor from the local maximum to zero and its further increase to the next local maximum can be abrupt or smooth. At a point of topological transition the last zero of the Seebeck coefficient is achieved. After that, power factor which increases at first abruptly, and then smoothly, reaches the last maximum, following which it starts vanishing in both models according to asymptotic laws of the kind $P \propto T^{-3}B^{-6}$ at $\tau = const$ and $P \propto T^{-3}B^{-7}$ at $\tau \propto v_z$. The last maxima in the case of acoustic phonon deformation potential scattering of electrons with previously stipulated problem parameters and $W_0/\zeta_{02D} = 1.5;2;2.5$, respectively, make $3.643 \cdot 10^{-6};1.822 \cdot 10^{-6};9.108 \cdot 10^{-7}$ W/(m·K^2), respectively. Similar maxima in the case of electron scattering on ionized impurities with previously stipulated problem parameters, $N = 50000$ and $W_0/\zeta_{02D} = 1.5;2;2.5$, respectively, make $7.151 \cdot 10^{-4};4.552 \cdot 10^{-4};3.917 \cdot 10^{-4}$ W/(m·K^2), respectively. Decrease in power factor maxima in both models with increase in the value of effective attractive interaction is attributable to narrowing of conduction miniband when passing into the ordered state. However, power factor decreases slower than conduction, by virtue of the fact that maximum value of the Seebeck coefficient increases with increase in the value of effective attractive interaction between electrons. Besides, relatively high power factor values in the case of electron scattering on ionized impurities, on the one hand, are due to a more abrupt dependence of longitudinal conductivity on the miniband width in the model $\tau \propto v_z$ as compared to the model $\tau = const$ with almost identical value of the Seebeck coefficient maximum in both models. On the other hand, they are due to the fact that the value $N = 50000$ with selected problem parameters corresponds to mean free path $l \approx 50$ μm at 3.5K, which complies with quite pure and perfect samples. This condition can be considered as a variant of performance requirements to thermoelectric material.

Chapter 35

LONGITUDINAL ELECTRIC CONDUCTIVITY OF LAYERED CRYSTALS IN THE ABSENCE OF A MAGNETIC FIELD UNDER CONDITIONS OF WEAK AND INTERMEDIATE DEGENERACY

From formula (16.3) with regard to formulae (16.4) and (16.6) follows the following formula for the electric conductivity of a layered crystal in the absence of a magnetic field:

$$\sigma_{zz}(0)=\frac{16\pi^2 e^2 m^* a}{h^4}\int_{all}\tau(x)|W'(x)|^2\left\{1+\exp\left[\frac{W(x)-\zeta}{kT}\right]\right\}^{-1}dx\,. \tag{35.1}$$

Using constant relaxation time for the case of elastic scattering on acoustic phonons in the form proposed in [58], namely:

$$\tau_0=\frac{a\rho s^2 h^3}{12\pi^3 m^* \Xi^2 kT}, \tag{35.2}$$

we finally get:

$$\sigma_{zz}(0)\equiv\sigma_{zz}=\frac{4e^2 a^2 \rho s^2}{3\pi h \Xi^2 kT}\int_{all}|W'(x)|^2\left\{1+\exp\left[\frac{W(x)-\zeta}{kT}\right]\right\}^{-1}dx\,. \tag{35.3}$$

As before, calculations of the temperature dependence of electric conductivity were made for the cases of a real layered crystal at $W(x)=\Delta(1-\cos x)$ and in the effective mass approximation. The results of these calculations are given in Figure 129. In this figure $\sigma_0=4e^2a^2\rho s^2\Delta/(3\pi h\Xi^2)$.

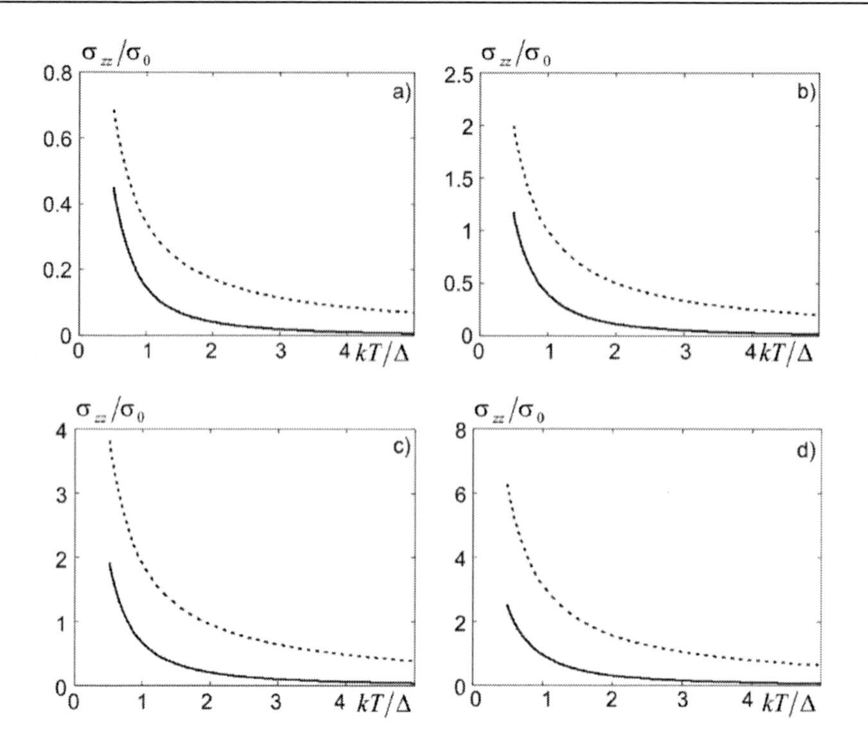

Figure 129. Temperature dependence of longitudinal electric conductivity of a layered crystal in the absence of a magnetic field at $0.5 \le kT/\Delta \le 5$.

From the figures it is evident that for all degrees of miniband filling and all temperatures considered, the electric conductivity of a real layered crystal is lower than in the effective mass approximation. This is caused, on the one hand, by restriction of the longitudinal velocity of electrons in a real layered crystal, on the other hand – by higher electron density of states in it as compared to the effective mass approximation. For instance, with previously specified problem parameters, $kT/\Delta = 0.5$ and $\zeta_0/\Delta = 0.5;1;1.5;2$, the longitudinal conductivity of a real layered crystal is $\sigma_{zz} = 577;1508;2454;3235$ S/m, respectively, and in the effective mass approximation – $\sigma_{zz} = 877;2560;4897;8042$ S/m, respectively. At $kT/\Delta = 0.5$ and $\zeta_0/\Delta = 0.5;1;1.5;2$ the longitudinal conductivity of a real layered crystal is $\sigma_{zz} = 8.283;24.32;46.08;72.96$ S/m, respectively, and in the effective mass approximation – $\sigma_{zz} = 87.04;256.00;490.24;793.6$ S/m, respectively. Thus, at $kT/\Delta = 0.1$ the longitudinal conductivity of a real layered crystal under conditions of weak and intermediate degeneracy and in the nondegenerate gas is factor of 1.5÷2.5 lower than in the effective mass approximation, and at $kT/\Delta = 5$ the longitudinal conductivity of a real layered crystal is an order of magnitude lower than in the effective mass approximation. Thus, under conditions of weak and intermediate degeneracy of electron gas or in its absence the longitudinal conductivity of a real layered crystal decreases with temperature faster than in the effective mass approximation.

Chapter 36

LONGITUDINAL MAGNETORESISTANCE OF LAYERED CRYSTALS UNDER WEAK AND INTERMEDIATE DEGENERACY

Let us consider the longitudinal magnetoresistance of layered crystals under weak and intermediate degeneracy of electron gas in model of constant relaxation time. Under these conditions, at $\pi^2 kT/\mu^* B >> 1$ one can ignore the oscillating part of longitudinal electric conductivity as compared to the monotone part (in the absence of degeneracy the oscillating part of longitudinal electric conductivity is equal to zero). Therefore, total longitudinal electric conductivity comprises only components determined by formulae (16.4) and (16.6). The results of calculations of the field dependence of longitudinal electric conductivity by these formulae are given in Figure 130. In this figure, $\sigma_0 = 4e^2 a^2 \rho s^2 \Delta / (3\pi h \Xi^2)$.

The figure demonstrates that in a layered crystal under weak and intermediate degeneracy of electron gas there is increase in longitudinal electric conductivity with increase in magnetic field, i.e. negative longitudinal magnetoresistance (NLMR). Its main reason is that increase in a quantizing magnetic field is in the certain sense equivalent to crystal cooling. Moreover, increase in longitudinal electric conductivity is brought about by growth of electron gas chemical potential due to the fact that free energy of a layered crystal electron subsystem in the presence of the Landau levels is larger than in the absence of said levels. For instance, with a change in magnetic field from zero to the level whereby $\mu^* B/\Delta = 1$ for values $\zeta_0/\Delta = 0.5;1;1.5;2$, increase in longitudinal electric conductivity is 1;2.4;3.7;3.9%, respectively. These values are low, but quite measurable. Calculations also show that in the effective mass approximation the relative increase in longitudinal electric conductivity is 1 to 2 orders of magnitude lower than in a layered crystal. So, it is not shown here. From the results obtained it is clear why high NLMR values are not observed in typical metals for which quasi-free electron approximation is valid. Thus, relatively high NLMR values with a weak or intermediate degeneracy of electron gas are largely due to the finite width of conduction miniband. Note that this result cannot be obtained in the framework of traditional approaches whereby the specific electron dispersion law and FS extension along the direction of a magnetic field are inessential.

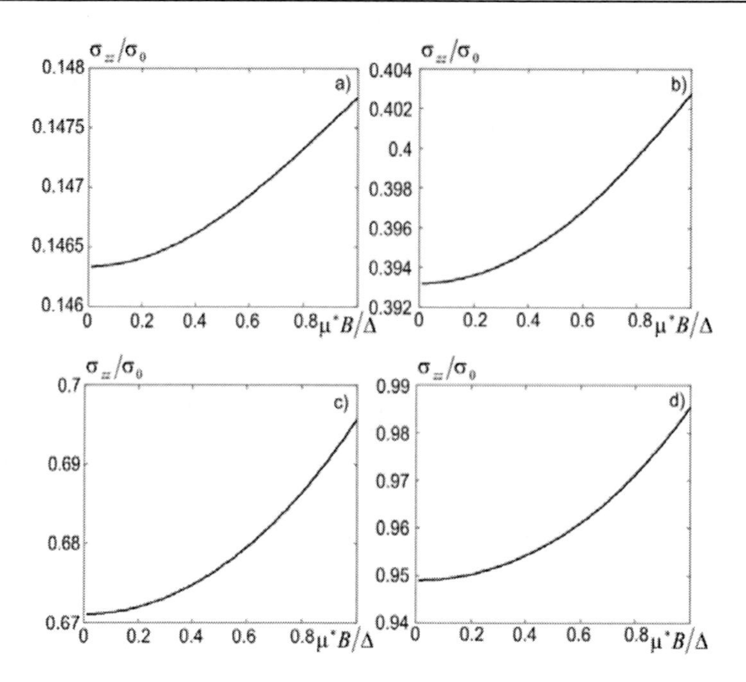

Figure 130. Field dependence of longitudinal electric conductivity of a layered crystal in model $\tau = const$ at $kT/\Delta = 1; 0 \le \mu^* B/\Delta \le 1$.

Chapter 37

LONGITUDINAL ELECTRIC CONDUCTIVITY OF LAYERED CRYSTALS IN THE ABSENCE OF MAGNETIC FIELD UNDER CONDITIONS OF WEAK AND INTERMEDIATE DEGENERACY WITH REGARD TO THE INFLUENCE OF FS CLOSENESS ON SCATTERING

Approximation of constant relaxation time proposed in [58] is valid for an open FS, but is too rough for a closed FS. If we do not use this approximation, but consider the relaxation time to be inversely proportional to electron density of states $g(\varepsilon)$, then in the case of a real layered crystal with a closed FS we will get the following formula for longitudinal conductivity of a layered crystal in the absence of a magnetic field with acoustic phonon scattering:

$$\sigma_{zz}(0) \equiv \sigma_{zz} = \frac{4e^2a^2\rho s^2\Delta}{3h\Xi^2 t}\int_0^{\pi}\left(0.25\frac{\sin 2x}{x^2} - 0.5\frac{\cos 2x}{x}\right)\left\{1 + \exp\left[\frac{1-\cos x - \gamma}{t}\right]\right\}^{-1} dx \cdot \qquad (37.1)$$

In the effective mass approximation this formula acquires the form:

$$\sigma_{zz}(0) \equiv \sigma_{zz} = \frac{8e^2a^2\rho s^2\Delta}{9h\Xi^2}\ln[\exp(\gamma/t)+1]. \qquad (37.2)$$

The results of calculations by formulae (37.1) and (37.2) are given in Figure 131. In this figure, $\sigma_0 = 4e^2a^2\rho s^2\Delta/(3\pi h\Xi^2)$.

The figure demonstrates that again, for all degrees of miniband filling and all temperatures considered, the electric conductivity of a real layered crystal is lower than in the effective mass approximation. Like in the model $\tau = const$, it is caused, on the one hand, by restriction of the longitudinal velocity of electrons in a real layered crystal, on the other hand – by higher electron density of states in it as compared to the effective mass approximation. For instance, with previously specified problem parameters, $kT/\Delta = 0.5$ and

$\zeta_0/\Delta = 0.5;1;1.5;2$, the longitudinal conductivity of a real layered crystal is $\sigma_{zz} = 1185;2620;3956;4032$ S/m, respectively, and in the effective mass approximation – $\sigma_{zz} = 1875;4756;7937;11460$ S/m, respectively. At $kT/\Delta = 0.5$ and $\zeta_0/\Delta = 0.5;1;1.5;2$ the longitudinal conductivity of a real layered crystal is $\sigma_{zz} = 9.821;28.16;51.20;80.64$ S/m, respectively, and in the effective mass approximation – $\sigma_{zz} = 65.28;189.44;358.40;582.40$ S/m, respectively. Thus, at $kT/\Delta = 0.5$ the longitudinal conductivity in the effective mass approximation under conditions of weak and intermediate degeneracy and in the nondegenerate gas is factor of 1.6÷2.8 higher than in a real layered crystal, and at $kT/\Delta = 5$ the longitudinal conductivity in the effective mass approximation is factor of 6.6÷7.2 higher than in a real layered crystal. Thus, under conditions of weak and intermediate degeneracy of electron gas or in its absence, in the model $\tau \propto 1/g(\varepsilon)$, just as in the model $\tau = const$, the longitudinal conductivity of a real layered crystal decreases with temperature faster than in the effective mass approximation. With regard to the influence of FS closeness on scattering, the longitudinal conductivity is higher than in the approximation of constant relaxation time, largely due to the fact that the relaxation time of low-energy electron states in the model $\tau \propto 1/g(\varepsilon)$ is higher than in the model $\tau = const$.

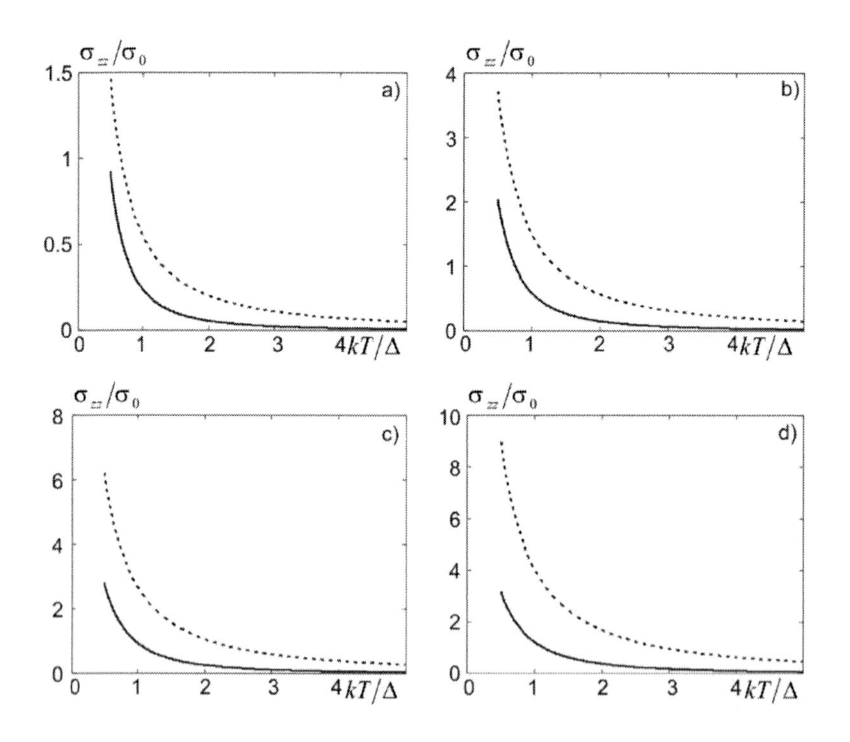

Figure 131. Temperature dependence of longitudinal electric conductivity of a layered crystal in the absence of magnetic field in the model $\tau \propto 1/g(\varepsilon)$ at $0.5 \le kT/\Delta \le 5$.

Chapter 38

LONGITUDINAL MAGNETORESISTANCE OF A LAYERED CRYSTAL UNDER CONDITIONS OF WEAK AND INTERMEDIATE DEGENERACY WITH REGARD TO THE INFLUENCE OF DENSITY OF STATES ON SCATTERING

We now consider the longitudinal magnetoresistance of a layered crystal under conditions of weak and intermediate degeneracy of electron gas with regard to the influence of the density of electron states on scattering. As before, we will consider the range of temperatures where acoustic phonon deformation potential scattering of electrons is predominant. We will use the model of relaxation time inversely proportional to the density of electron states in the framework of which $\tau \propto 1/g(\varepsilon)$. In this model, the longitudinal conductivity of a real layered crystal is determined as:

$$\sigma_{zz}(0) \equiv \sigma_{zz} = \frac{4e^2a^2\rho s^2\Delta}{3h\Xi^2 t}\left\{ \int\limits_0^{\arccos(1-\gamma)} \left(0.25\frac{\sin 2x}{x^2} - 0.5\frac{\cos 2x}{x}\right)dx + \sum_{l=1}^{\infty}(-1)^l \frac{bl/t}{\sinh(bl/t)} \times \right.$$
$$\times \left[\int\limits_0^{\arccos(1-\gamma)} \left(0.25\frac{\sin 2x}{x^2} - 0.5\frac{\cos 2x}{x}\right)\exp\left(l\frac{1-\cos x-\gamma}{t}\right)dx - \right.$$
$$\left.\left. - \int\limits_{\arccos(1-\gamma)}^{\pi} \left(0.25\frac{\sin 2x}{x^2} - 0.5\frac{\cos 2x}{x}\right)\exp\left(l\frac{\gamma-1+\cos x}{t}\right)dx \right]\right\}. \quad (38.1)$$

In the effective mass approximation this formula acquires the form:

$$\sigma_{zz}(0) \equiv \sigma_{zz} = \frac{8e^2a^2\rho s^2\Delta}{9h\Xi^2 t}\left\{0.5\gamma[1+\mathrm{sgn}(\gamma)] + b\sum_{l=1}^{\infty}(-1)^{l-1}\frac{\exp(-lt^{-1}|\gamma|)}{\sinh(bl/t)}\right\}. \quad (38.2)$$

The results of calculations of longitudinal magnetoresistance of a layered crystal by these formulae are given in Figures 131 and 132. In these figures, $\sigma_0 = 4e^2a^2\rho s^2\Delta/(3\pi h\Xi^2)$.

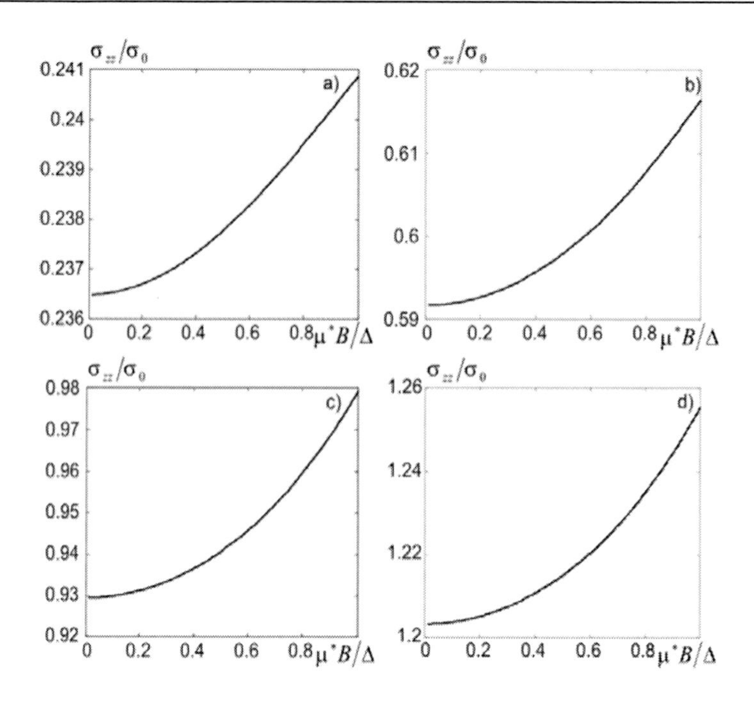

Figure 132. Field dependence of total conductivity of a real layered crystal in the model $\tau \propto 1/g(\varepsilon)$ under conditions of weak and intermediate degeneracy at $kT/\Delta = 1$ and $0 \le \mu^* B/\Delta \le 1$.

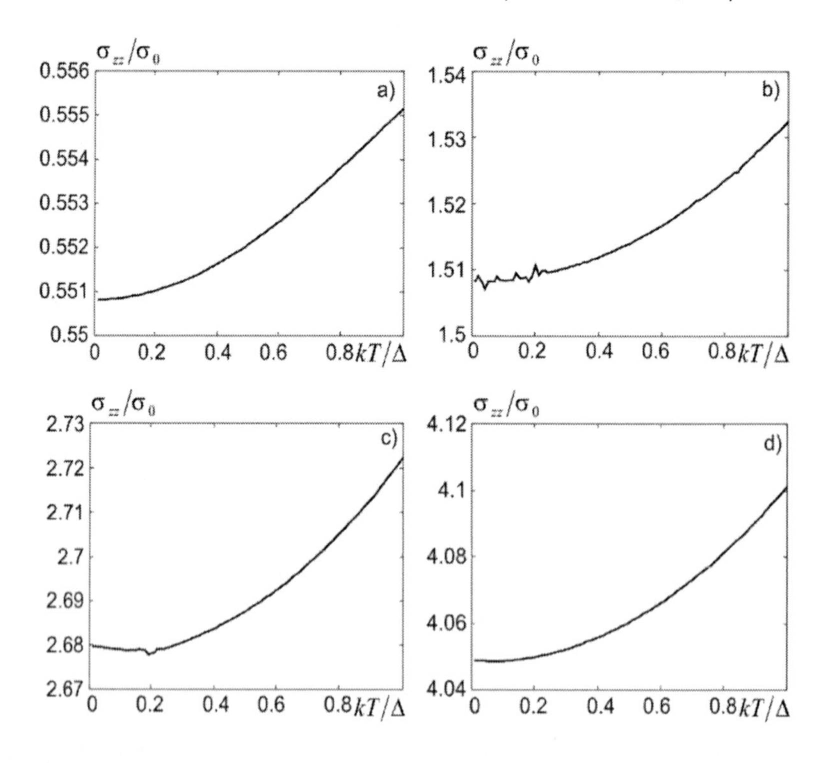

Figure 133. Field dependence of total conductivity in the effective mass approximation under conditions of weak and intermediate degeneracy at $kT/\Delta = 1$ and $0 \le \mu^* B/\Delta \le 1$.

The figures demonstrate that in a real layered crystal with a closed FS in the model $\tau \propto 1/g(\varepsilon)$, as well as in the model $\tau = const$, for all degrees of miniband filling and over the entire range of magnetic fields considered the longitudinal electric conductivity increases with magnetic field growth, i.e. there is a stable NLMR. With increase in the degree of miniband filling in a real layered crystal the absolute conductivity value is increased. However, the NLMR value is a nonmonotonic function of the degree of miniband filling. At $\mu^* B/\Delta = 1$ and $\zeta_0/\Delta = 0.5;1;1.5;2$, respectively, the NLMR of a real layered crystal is 1.8;4.4;5.1;2.9%, respectively. However, in the effective mass approximation at $\gamma = 0.5$ there are weak oscillations of longitudinal electric conductivity. At $\zeta_0/\Delta = 1.5$ on the field dependence of longitudinal electric conductivity there is a little interval of electric conductivity drop, i.e. "habitual" positive longitudinal magnetoresistance. Finally, at $\zeta_0/\Delta = 2$ the electric conductivity oscillations disappear, and a stable NLMR again happens to be the case. These phenomena, just as the sign inversion of the Langevin diamagnetic susceptibility, are attributable to two competing processes of charge carriers' thermal transfer, or activation. When thermal transfer or activation of charge carriers from the lower Landau levels to chemical potential level is dominant, then longitudinal conductivity is increased with magnetic field growth and NLMR takes place. However, when thermal transfer or activation of charge carriers from chemical potential level to the upper Landau levels is dominant, then the conductivity is reduced with magnetic field growth, and we get a "habitual" positive magnetoresistance. The ratio between these two processes is affected not only by crystal band structure, but also by temperature, magnetic field value and the way in which charge carrier scattering is simulated. In so doing, magnetic field affects the ratio between said processes both through the ratio of average thermal energy to the distance between the Landau levels, and through the magnetic field dependence of chemical potential. In particular, it is clear that as long as electron gas chemical potential, both for a real layered crystal and in the effective mass approximation, increases with magnetic field growth, then longitudinal magnetoresistance in general must be negative. However, in the effective mass approximation with equal degree of miniband filling the electron gas chemical potential with all magnetic field inductions is higher than for a real layered crystal. This difference is the larger the lower is the degree of miniband filling. That is the main reason why in the effective mass approximation there are electric conductivity oscillations and a "habitual" positive longitudinal magnetoresistance. At $\mu^* B/\Delta = 1$ and $\zeta_0/\Delta = 0.5;1;1.5;2$, respectively, the NLMR in the effective mass approximation is 0.76;1.5;1.6;1.3%, respectively, which is factor of 2.2÷3.2 lower than in the case of a real layered crystal. However, the pattern of NLMR change with increase in the degree of miniband filling for a real layered crystal and in the effective mass approximation is identical.

Chapter 39

TEMPERATURE-DEPENDENT KAPITSA EFFECT

Earlier we have considered the longitudinal Kapitsa effect related to charge carrier condensation in the lowest Landau subband with the number $n = 0$. We now consider another opportunity for the origin of this effect which is related to a linear dependence of scattering probability on magnetic field induction [59]. Such scattering probability is typical of induced acoustic phonon deformation potential scattering of electrons under suppressed transitions between the Landau subbands. These transitions can be considered to be suppressed on fulfilment of conditions $\Delta \leq \mu^* B$ and $kT \leq \mu^* B$. Assuming, for instance, that $m^* = 0.1 m_0$ and $T = 88$ K, these conditions are fulfilled in the fields with inductions $B \geq 7$ T. Whereas Kapitsa, studying magnetoresistance of metals as a function of magnetic field, worked in the fields with inductions up to 30T. If the conditions for suppression of transitions between the Landau subbands are fulfilled, then, as is shown in [36], in the calculation of longitudinal electric conductivity one can introduce the relaxation time and consider it to be inversely proportional to the density of states in a magnetic field per one Landau subband.

The respective proportionality factor was calculated in [36] for 0 – 0 intraband transitions. However, in the case of acoustic phonon deformation potential scattering of electrons this factor will be equal for all $n - n$ transitions in two extreme cases: induced and spontaneous scattering. Which of the mechanisms will be dominant depends on the value of dimensionless parameter $\kappa = 2\left(\pi k T a_B / h s_{\parallel}\right)^2$, $a_B = \left(h / 2\pi e B\right)^{1/2}$ – magnetic length, $s_{\parallel}$ – sound velocity in layers plane, the rest of designations have been explained above or generally accepted. When $\kappa >> 1$, induced scattering is dominant, whereas when $\kappa << 1$ – spontaneous. If, for instance, $s_{\parallel} = 5 \cdot 10^3$ m/s, $a = 1$ nm, then it turns out that in the region of nitrogen temperatures induced scattering is dominant. Then the nonoscillating part of electric conductivity of a layered crystal can be defined by the formula:

$$\sigma_{zz}(B)=\frac{16\pi\tau_0 e^2 m^* a\overline{W}}{h^4 kT|\mu^* B|}\left\{\int\limits_{W(x)\le\zeta} dx|W'(x)|^3+\sum_{l=1}^{\infty}(-1)^l h_l^\sigma\left[\int\limits_{W(x)\le\zeta} dx|W'(x)|^3\exp\left(l\frac{W(x)-\zeta}{kT}\right)-\right.\right. \tag{39.1}$$
$$\left.\left.-\int\limits_{W(x)\ge\zeta} dx|W'(x)|^3\exp\left(l\frac{\zeta-W(x)}{kT}\right)\right]\right\}$$

In this formula, τ_0 – certain crystal constant characterizing scattering intensity, $\overline{W}=\Delta$ – miniband half-width. Integration in formula (39.1) is performed only with respect to positive x values. Using this formula, we will consider the Kapitsa effect at $W(x)=\Delta(1-\cos x)$. In so doing, using the relationships from [36], we will express τ_0 through deformation potential constant Ξ, crystal density ρ and sound velocity in layers plane $s_{||}$.

First of all, analyzing formula (39.1) we note that dependence of crystal resistivity on magnetic field induction will be linear in a strongly degenerate case, when $\zeta/\Delta >> 1$, since corrections to a term linear in *B* are exponentially small in this case. This yields the following final formula for ρ_{zz}:

$$\rho_{zz}=\frac{3\pi h\Xi^2 kT|\mu^* B|}{e^2\Delta^3\rho s_{||}^2 a^2}. \tag{39.2}$$

The second case is realized for such temperatures and electron concentrations when $\zeta=\Delta$, i.e. the chemical potential level lies in the middle of allowed miniband. Then the value ρ_{zz} is obtained by doubling the expression (39.2). The sum over l in formula (39.1) in this case is identically zero. The third case corresponds to nondegenerate electron gas, when $\zeta<0$ and $|\zeta|/kT>>1$. In this case, with additional account of the Landau quantization effect on chemical potential, we get the following final formula for longitudinal magnetoresistance:

$$\rho_{zz}=\frac{\pi\Xi^2 B I_0(\Delta/kT)}{4n_0 a^3 e\rho s_{||}^2}\left(\Delta\cosh\frac{\Delta}{kT}-kT\sinh\frac{\Delta}{kT}\right)^{-1}, \tag{39.3}$$

where $I_0(x)$ is imaginary Bessel function of the first kind. Transfer to the effective mass approximation in this formula is matched by a larger value of Δ/kT relationship, and then, with regard to the asymptotic kind of the Bessel function $I_0(x)$ we get the following asymptotic formula of magnetoresistance:

$$\rho_{zz}\propto B\sqrt{T}. \tag{39.4}$$

This formula, in terms of temperature dependence, is similar to that which follows (though, for the resistance in the absence of a magnetic field) from the classical Drude model [60]. But this model, for classical statistics and square-law dispersion law, should yield correct results, if the geometry of current carrier scattering is approximately one-dimensional, which is the case in a strong quantizing magnetic field [61]. An auxiliary condition for the validity of using formula (39.1) for superlattice crystals is $hs_{\|}/2\pi a_B << \Delta$, which, for instance, at $s_{\|} = 5\cdot 10^3$ m/s and $\Delta = 0.01$ eV can be violated only in the fields with inductions over 6100T. Note that if instead of electron-phonon collisions we would have considered collisions with impurities and (or) point defects as scattering mechanism, then formula (273) would only change so that the multiplier before the expression in braces would cease to depend on magnetic field and temperature, and in strong fields the longitudinal resistance would tend to saturation. Therefore, on exposure to both scattering mechanisms, the magnetic field dependence of resistance would be of the form

$$\rho_{zz} = \rho_i + KB\,, \tag{39.5}$$

where ρ_i – part of resistance due to impurities, K – the Kapitsa coefficient.

However, in a weak field, when relaxation time can be considered to be magnetic field independent, with all scattering mechanisms we get a square-law field dependence of longitudinal magnetoresistance, as long as even in magnetic field multiplier h_l^{σ} is caused by the thermodynamic properties of electron gas in a magnetic field, rather than by subtle details of scattering mechanisms.

Finally, let us summarize the conditions for the Kapitsa effect observability and estimate the Kapitsa coefficient for some materials. Note that if $\Delta = 0.01$ eV, then at $m^* = 0.1m_0$ for the observation of the Kapitsa effect we need magnetic fields with inductions over 10T, and if $m^* = 0.01m_0$, the fields with inductions over 1T. The temperature range in this case can include from helium to nitrogen, and in some cases, i.e. with particularly small transverse effective masses of charge carriers, even room temperatures. Then, in strongly degenerate semiconductors and semimetals the Kapitsa coefficient can be determined by formula (39.2), and, for instance, at

$$T = 4\text{K}, \Xi = 1\text{eV}, m^* = 0.01m_0, \Delta = 0.01\text{eV}, \rho = 5\cdot 10^3\text{kg/m}^3, a = 10\text{nm}, s_{\|} = 5\cdot 10^3\text{m/s}$$

it will make $1.56\cdot 10^{-9}$ Ω·m/T, which is matched by charge carrier concentration $n_0 = 4.16\cdot 10^{16}$ cm^{-3} or higher. In the nondegenerate layered semiconductors the Kapitsa coefficient can be determined by formula (39.3), and with the above specified band parameters, the deformation potential and temperature 88K it will make $1.37\cdot 10^{-6}$Ω·m/T. Note that knowing the band parameters of the crystal, from the investigations of the longitudinal Kapitsa effect one can establish the amplitude of deformation potential and, hence, the value of effective interaction which, according to Bardeen-Cooper-Schrieffer theory, is responsible for a superconducting transition. Thus, at least for the longitudinal temperature-dependent effect, Kapitsa's assumption on its relation to superconductivity is confirmed. However, for the transverse effect in its traditional interpretation such a relation is absent.

Moreover, from the calculated results it follows that if crystal FS is a corrugated circular (elliptical) cylinder or an ellipsoid, then the conditions for the resulting formulae validity and, hence, the longitudinal temperature-dependent Kapitsa effect observability are best assured with orientation of the electric and quantizing magnetic fields along the cylinder axis or the longest ellipsoid axis.

Chapter 40

THE KAPITSA EFFECT IN CHARGE-ORDERED LAYERED CRYSTALS

The longitudinal conductivity of a charge-ordered layered crystal under conditions of the Kapitsa effect observability is determined by formula (39.1) with substitution of $W(x) = \pm\sqrt{\Delta^2 \cos^2 x + W_0^2\delta^2}$, $\overline{W} = \sqrt{\Delta^2 + W_0^2\delta^2} - W_0\delta$.

Conditions of the magnetic field independence of parameter τ_0 , and hence, of the Kapitsa longitudinal effect observability in charge-ordered layered crystals, are as follows:

$$\mu^* B \geq kT , \tag{40.1}$$

$$\mu^* B \geq \sqrt{\Delta^2 + W_0^2\delta^2} - W_0\delta , \tag{40.2}$$

$$2\pi^2\left(kTa_B / hs_{\parallel}\right) >> 1 , \tag{40.3}$$

$$hs_{\parallel}a_B / 2\pi << \sqrt{\Delta^2 + W_0^2\delta^2} - W_0\delta . \tag{40.4}$$

If $T = 10\text{K}$, $m^* = 0.1m_0$, $\Delta = 0.01\text{eV}$, $W_0 = 0.04\text{eV}$, $s_{\parallel} = 5\cdot 10^3$ m/s, it turns out that conditions (40.1) and (40.2) expressing smallness of contribution of interband transitions to longitudinal conductivity, are fulfilled in the fields with inductions 0.86T and more, whereas conditions (40.3) and (40.4), expressing prevalence of induced quasi-elastic acoustic phonon scattering of electrons are not violated even in the fields with inductions up to 1000T.

As before, we will be interested in the cases when the magnetic field dependence of resistivity can be reduced to linear. The first such case is realized when $\zeta > \sqrt{\Delta^2 + W_0^2\delta^2}$. Then, by analogy with formula (39.2), we get the following formula for the longitudinal magnetoresistance [62]:

$$\rho_{zz}=\frac{3\pi h\Xi^2 kT\left|\mu^* B\right|}{4e^2\rho s_{\|}a^2\left(\sqrt{\Delta^2+W_0^2\delta^2}-W_0\delta\right)^3}. \tag{40.5}$$

The second case is realized when $W_0/\zeta_{02D}=2$. Then, in the entire subcritical temperature region, under weak and intermediate degeneracy, the level of electron gas chemical potential even in the presence of a magnetic field lies strictly in the middle of pseudo-gap between the minibands. Therefore, in the second case

$$\rho_{zz}=\frac{3\pi h\Xi^2 kT\left|\mu^* B\right|}{2e^2\rho s_{\|}a^2\left(\sqrt{\Delta^2+W_0^2\delta^2}-W_0\delta\right)^3}. \tag{40.6}$$

In this case, as long as the order parameter δ, as it follows from Figure 18b, depends only slightly on the magnetic field, and hence, the magnetic field dependence of resistivity is essentially near linear, it is appropriate to cite a modelling numerical example of estimating the Kapitsa coefficient. Assuming $kT/\Delta=0.6, \zeta_{02D}/\Delta=2, W_0/\zeta_{02D}=2, \Delta=0.01\,\mathrm{eV}$, $\rho=5000$ kg/m^3, $s_{\|}=5000$ m/s, $\Xi=10\,\mathrm{eV}$, $a=10$ nm, $m^*=0.1m_0$, and, moreover, in conformity with Figure 18b, on the average $\delta\approx 0.925$, for the Kapitsa coefficient K we will get the value 0.168Ω·m/T. Thus, it turns out that charge ordering is useful in observation of the Kapitsa effect. As regards the above cited numerical example, the most meticulous readers may ask to what degree the sufficiently high value of acoustic phonon deformation potential constant assumed in it agrees with the low value of effective attractive interaction W_0 leading to charge ordering. However, it should be taken into account that interaction W_0 is the result of competence between electron-phonon interaction and the Coulomb electron repulsion, and hence, even with a large value of deformation potential it can be low.

Finally, the third case is realized with rather high values of effective attractive interaction $W_0/\zeta_{02D}>>1$, or with low electron concentrations, when chemical potential level lies below the lower conduction miniband, and electron gas can be considered as nondegenerate. In this case, for the longitudinal magnetoresistance the following formula is valid:

$$\begin{aligned}\rho_{zz}=\frac{\Xi^2 IB}{4n_0a^3e\rho s^2}&\left\{\frac{1}{2kT}\left[\left(\Delta_\delta^2+W_0^2\delta^2\right)\left(\sinh\frac{\Delta_\delta}{kT}-\sinh\frac{W_0\delta}{kT}\right)+W_0^2\delta^2\sinh\frac{W_0\delta}{kT}-\right.\right.\\ &\left.-\Delta_\delta^2\sinh\frac{\Delta_\delta}{kT}\right]+\Delta_\delta\cosh\frac{\Delta_\delta}{kT}-W_0\delta\cosh\frac{W_0\delta}{kT}+kT\left(\sinh\frac{W_0\delta}{kT}-\sinh\frac{\Delta_\delta}{kT}\right)+\\ &+\frac{W_0^2\delta^2\Delta_\delta^2}{(kT)^3}\left(\mathrm{Shi}\frac{W_0\delta}{kT}-\mathrm{Shi}\frac{\Delta_\delta}{kT}\right)+\\ &\left.+\frac{1}{2(kT)^2}\left(W_0^2\delta^2\Delta_\delta\exp\frac{\Delta_\delta}{kT}-W_0\delta\Delta_\delta^2\exp\frac{W_0\delta}{kT}\right)\right\}^{-1},\end{aligned} \tag{40.7}$$

where

$$I = \int_0^{\pi} \cosh \frac{\sqrt{\Delta^2 \cos^2 x + W_0^2 \delta^2}}{kT} dx , \tag{40.8}$$

$\mathrm{Shi}(x)$ is integral hyperbolic sine.

The order parameter δ in this case is found from the equation:

$$\delta = W_0 \delta I^{-1} \int_0^{\pi} \left(\sqrt{\Delta^2 \cos^2 x + W_0^2 \delta^2} \right)^{-1} \sinh \left(\frac{\sqrt{\Delta^2 \cos^2 x + W_0^2 \delta^2}}{kT} \right) dx . \tag{40.9}$$

One can see from the equation that the order parameter in this case depends only on crystal parameters and temperature. Therefore, the magnetic field dependence of charge-ordered crystal magnetoresistance in the absence of electron gas degeneracy reflected by formula (284) is really linear.

Note also that our earlier temperature dependences of the basic thermodynamic characteristics of electron gas in charge-ordered crystals at least qualitatively resemble similar dependences for superconductors including, for instance, dichalcogenides of transient metals. Thus, in these substances, formation of instability of charge-density wave type appears to be related to superconducting transition, which comes as no surprise, since the primary cause for both effects is the effective attraction between electrons due to electron-phonon interaction. Therefore, crystal electron spectrum anisotropy caused by layered structure does not interfere with the Cooper effect, but is instrumental in it.

Chapter 41

TEMPERATURE DEPENDENCE OF LONGITUDINAL ELECTRIC CONDUCTIVITY OF LAYERED CHARGE-ORDERED CRYSTALS IN THE ABSENCE OF A MAGNETIC FIELD

At high temperatures which, however, are lower than the critical temperature of second order phase transition, the longitudinal electric conductivity of a layered charge-ordered crystal can be determined by the formula:

$$\sigma_{zz}(0) \equiv \sigma_{zz} = \frac{4e^2a^2\rho s^2\Delta}{3\pi h\Xi^2 t}\int_{w\delta}^{\sqrt{w^2\delta^2+1}} y^{-1}\sqrt{\left(y^2-w^2\delta^2\right)\left(w^2\delta^2+1-y^2\right)}\left\{\left[1+\exp\left(\frac{y-\gamma}{t}\right)\right]^{-1}+\right. \\ \left. +\left[1+\exp\left(-\frac{y+\gamma}{t}\right)\right]^{-1}\right\}dy \quad (41.1)$$

The results of calculations of the temperature dependences of longitudinal electric conductivity of a layered crystal by this formula are given in Figures 134 and 135. In these figures $\sigma_0 = 4e^2a^2\rho s^2\Delta/\left(3\pi h\Xi^2\right)$.

The figures demonstrate that in the framework of the assumed relaxation time model corresponding to predominance of acoustic phonon deformation potential the longitudinal electric conductivity of a layered crystal on the whole, as could be expected, drops with a rise in temperature both in the disordered and ordered phase. In so doing, in the low-temperature region the longitudinal electric conductivity of a layered crystal in the ordered phase is 1 – 2 orders of magnitude lower than in the disordered phase and decreases with increasing the effective interaction. It is caused by conduction miniband narrowing in going to the charge-ordered state. However, in the ordered state on the temperature dependence of conductivity there are additional local minimum and local maximum of conductivity. The physical meaning of these extra features of conductivity is as follows. On the one hand, conductivity decrease with a rise in temperature is mainly governed by increase in the intensity of acoustic phonon deformation potential scattering of electrons due to increase in the number of these phonons. On the other hand, temperature rise should lead to growth of miniband width owing to gradual destruction of charge ordering. Therefore, at sufficiently low temperatures, while the order parameter is close to unity, and its temperature dependence is weak, the longitudinal electric conductivity decreases with a rise in temperature. However, at higher temperatures, when the order parameter starts dropping drastically with a rise in temperature, conduction miniband is suddenly expanded, owing to which the electric conductivity first drops with a rise in temperature considerably slower, and then, on reaching the minimum, it starts to grow. This growth continues up to phase transition temperature, following which the longitudinal electric conductivity of crystal drops again. Moreover, at point of phase transition on the temperature dependence of electric conductivity there is a kink, which is typical of second order phase transitions. In the process, as the value of effective attractive interaction between electrons increases, the additional maximum and minimum of conductivity are displaced toward the area of higher temperatures and their values are reduced.

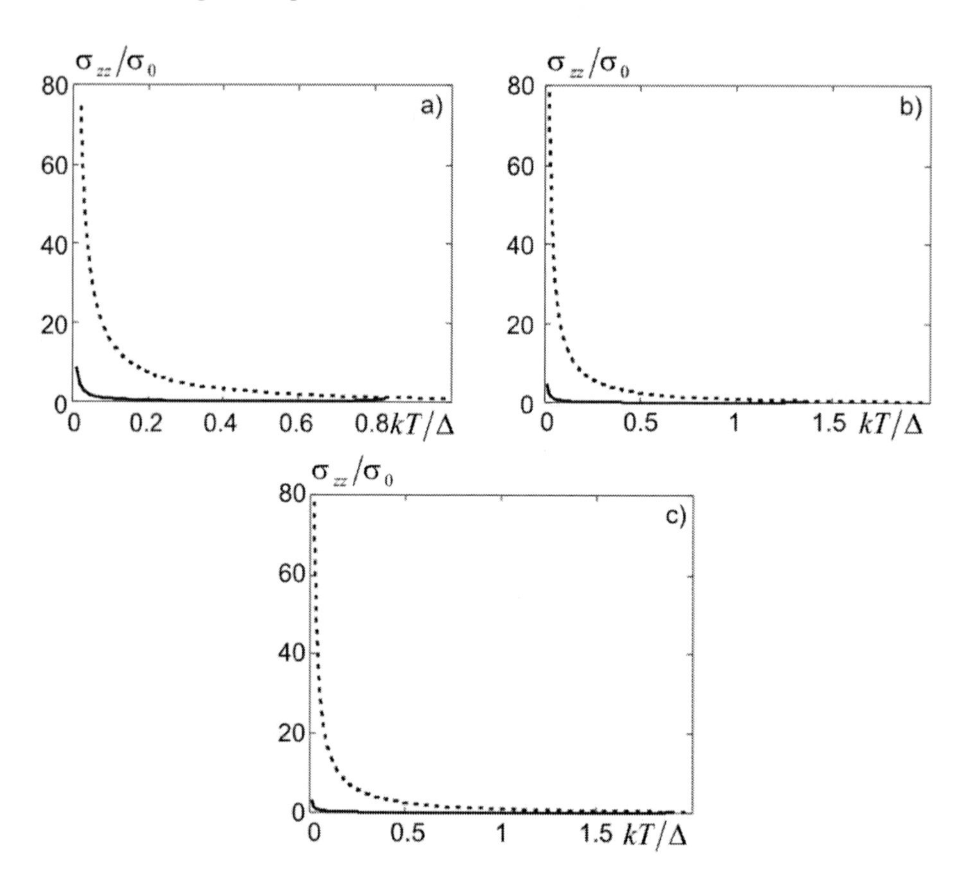

Figure 134. Temperature dependences of longitudinal electric conductivity of a layered crystal at:
a) $\zeta_{02D}/\Delta = 1; W_0/\zeta_{02D} = 1.5; 0.01 \le kT/\Delta \le 1$;
b) $\zeta_{02D}/\Delta = 1; W_0/\zeta_{02D} = 2; 0.01 \le kT/\Delta \le 2$;
c) $\zeta_{02D}/\Delta = 1; W_0/\zeta_{02D} = 2.5; 0.01 \le kT/\Delta \le 2$.
The solid curves are for the ordered state, the dashed curves - for the disordered state.

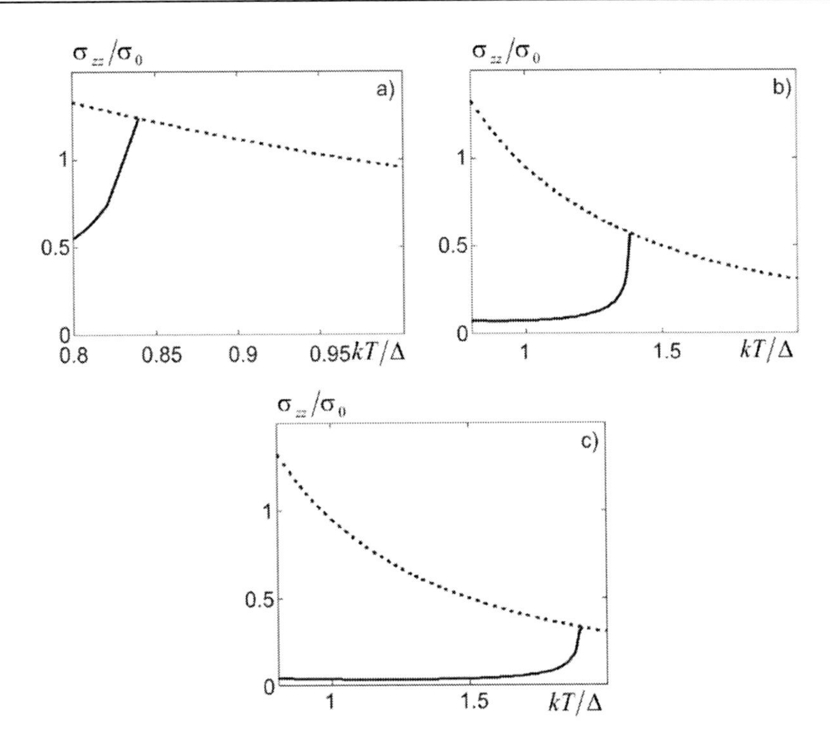

Figure 135. Temperature dependences of longitudinal electric conductivity of a layered crystal at $\zeta_{02D}/\Delta = 1, 0.8 \le kT/\Delta \le 2$: a) $W_0/\zeta_{02D} = 1.5$; b) $W_0/\zeta_{02D} = 2$; c) $W_0/\zeta_{02D} = 2.5$. The solid curves are for the ordered state, the dashed curves – for the disordered state.

Chapter 42

MAGNETORESISTANCE INVERSION IN CHARGE-ORDERED LAYERED CRYSTALS AT $\tau = const$

We now consider the effect of inversion of the monotonous part of longitudinal magnetoresistance in charge-ordered layered crystals in the approximation of constant relaxation time, assuming that acoustic phonon deformation potential scattering of charge carriers is predominant. Then full longitudinal electric conductivity of a charge-ordered layered crystal can be determined by the formula:

$$\sigma_{zz} = \frac{4e^2a^2\rho s^2\Delta}{3h\Xi^2 t}\Bigg\{ \int\limits_{\gamma+y\ge 0} y^{-1}\sqrt{\left(1+w^2\delta^2-y^2\right)\left(y^2-w^2\delta^2\right)} + \int\limits_{\gamma-y\ge 0} y^{-1}\sqrt{\left(1+w^2\delta^2-y^2\right)\left(y^2-w^2\delta^2\right)} +$$
$$+\sum_{l=1}^{\infty}(-1)^l \frac{bl/t}{\sinh(bl/t)}\Bigg[\int\limits_{\gamma+y\ge 0} y^{-1}\sqrt{\left(1+w^2\delta^2-y^2\right)\left(y^2-w^2\delta^2\right)}\exp\left(-l\frac{\gamma+y}{t}\right)dy -$$
$$- \int\limits_{\gamma+y\le 0} y^{-1}\sqrt{\left(1+w^2\delta^2-y^2\right)\left(y^2-w^2\delta^2\right)}\exp\left(l\frac{\gamma+y}{t}\right)dy +$$
$$+ \int\limits_{\gamma-y\ge 0} y^{-1}\sqrt{\left(1+w^2\delta^2-y^2\right)\left(y^2-w^2\delta^2\right)}\exp\left(-l\frac{\gamma-y}{t}\right)dy -$$
$$- \int\limits_{\gamma-y\le 0} y^{-1}\sqrt{\left(1+w^2\delta^2-y^2\right)\left(y^2-w^2\delta^2\right)}\exp\left(l\frac{\gamma-y}{t}\right)dy \Bigg]\Bigg\} \quad . \quad (42.1)$$

The results of calculation of crystal longitudinal electric conductivity by this formula are given in Figure 136. In this figure, as before, $\sigma_0 = 4e^2a^2\rho s^2\Delta/\left(3\pi h\Xi^2\right)$.

The figure demonstrates that the monotonous part of longitudinal magnetoresistance of a layered crystal undergoes inversion depending on the value of effective interaction. Thus, at $W_0/\zeta_{02D} = 1.5$ in the subcritical temperature range the level of electron gas chemical potential lies above the middle of a pseudogap between the minibands and drops with magnetic field increase. On the other hand, the order parameter grows with magnetic field increase. The two processes bring about conductivity reduction with magnetic field increase, hence, the longitudinal magnetoresistance at $W_0/\zeta_{02D} = 1.5$ should be "habitual", i.e.

positive. Indeed, with increase in the ratio $\mu^* B/\Delta$ from 0 to 5, the longitudinal conductivity of a layered crystal at $W_0/\zeta_{02D} = 1.5$ and $kT/\Delta = 0.6$ drops by a factor of 1.72.

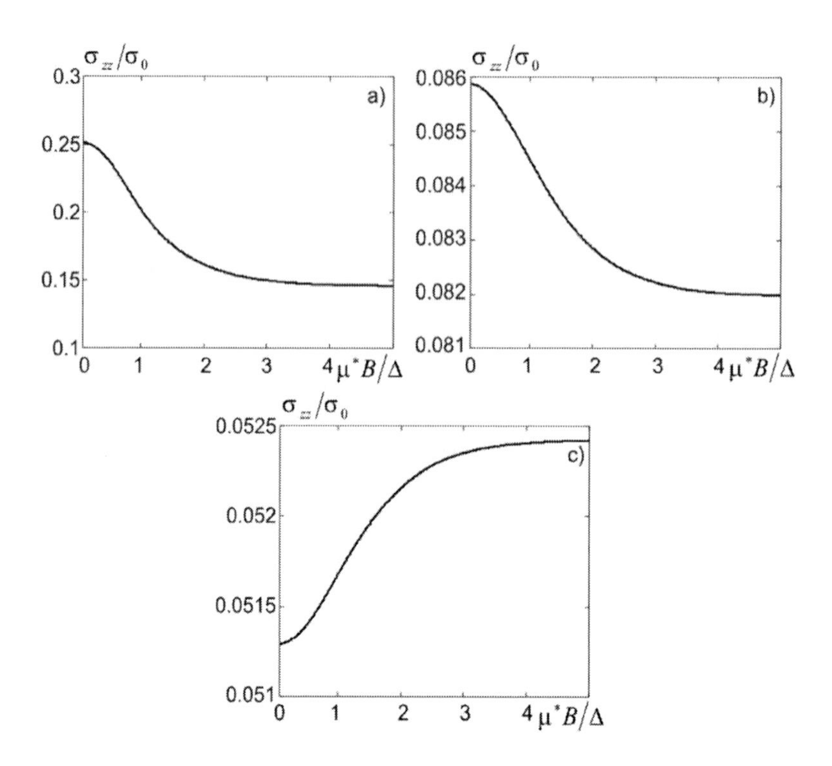

Figure 136. Field dependences of the nonoscillating part of longitudinal electric conductivity of a layered crystal at $\zeta_{02D}/\Delta = 1; kT/\Delta = 0.6; 0 \le \mu^* B/\Delta \le 5$ and: a) $W_0/\zeta_{02D} = 1.5$; b) $W_0/\zeta_{02D} = 2$; c) $W_0/\zeta_{02D} = 2.5$.

At $W_0/\zeta_{02D} = 2$ the level of electron gas chemical potential in the entire subcritical temperature range lies strictly in the middle of a pseudo-gap between the minibands and is magnetic field independent. Therefore, if the order parameter were magnetic field independent, the monotonous part of the longitudinal magnetoresistance would be equal to zero. However, at $W_0/\zeta_{02D} = 2$ the order parameter is weakly increased with magnetic field increase. So, due to narrowing of conduction minibands, the longitudinal conductivity of a layered crystal again drops with magnetic field increase, though considerably slower. Thus, with increase in the ratio $\mu^* B/\Delta$ from 0 to 5, the longitudinal conductivity of a layered crystal at $W_0/\zeta_{02D} = 2$ and $kT/\Delta = 0.6$ is decreased by 4.9%.

At $W_0/\zeta_{02D} = 2.5$ the level of electron gas chemical potential in the entire subcritical temperature range lies below the middle of a pseudo-gap between the minibands and grows with increase in magnetic field. The order parameter at $W_0/\zeta_{02D} = 2$ weakly grows with magnetic field increase. Therefore, due to growth of electron gas chemical potential level, the longitudinal conductivity of a layered crystal should grow with magnetic field increase, and due to narrowing of conduction minibands it should drop. But as long as the dominating contribution to change of crystal longitudinal conductivity as a function of a magnetic field is made in this case by the former process, with increase in the ratio $\mu^* B/\Delta$ from 0 to 5, the longitudinal conductivity of a layered crystal at $W_0/\zeta_{02D} = 2.5$ and $kT/\Delta = 0.6$ is increased by 2.1%, i.e. NLMR takes place. The relatively low value of NLMR in this case is attributable to the fact that the larger the effective attractive interaction leading to charge ordering, the weaker the magnetic field dependence of the order parameter and the chemical potential.

Chapter 43

Longitudinal Electric Conductivity of Charge-Ordered Layered Crystals in the Absence of a Magnetic Field under Conditions of Weak and Intermediate Degeneracy with Regard to FS Closeness Impact on Scattering

Assuming the relaxation time to be inversely proportional to the density of electron states $g(\varepsilon)$, in the case of a charge-ordered layered crystal with closed FS the formula of longitudinal electric conductivity in the absence of a magnetic field with scattering on acoustic phonons is as follows:

$$\sigma_{zz}(0) \equiv \sigma_{zz} = \frac{e^2 a^2 \rho s^2 \Delta}{3h\Xi^2 t} \int_0^{\pi/2} \left(\frac{\sin^2 2x}{x\left(w^2\delta^2 + \cos^2 x\right)} - \frac{1}{x^2} \int_0^x \frac{\sin^2 2z dz}{\left(w^2\delta^2 + \cos^2 z\right)} \right) \times$$
$$\times \left\{ \left[1 + \exp\left(\frac{-\sqrt{w^2\delta^2 + \cos^2 x} - \gamma}{t} \right) \right]^{-1} + \left[1 + \exp\left(\frac{\sqrt{w^2\delta^2 + \cos^2 x} - \gamma}{t} \right) \right]^{-1} \right\} dx \quad . \quad (43.1)$$

The results of calculations by formula (43.1) are given in Figures 137 and 138. In these figures, $\sigma_0 = 4e^2 a^2 \rho s^2 \Delta / \left(3\pi h \Xi^2\right)$.

The figures demonstrate that within the framework of model $\tau \propto 1/g(\varepsilon)$, the longitudinal electric conductivity of a layered crystal as a whole also drops with a rise in temperature both in the disordered and ordered phases. In so doing, in the low-temperature region the longitudinal electric conductivity of a layered crystal in the ordered phase is 1 – 2 orders lower than in the disordered phase and is reduced with effective interaction increase. As in the model of constant relaxation time, this is due to narrowing of conduction miniband when passing to the charge-ordered state. As well as in the model of constant relaxation time, in the ordered state on the temperature dependence of conductivity there exist additional local minimum and local maximum of conductivity. The physical meaning of these additional conductivity features is the same as in the model of constant relaxation time. Moreover, as in the model of constant relaxation time, at phase-transition point on the temperature dependence of electric conductivity there is a break typical of second order phase transitions. In so doing, with increase in effective attractive interaction between electrons, the additional maximum and minimum are displaced toward the higher temperature range and their values are reduced. Thus, the above peculiarities of the longitudinal conductivity behavior are attributable, on the one hand, to a more intensive acoustic phonon deformation potential scattering of electrons with a rise in temperature, and, on the other hand, to peculiarities of electron gas thermodynamics in the presence of interlayer charge ordering. However, at all temperatures in the model $\tau \propto 1/g(\varepsilon)$ the longitudinal conductivity of a layered charge-ordered crystal is approximately 2-3 times higher than in the model $\tau = const$. This difference is due to the fact that in the model $\tau = const$ the contribution of low-energy electron states to longitudinal conductivity is reduced.

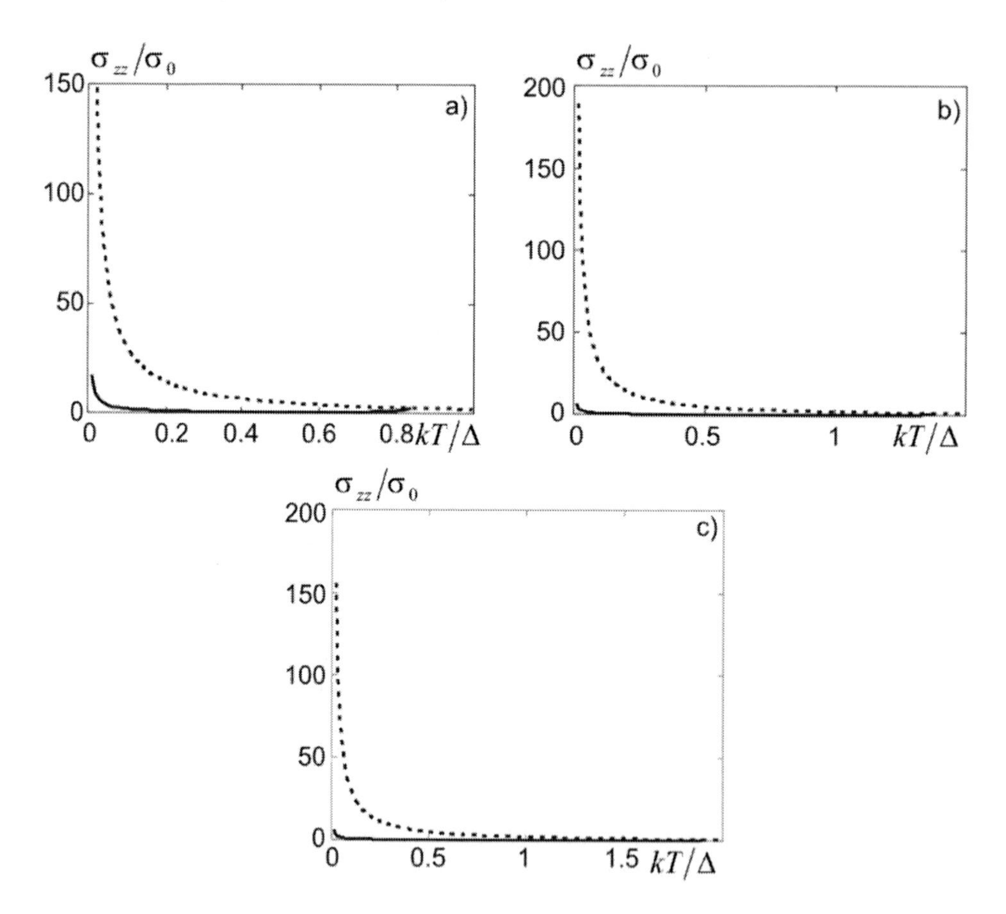

Figure 137. Temperature dependence of longitudinal conductivity of a layered charge-ordered crystal in the absence of a magnetic field in the model $\tau \propto 1/g(\varepsilon)$ over a wide range of temperatures at $\zeta_{02D}/\Delta = 1$ and: a) $W_0/\zeta_{02D} = 1.5$; b) $W_0/\zeta_{02D} = 2$; c) $W_0/\zeta_{02D} = 2.5$. Dashed curves correspond to the disordered state.

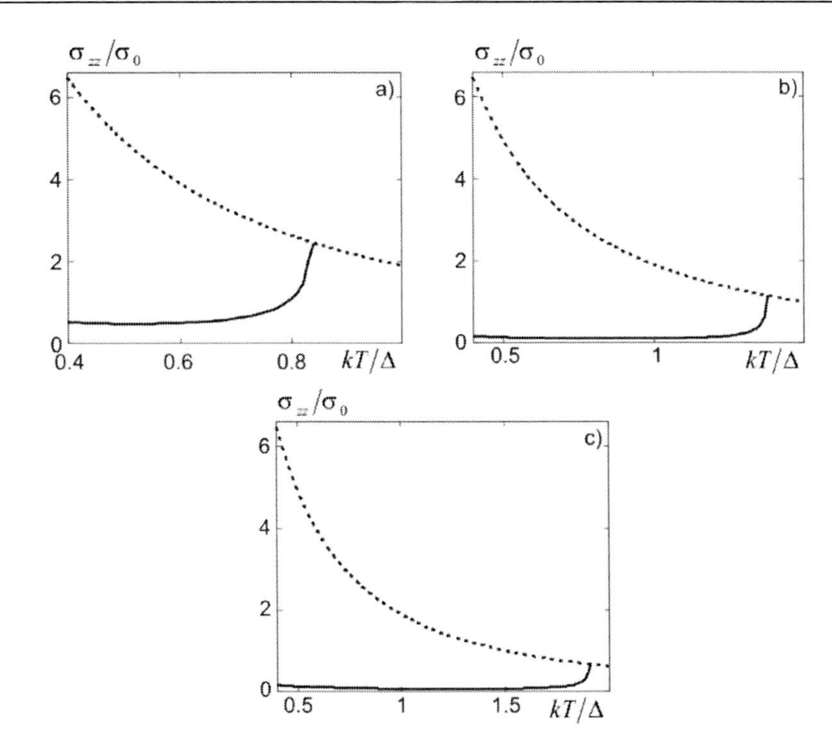

Figure 138. Temperature dependence of longitudinal electric conductivity of a layered charge-ordered crystal near phase-transition point in the absence of a magnetic field in the model $\tau \propto 1/g(\varepsilon)$ over a wide range of temperatures at $\zeta_{02D}/\Delta = 1$ and: a) $W_0/\zeta_{02D} = 1.5$; b) $W_0/\zeta_{02D} = 2$; c) $W_0/\zeta_{02D} = 2.5$. Dashed curves correspond to the disordered state.

Chapter 44

MAGNETORESISTANCE INVERSION IN THE CHARGE-ORDERED LAYERED CRYSTALS AT $\tau \propto 1/g(\varepsilon)$

We now consider the effect of inversion of the monotonous part of longitudinal magnetoresistance in the charge-ordered layered crystals with regard to energy dependence of relaxation time according to model $\tau \propto 1/g(\varepsilon)$. Then full longitudinal electric conductivity of a charge-ordered layered crystal can be determined by the formula:

$$\sigma_{zz} = \frac{4e^2a^2\rho s^2\Delta}{3h\Xi^2 t}\left\{ \left[\int\limits_{\gamma+\sqrt{w^2\delta^2+\cos^2 x}\geq 0, x\leq\pi/2} \left(\frac{\sin^2 2x}{x\left(w^2\delta^2+\cos^2 x\right)} - \frac{1}{x^2}\int\limits_0^x \frac{\sin^2 2z dz}{\left(w^2\delta^2+\cos^2 z\right)}\right)dx + \right.\right.$$
$$+ \int\limits_{\gamma-\sqrt{w^2\delta^2+\cos^2 x}\geq 0, x\leq\pi/2} \left(\frac{\sin^2 2x}{x\left(w^2\delta^2+\cos^2 x\right)} - \frac{1}{x^2}\int\limits_0^x \frac{\sin^2 2z dz}{\left(w^2\delta^2+\cos^2 z\right)}\right)dx +$$
$$+\sum_{l=1}^{\infty}(-1)^l \frac{bl/t}{\sinh(bl/t)}\left[\int\limits_{\gamma+\sqrt{w^2\delta^2+\cos^2 x}\geq 0, x\leq\pi/2} \left(\frac{\sin^2 2x}{x\left(w^2\delta^2+\cos^2 x\right)} - \frac{1}{x^2}\int\limits_0^x \frac{\sin^2 2z dz}{\left(w^2\delta^2+\cos^2 z\right)}\right)\times\right.$$
$$\times \exp\left(-l\frac{\gamma+\sqrt{w^2\delta^2+\cos^2 x}}{t}\right)dx -$$
$$- \int\limits_{\gamma+\sqrt{w^2\delta^2+\cos^2 x}\leq 0, x\leq\pi/2} \left(\frac{\sin^2 2x}{x\left(w^2\delta^2+\cos^2 x\right)} - \frac{1}{x^2}\int\limits_0^x \frac{\sin^2 2z dz}{\left(w^2\delta^2+\cos^2 z\right)}\right)\exp\left(l\frac{\gamma+\sqrt{w^2\delta^2+\cos^2 x}}{t}\right)dx +$$
$$+ \int\limits_{\gamma-\sqrt{w^2\delta^2+\cos^2 x}\geq 0, x\leq\pi/2} \left(\frac{\sin^2 2x}{x\left(w^2\delta^2+\cos^2 x\right)} - \frac{1}{x^2}\int\limits_0^x \frac{\sin^2 2z dz}{\left(w^2\delta^2+\cos^2 z\right)}\right)\exp\left(-l\frac{\gamma-\sqrt{w^2\delta^2+\cos^2 x}}{t}\right)dx -$$
$$\left.\left.- \int\limits_{\gamma-\sqrt{w^2\delta^2+\cos^2 x}\leq 0, x\leq\pi/2} \left(\frac{\sin^2 2x}{x\left(w^2\delta^2+\cos^2 x\right)} - \frac{1}{x^2}\int\limits_0^x \frac{\sin^2 2z dz}{\left(w^2\delta^2+\cos^2 z\right)}\right)\exp\left(l\frac{\gamma-\sqrt{w^2\delta^2+\cos^2 x}}{t}\right)dx\right]\right\} \quad (44.1)$$

The results of calculations of crystal longitudinal electric conductivity by this formula are given in Figure 139. In this figure, as before, $\sigma_0 = 4e^2a^2\rho s^2\Delta/\left(3\pi h\Xi^2\right)$.

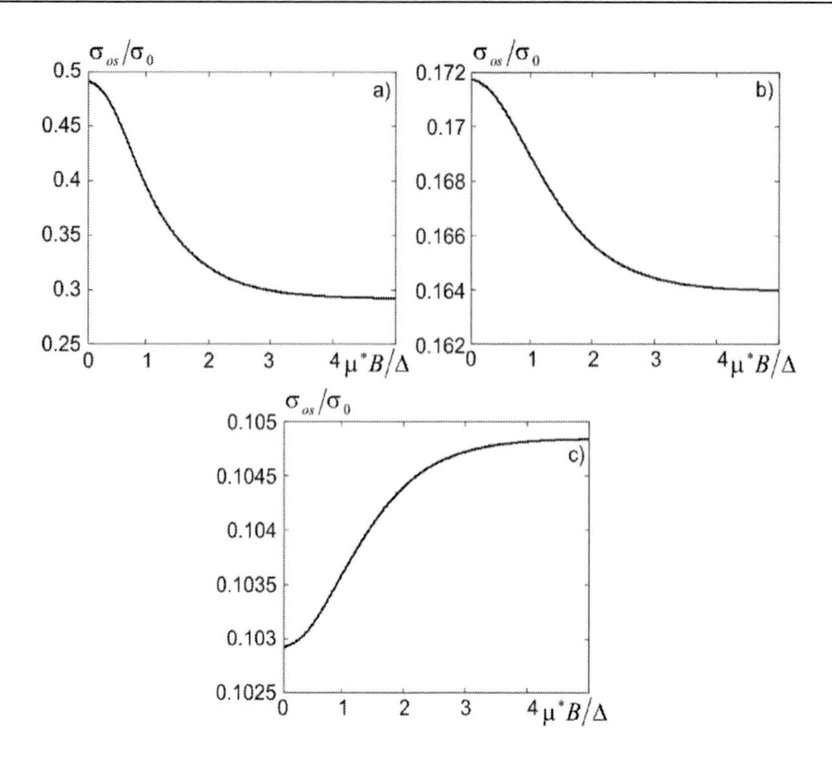

Figure 139. Field dependences of the nonoscillating part of longitudinal electric conductivity of a layered crystal at $\zeta_{02D}/\Delta = 1; kT/\Delta = 0.6; 0 \le \mu^* B/\Delta \le 5$ and: a) $W_0/\zeta_{02D} = 1.5$; b) $W_0/\zeta_{02D} = 2$; c) $W_0/\zeta_{02D} = 2.5$.

The figure demonstrates that the monotonous part of longitudinal magnetoresistance of a layered crystal, just as in the framework of the model $\tau = const$, undergoes inversion depending on the value of effective interaction [63]. As well as in the model $\tau = const$, the longitudinal magnetoresistance at $W_0/\zeta_{02D} = 1.5$ is "habitual", i.e. positive. In the framework of the model $\tau \propto 1/g(\varepsilon)$ with increase in the ratio $\mu^* B/\Delta$ from 0 to 5, the longitudinal conductivity of a layered crystal at $W_0/\zeta_{02D} = 1.5$ and $kT/\Delta = 0.6$ drops by a factor of 1.68.

With the same increase of this ratio, but at $W_0/\zeta_{02D} = 2$ the longitudinal conductivity of a layered crystal at $kT/\Delta = 0.6$ drops by 4.9%, i.e. there is also a positive longitudinal magnetoresistance, though of considerably lower value. However, with the same increase in this ratio, but at $W_0/\zeta_{02D} = 2.5$ the longitudinal conductivity of a layered crystal at $kT/\Delta = 0.6$ is increased by 1.9%, i.e. a negative longitudinal magnetoresistance takes place.

In connection with these results it is interesting to note that *a relative change* in the longitudinal resistance of a layered charge-ordered crystal in a magnetic field for all considered values of effective attractive interaction in the model $\tau \propto 1/g(\varepsilon)$ is virtually the same as in the model $\tau = const$, from which it can be concluded that this change is mainly governed by electron gas thermodynamics in a quantizing magnetic field in the presence of interlayer charge ordering. However, *the absolute value* of electric conductivity in the model $\tau \propto 1/g(\varepsilon)$ for all considered values of effective interaction is about 2 times greater than in the model $\tau = const$. Electric conductivity increase in this case, as before, is attributable to increased contribution of low-energy states in the model $\tau \propto 1/g(\varepsilon)$ as compared to the model $\tau = const$.

Chapter 45

GENERAL FORMULA FOR THE SEEBECK COEFFICIENT OF A LAYERED CRYSTAL UNDER CONDITIONS OF WEAK AND INTERMEDIATE DEGENERACY

Using Eq.(5.3), we obtain the following expression for the longitudinal Seebeck coefficient of a layered crystal under conditions of weak and intermediate degeneracy:

$$\alpha_{zz} = \alpha_0 \frac{A^{(\alpha)} + B^{(\alpha)}}{C^{(\alpha)} + D^{(\alpha)}}, \tag{45.1}$$

where:

$$A^{(\alpha)} = \left(\frac{\mu^* B}{kT}\right)^2 \sum_{l=1}^{\infty} \frac{(-1)^l l \coth(\mu^* Bl/kT)}{\sinh(\mu^* Bl/kT)} \left\{ \int\limits_{W(x)\le\zeta} \tau(x) W'^2(x) \exp\left[l \frac{W(x)-\zeta}{kT}\right] + \right. \\ \left. + \int\limits_{W(x)\ge\zeta} \tau(x) W'^2(x) \exp\left[l \frac{\zeta - W(x)}{kT}\right] dx \right\}, \tag{45.2}$$

$$B^{(\alpha)} = \frac{\mu^* B}{(kT)^2} \sum_{l=1}^{\infty} \frac{(-1)^{l-1} l}{\sinh(\mu^* Bl/kT)} \left\{ \int\limits_{W(x)\le\zeta} \tau(x) [W(x)-\zeta] W'^2(x) \exp\left[l \frac{W(x)-\zeta}{kT}\right] dx + \right. \\ \left. + \int\limits_{W(x)\ge\zeta} \tau(x) [\zeta - W(x)] W'^2(x) \exp\left[l \frac{\zeta - W(x)}{kT}\right] dx \right\}, \tag{45.3}$$

$$C^{(\alpha)} = \int\limits_{W(x)\le\zeta} \tau(x) W'^2(x) dx, \tag{45.4}$$

$$D^{(\alpha)} = \sum_{l=1}^{\infty} \frac{(-1)^l \mu^* Bl}{kT \sinh(\mu^* Bl/kT)} \left\{ \int_{W(x)\le\zeta} \tau(x) W'^2(x) \exp\left[l \frac{W(x)-\zeta}{kT} \right] - \right. \quad . \tag{45.5}$$
$$\left. - \int_{W(x)\ge\zeta} \tau(x) W'^2(x) \exp\left[l \frac{\zeta - W(x)}{kT} \right] dx \right\}$$

From here on we shall use these formulae for the analysis of the temperature and field dependences of the Seebeck coefficient under conditions of weak and intermediate degeneracy, both for a real layered crystal and in the effective mass approximation, as well as in the presence of interlayer charge ordering.

Chapter 46

TEMPERATURE DEPENDENCE OF THE SEEBECK COEFFICIENT IN A LAYERED CRYSTAL IN THE ABSENCE OF A MAGNETIC FIELD IN THE APPROXIMATION OF CONSTANT RELAXATION TIME

Now we consider this dependence in the approximation of constant relaxation time. The Seebeck coefficient of a real layered crystal in this case can be written as follows:

$$\alpha_{zz} = -\alpha_0 \frac{\int_0^{\pi} \sin^2 x \left\{ \ln\left[1+\exp\left(\frac{\gamma-1+\cos x}{t}\right)\right] + \frac{(1-\cos x-\gamma)}{t}\left[1+\exp\left(\frac{1-\cos x-\gamma}{t}\right)\right]^{-1} \right\} dx}{\int_0^{\pi} \sin^2 x \left[1+\exp\left(\frac{1-\cos x-\gamma}{t}\right)\right]^{-1} dx}. \quad (46.1)$$

In the effective mass approximation this formula acquires the form:

$$\alpha_{zz} = -\alpha_0 \frac{\int_0^{\infty} x^2 \left\{ \ln\left[1+\exp\left(\frac{\gamma-0.5x^2}{t}\right)\right] + \frac{(0.5x^2-\gamma)}{t}\left[1+\exp\left(\frac{0.5x^2-\gamma}{t}\right)\right]^{-1} \right\} dx}{\int_0^{\infty} x^2 \left[1+\exp\left(\frac{0.5x^2-\gamma}{t}\right)\right]^{-1} dx}. \quad (46.2)$$

The results of calculations of the temperature dependences of the Seebeck coefficient by these formulae are given in Figure 140.

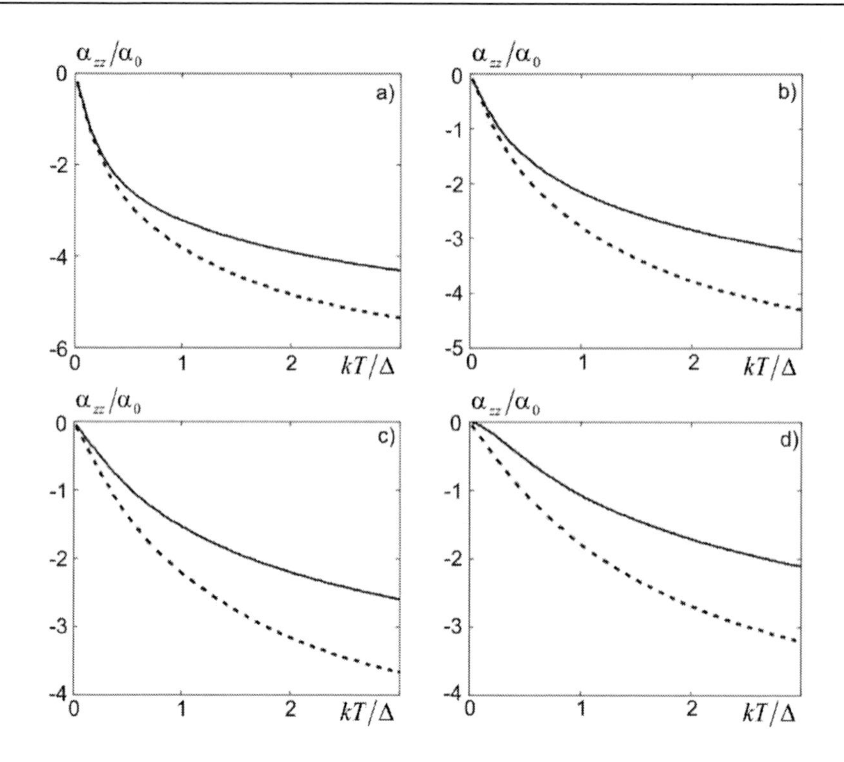

Figure 140. Temperature dependence of the Seebeck coefficient of a layered crystal at $0 \le kT/\Delta \le 3$.

The figure demonstrates that for all degrees of miniband filling, as it must be, the Seebeck coefficient is equal to zero at $T = 0$ and gradually increases with a rise in temperature. In so doing, as could be expected, the Seebeck coefficient, with a reduction of miniband filling degree, increases with temperature quicker and reaches high values. This occurs because with a reduction of miniband filling degree, the degree of electron gas degeneracy is reduced, owing to which the role of charge carriers with the energies lower than chemical potential, "harmful" from the viewpoint of the Seebeck coefficient modulus, is reduced. Moreover, for all degrees of miniband filling, with a rise in temperature, the Seebeck coefficient moduli calculated for a real layered crystal, become increasingly different from the Seebeck coefficient moduli calculated in the effective mass approximation. This occurs because in the effective mass approximation the energy of longitudinal motion of charge carriers, in principle, can assume any value from zero to infinity, whereas, as in the case of a real layered crystal, this energy is restricted by miniband width 2Δ . Thus, in the case of a real layered crystal the contribution of low-energy carriers to the Seebeck coefficient is greater than in the effective mass approximation, and, hence, its modulus is smaller. With a rise in temperature, for each degree of miniband filling, not only are the Seebeck coefficient moduli increased, calculated both for a real layered crystal and in the effective mass approximation, but also their absolute difference. The reason is that the finite width of a miniband in a real layered crystal restricts considerably the thermal spread of carriers in energies and, hence, the Seebeck coefficient growth with temperature. With increase in miniband filling degree, not only are the Seebeck coefficient moduli decreased calculated both for a real layered crystal and in the effective mass approximation, but also their relative difference. This occurs because the closer is electron gas chemical potential level to miniband ceiling, to a greater degree in the case of a real layered crystal the value of the Seebeck coefficient modulus and the character of its temperature dependence is affected by the singularity of the density of electron states at $\varepsilon = 2\Delta$, whereas in the effective mass approximation such a singularity is absent. Therefore, for instance, at $kT/\Delta = 0.3$ and $\zeta_0/\Delta = 0.5;1;1.5;2$ the Seebeck coefficient modulus of a real layered crystal is $372;280;224;182$ μV/K, respectively, and in the effective mass approximation – $461;371;316;276$ μV/K, respectively, and it is 24;33;41 and 52%, respectively, greater than for a real layered crystal. At $\Delta = 0.01$ eV the value $kT/\Delta = 0.3$ is matched by temperature 349K. Electron concentrations corresponding to said values of the Seebeck coefficient, both for a real layered crystal and in the effective mass approximation, with previously stipulated problem parameters are equal to $4.54 \cdot 10^{16}$; $1.33 \cdot 10^{17}$; $2.54 \cdot 10^{17}$; $4.16 \cdot 10^{17} cm^{-3}$. Thus, it turns out that even with a closed FS the layered structure effects produce a considerable effect on the Seebeck coefficient. Therefore, when developing novel thermoelectric materials and improving the properties of traditional thermoelectric materials, account must be taken of the influence of layered structure effects on the Seebeck coefficient, power factor and thermoelectric figure of merit of these materials.

Chapter 47

MAGNETIC FIELD DEPENDENCE OF THE SEEBECK COEFFICIENT OF A LAYERED CRYSTAL UNDER CONDITIONS OF WEAK AND INTERMEDIATE DEGENERACY IN THE APPROXIMATION OF CONSTANT RELAXATION TIME

Under conditions of weak and intermediate degeneracy in the presence of a longitudinal quantizing magnetic field the coefficients $A^{(\alpha)}, B^{(\alpha)}, C^{(\alpha)}, D^{(\alpha)}$ for a real layered crystal may be represented in the following dimensionless form:

$$A^{(\alpha)} = (b/t)^2 \sum_{l=1}^{\infty} \frac{(-1)^l l \coth(bl/t)}{\sinh(bl/t)} \left\{ \int_0^{\gamma} \sqrt{2y - y^2} \exp\left[lt^{-1}(y-\gamma)\right] dy + \int_{\gamma}^{2} \sqrt{2y - y^2} \exp\left[lt^{-1}(\gamma - y)\right] dy \right\}, \quad (47.1)$$

$$B^{(\alpha)} = \left(b/t^2\right) \sum_{l=1}^{\infty} \frac{(-1)^{l-1} l}{\sinh(bl/t)} \left\{ \int_0^{\gamma} \sqrt{2y - y^2} (y-\gamma) \exp\left[lt^{-1}(y-\gamma)\right] dy + \right.$$
$$\left. + \int_{\gamma}^{2} \sqrt{2y - y^2} (\gamma - y) \exp\left[lt^{-1}(\gamma - y)\right] dy \right\}, \quad (47.2)$$

$$C^{(\alpha)} = \int_0^{\gamma} \sqrt{2y - y^2}\, dy, \quad (47.3)$$

$$D^{(\alpha)} = \sum_{l=1}^{\infty} \frac{(-1)^l bl}{t \sinh(bl/t)} \left\{ \int_0^{\gamma} \sqrt{2y - y^2} \exp\left[lt^{-1}(y-\gamma)\right] dy - \int_{\gamma}^{2} \sqrt{2y - y^2} \exp\left[lt^{-1}(\gamma - y)\right] dy \right\}. \quad (47.4)$$

In the effective mass approximation these formulae acquire the form:

$$A^{(\alpha)} = (b/t)^2 \sum_{l=1}^{\infty} \frac{(-1)^{l-1} l \coth(bl/t)}{\sinh(bl/t)} \left\{ \int_0^{\gamma} \sqrt{2y} \exp\left[lt^{-1}(y-\gamma)\right] dy + \int_{\gamma}^{\infty} \sqrt{2y} \exp\left[lt^{-1}(\gamma-y)\right] dy \right\}, \quad (47.5)$$

$$B^{(\alpha)} = (b/t^2) \sum_{l=1}^{\infty} \frac{(-1)^{l-1} l}{\sinh(bl/t)} \left\{ \int_0^{\gamma} \sqrt{2y}(y-\gamma) \exp\left[lt^{-1}(y-\gamma)\right] dy + \int_{\gamma}^{\infty} \sqrt{2y}(\gamma-y) \exp\left[lt^{-1}(\gamma-y)\right] dy \right\}, \quad (47.6)$$

$$C^{(\alpha)} = (2\gamma)^{3/2}/3, \quad (47.7)$$

$$D^{(\alpha)} = \sum_{l=1}^{\infty} \frac{(-1)^l bl}{t \sinh(bl/t)} \left\{ \int_0^{\gamma} \sqrt{2y} \exp\left[lt^{-1}(y-\gamma)\right] dy - \int_{\gamma}^{\infty} \sqrt{2y} \exp\left[lt^{-1}(\gamma-y)\right] dy \right\}. \quad (47.8)$$

The results of calculations of the field dependences of the Seebeck coefficient by these formulae are given in Figure 141.

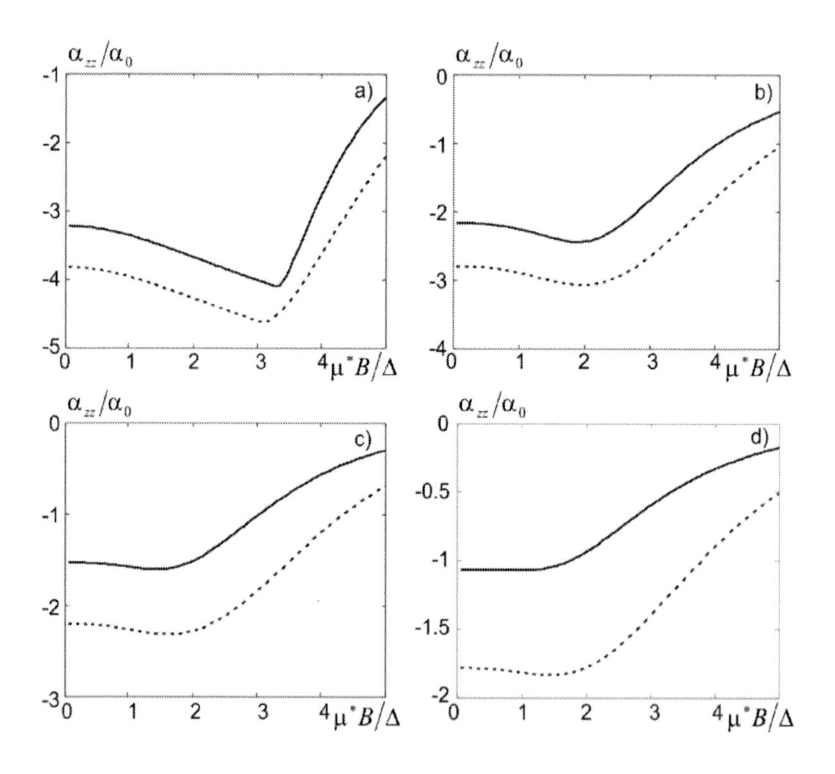

Figure 141. Field dependences of the Seebeck coefficient of a layered crystal under conditions of weak and intermediate degeneracy at $kT/\Delta = 1; 0 \le \mu^* B/\Delta \le 5$.

The figure demonstrates that for all degrees of miniband filling the Seebeck coefficient remains negative, i.e. purely "electron", since we suppose that free charge carriers are electrons. For all degrees of miniband filling the Seebeck coefficient modulus in the effective mass approximation is larger than in a real layered crystal. This occurs because in the effective mass approximation the width of a miniband describing the interlayer motion of electrons is assumed to be infinite, whereas in a real layered crystal it is essentially finite. Therefore, in a real layered crystal the reducing role of "harmful" charge carriers with the energies lower than chemical potential level is considerably larger than in the effective mass approximation. Hence, with increase in miniband filling degree, i.e., in fact, the degeneracy degree, the Seebeck coefficient modulus drops. Moreover, with increase in degeneracy degree, the number of "harmful" charge carriers with the energies lower than chemical potential level is increased. However, as long as on the whole the energy of charge carriers in the presence of the Landau levels is larger than in their absence, at first, while the electron gas chemical potential with magnetic field increase is slowly increased, and thermal spread of current carriers between the Landau subbands is not too difficult, the Seebeck coefficient modulus with magnetic field increase is increased. However, as mentioned above, magnetic field increase is in a sense equivalent to crystal cooling. Therefore, when the distance between the Landau subbands becomes comparable both to miniband width and the energy of thermal motion, the spread of charge carriers in energies becomes difficult. At the same time, the electron gas chemical potential starts increasing with magnetic field increase considerably faster, and the reducing role of "harmful" charge carriers is increased considerably. Exactly for this reason the Seebeck coefficient modulus at first starts increasing slower and with a certain magnetic field induction reaches its maximum, and then starts decreasing. With increase in miniband filling degree, the Seebeck coefficient modulus maximum becomes less pronounced. In so doing, in the effective mass approximation this maximum is more pronounced than for a real layered crystal, therefore, in the case of a transient FS, i.e. at $\zeta_0/\Delta = 2$, in a real layered crystal this maximum is practically smoothed over, whereas in the effective mass approximation it remains detectable. At $kT/\Delta = 1$ and $\zeta_0/\Delta = 0.5;1;1.5;2$, respectively, for a real layered crystal the Seebeck coefficient modulus maxima are reached at $\mu^* B/\Delta = 3.3;1.9;1.4;0.85$, respectively, and make 353;210;138;92μV/K, respectively. In the effective mass approximation, with the same electron concentration and temperature, these maxima are achieved at $\mu^* B/\Delta = 3.1;1.95;1.6;1.45$ and make 399;264;200;158μV/K, respectively, i.e. they are 13, 26,45 and 71%, respectively, larger than in a real layered crystal. At $m^* = 0.01m_0; \Delta = 0.01\text{eV}$ the magnetic field inductions whereby maxima are achieved, make for a real layered crystal 5.69;3.28;2.42;1.47T, respectively, and in the effective mass approximation – 5.35;3.37;2.76;2.50T, respectively. So, with regard to NLMR, application of a magnetic field parallel to temperature gradient can be considered as one of the methods for properties control of at least some of thermoelectric materials.

Chapter 48

Temperature Dependence of the Seebeck Coefficient in a Layered Crystal with Regard to Energy Dependence of Relaxation Time

Constant relaxation time approximation, not infrequently used to describe scattering of electrons on acoustic phonon deformation potential and on chaotically arranged charged impurity potential in a layered crystal is correct for an open FS, but is too rough for a transient and closed FS. However, for the two mentioned scattering mechanisms an approximation is often used in terms of which the relaxation time of charge carriers is inversely proportional to the density of their states, for instance, in the absence of a magnetic field. Then in the case of an open FS the relaxation time degenerates into constant one, and in the case of a closed FS it is proportional to the extension of FS (or constant-energy surface in general) along the superlattice axis coinciding in that event with the temperature gradient and the magnetic field direction. Therefore, the Seebeck coefficient for a real layered crystal is equal to:

$$\alpha_{zz} = -\alpha_0 \frac{\int_0^{\pi} \varphi(x)\left\{\ln\left[1+\exp\left(\frac{\gamma-1+\cos x}{t}\right)\right]+\frac{(1-\cos x-\gamma)}{t}\left[1+\exp\left(\frac{1-\cos x-\gamma}{t}\right)\right]^{-1}\right\}dx}{\int_0^{\pi} \varphi(x)\left[1+\exp\left(\frac{1-\cos x-\gamma}{t}\right)\right]^{-1} dx}, \tag{48.1}$$

where

$$\varphi(x) = x^{-2}\sin(2x) - 2x^{-1}\cos 2x\,. \tag{48.2}$$

In the effective mass approximation this formula acquires the form:

$$\alpha_{zz} = -\alpha_0 \frac{\int\limits_0^\infty x\left\{\ln\left[1+\exp\left(\frac{\gamma-0.5x^2}{t}\right)\right]+\frac{\left(0.5x^2-\gamma\right)}{t}\left[1+\exp\left(\frac{0.5x^2-\gamma}{t}\right)\right]^{-1}\right\}dx}{\int\limits_0^\infty x\left[1+\exp\left(\frac{0.5x^2-\gamma}{t}\right)\right]^{-1}dx}. \quad (48.3)$$

The results of calculations of the temperature dependence of the Seebeck coefficient by these formulae are given in Figure 142.

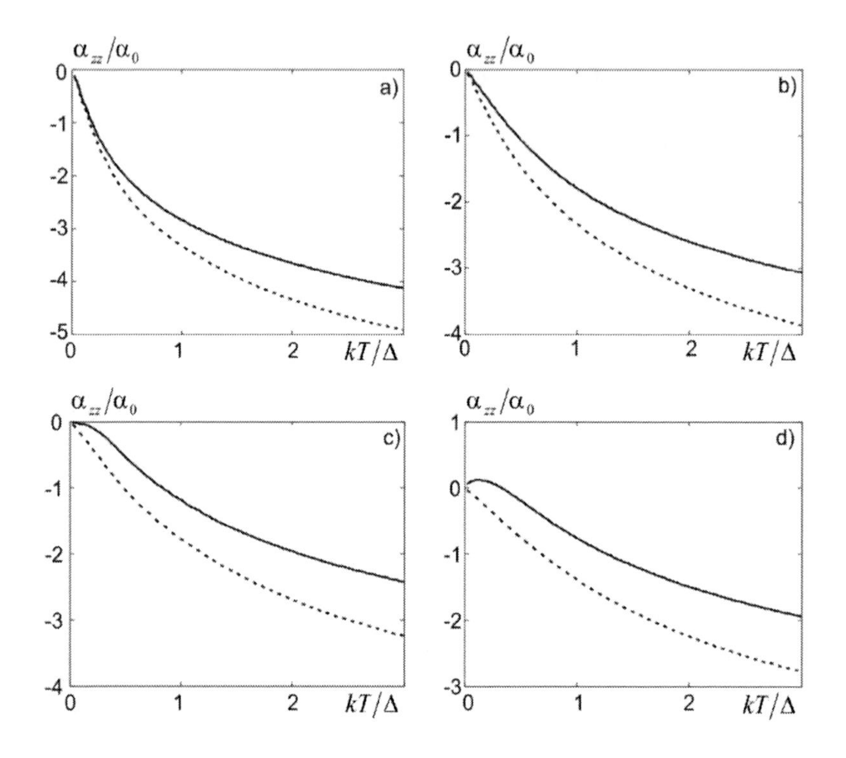

Figure 142. Temperature dependences of the Seebeck coefficient in the model $\tau \propto 1/g(\varepsilon)$ at $0 \le kT/\Delta \le 3$.

The figure demonstrates that in the model $\tau \propto 1/g(\varepsilon)$, as in the model $\tau = const$, with concentrations corresponding to the ratios $0.5 \le \zeta_0/\Delta \le 1.5$ the Seebeck coefficient for a real layered crystal, as well as in the effective mass approximation remains purely "electron", i.e. negative, but at $\zeta_0/\Delta = 2$ in the region of helium temperatures the Seebeck coefficient polarity is changed, i.e. it becomes a "hole" one. Moreover, at $kT/\Delta = 3$, i.e. at 348K, assuming that $\Delta = 0.01\text{eV}$, in the model $\tau \propto 1/g(\varepsilon)$ the Seebeck coefficient is somewhat lower than in the model $\tau = const$. Thus, for a real layered crystal at $\zeta_0/\Delta = 0.5;1;1.5;2$, respectively, the Seebeck coefficient modulus is 356;264;209;167μV/K, respectively, and in the effective mass approximation - 424;333;279;238 μV/K, respectively. Hence, in the model $\tau \propto 1/g(\varepsilon)$ the Seebeck coefficient modulus in the effective mass approximation is higher by 19, 26, 34 and 43%, respectively, than for a real layered crystal. As before, these differences are mainly attributable to the fact that with a finite width of miniband the thermal umklapp of charge carriers to the levels exceeding considerably the level of chemical potential, and thus the energy filtration of "harmful" charge carriers, is difficult. Moreover, it appears that with considered degrees of miniband filling for a real layered crystal the Seebeck coefficient modulus in the model $\tau = 1/g(\varepsilon)$ is lower by 4.3, 5.7, 6.7 and 8.2%, respectively, than in the model $\tau = const$, and in the effective mass approximation - by 8, 10.2, 11.7 and 13.8%, respectively, than in the model $\tau = const$. This slight but measurable (with appropriate equipment accuracy) impact is explained by the fact that in the model $\tau \propto 1/g(\varepsilon)$ the contribution of "harmful" charge carriers with low energies to the Seebeck coefficient is larger due to their larger relaxation time. For the same reason, certain effect on the reduction of the Seebeck coefficient modulus is produced by the fact that longitudinal conductivity calculated for a real layered crystal, as well as in the effective mass approximation, in the model $\tau \propto 1/g(\varepsilon)$ is higher than in the model $\tau = const$.

We now dwell on the reason for a change in the Seebeck coefficient sign in the model $\tau \propto 1/g(\varepsilon)$ at $\zeta_0/\Delta = 2$, i.e. in the case of a transient FS in the region of helium temperatures. At $T = 0$ the Seebeck coefficient from general thermodynamic considerations is equal to zero. With a rise in temperature, the electron gas chemical potential drops. Therefore, with location of chemical potential level close to miniband ceiling the major mass of charge carriers are "harmful" carriers with the energies lower than chemical potential. At the same time, due to singularity of the density of states $g(\varepsilon)$ at $\varepsilon = 2\Delta$ the relaxation time of charge carriers with the energies close to miniband ceiling is rather small, but is drastically increased with a reduction in chemical potential. Therefore, "harmful" charge carriers make a decisive contribution to the Seebeck coefficient which in the case of electrons becomes "hole", i.e. positive. However, with further rise in temperature, by virtue of degeneracy removal the contribution of "useful" charge carriers with the energies higher than chemical potential is larger and the Seebeck coefficient is also larger. It first achieves the local maximum of modulus, remaining positive, and then, decreasing in modulus, it vanishes, following which it becomes negative, i.e. "electron", and again increases in modulus with further rise in temperature. The maximum of the Seebeck coefficient modulus in the positive "hole" region makes 10.4μV/K and is achieved at $kT/\Delta = 0.12$, i.e. at a temperature of 13.9K, and conversion of this coefficient into zero occurs approximately at $kT/\Delta = 0.33$, i.e. at a temperature of 38.3K. Therefore, with selected problem parameters this temperature can be called the inversion temperature of the Seebeck coefficient.

Chapter 49

MAGNETIC FIELD DEPENDENCE OF THE SEEBECK COEFFICIENT OF A LAYERED CRYSTAL UNDER CONDITIONS OF WEAK AND INTERMEDIATE DEGENERACY IN THE APPROXIMATION OF ENERGY-DEPENDENT RELAXATION TIME

Under conditions of weak and intermediate degeneracy in the presence of a longitudinal quantizing magnetic field the coefficients $A^{(\alpha)}, B^{(\alpha)}, C^{(\alpha)}, D^{(\alpha)}$ for a real layered crystal in the model $\tau \propto 1/g(\varepsilon)$ may be represented in the following dimensionless form:

$$A^{(\alpha)} = (b/t)^2 \sum_{l=1}^{\infty} \frac{(-1)^l l \coth(bl/t)}{\sinh(bl/t)} \left\{ \int_0^{\arccos(1-\gamma)} \varphi(x) \exp\left[lt^{-1}(1-\cos x-\gamma)\right] dx + \int_{\arccos(1-\gamma)}^{\pi} \varphi(x) \exp\left[lt^{-1}(\gamma-1+\cos x)\right] dx \right\}, \tag{49.1}$$

$$B^{(\alpha)} = (b/t^2) \sum_{l=1}^{\infty} \frac{(-1)^{l-1} l}{\sinh(bl/t)} \left\{ \int_0^{\arccos(1-\gamma)} \varphi(x)(1-\cos x-\gamma) \exp\left[lt^{-1}(1-\cos x-\gamma)\right] dx + \int_{\arccos(1-\gamma)}^{\pi} \varphi(x)(\gamma-1+\cos x) \exp\left[lt^{-1}(\gamma-1+\cos x)\right] dx \right\}, \tag{49.2}$$

$$C^{(\alpha)} = \int_0^{\arccos(1-\gamma)} \varphi(x) dx, \tag{49.3}$$

$$D^{(\alpha)} = \sum_{l=1}^{\infty} \frac{(-1)^l bl}{t \sinh(bl/t)} \left\{ \int_0^{\arccos(1-\gamma)} \varphi(x) \exp\left[lt^{-1}(1-\cos x-\gamma)\right] dx - \int_{\arccos(1-\gamma)}^{\pi} \varphi(x) \exp\left[lt^{-1}(\gamma-1+\cos x)\right] dx \right\}. \tag{49.4}$$

In the effective mass approximation these formulae acquire the form:

$$A^{(\alpha)} = (b/t)^2 \sum_{l=1}^{\infty} \frac{(-1)^{l-1} l \coth(bl/t)}{\sinh(bl/t)} \left\{ \int_0^{\sqrt{2\gamma}} x \exp\left[lt^{-1}\left(0.5x^2 - \gamma\right)\right] dx + \right. \\ \left. + \int_{\sqrt{2\gamma}}^{\infty} x \exp\left[lt^{-1}\left(\gamma - 0.5x^2\right)\right] dx \right\} . \quad (49.5)$$

$$B^{(\alpha)} = \left(b/t^2\right) \sum_{l=1}^{\infty} \frac{(-1)^{l-1} l}{\sinh(bl/t)} \left\{ \int_0^{\sqrt{2\gamma}} x\left(0.5x^2 - \gamma\right) \exp\left[lt^{-1}\left(0.5x^2 - \gamma\right)\right] dx + \right. \\ \left. + \int_{\sqrt{2\gamma}}^{\infty} x\left(\gamma - 0.5x^2\right) \exp\left[lt^{-1}\left(\gamma - 0.5x^2\right)\right] dx \right\} , \quad (49.6)$$

$$C^{(\alpha)} = \gamma , \quad (49.7)$$

$$D^{(\alpha)} = \sum_{l=1}^{\infty} \frac{(-1)^{l} bl}{t \sinh(bl/t)} \left\{ \int_0^{\sqrt{2\gamma}} x \exp\left[lt^{-1}\left(0.5x^2 - \gamma\right)\right] dy - \int_{\sqrt{2\gamma}}^{\infty} x \exp\left[lt^{-1}\left(\gamma - 0.5x^2\right)\right] dx \right\} . \quad (49.8)$$

The results of calculations of the field dependences of the Seebeck coefficient by these formulae are given in Figure 143.

The figure demonstrates that in the model $\tau \propto 1/g(\varepsilon)$, as well, for all degrees of miniband filling the Seebeck coefficient remains negative, i.e. purely "electron", since we suppose that free charge carriers are electrons. For all degrees of miniband filling the Seebeck coefficient modulus in the effective mass approximation is larger than in a real layered crystal. This occurs because in the effective mass approximation the width of a miniband describing the interlayer motion of electrons is assumed to be infinite, whereas in a real layered crystal it is essentially finite. Therefore, in the model $\tau \propto 1/g(\varepsilon)$, as well as in the model $\tau = const$, in a real layered crystal the reducing role of "harmful" charge carriers with the energies lower than chemical potential level is considerably larger than in the effective mass approximation. Hence, with increase in miniband filling degree, i.e., in fact, the degeneracy degree, the Seebeck coefficient modulus drops. Moreover, with increase in degeneracy degree, the number of "harmful" charge carriers with the energies lower than chemical potential level is increased.

However, as long as on the whole the energy of charge carriers in the presence of the Landau levels is larger than in their absence, in the model $\tau \propto 1/g(\varepsilon)$, as well as in the model $\tau = const$, at first, while the electron gas chemical potential grows slowly with magnetic field increase, and thermal spread of current carriers between the Landau subbands is not too difficult, the Seebeck coefficient modulus increases with magnetic field increase.

However, as mentioned above, magnetic field increase is in a sense equivalent to crystal cooling.

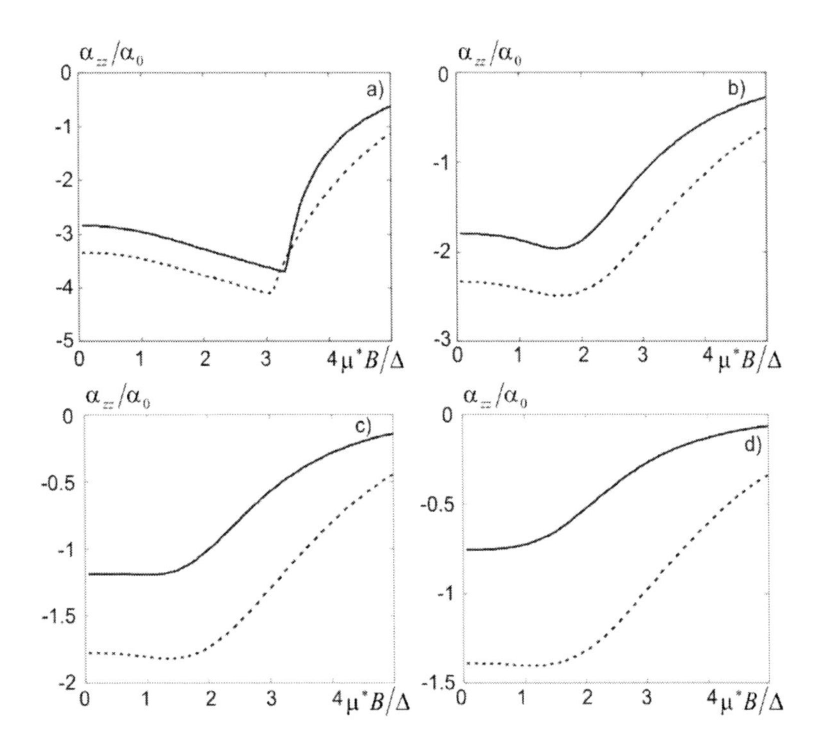

Figure 143. Field dependences of the Seebeck coefficient of a layered crystal under conditions of weak and intermediate degeneracy at $kT/\Delta = 1; 0 \le \mu^* B/\Delta \le 5$.

Therefore, when the distance between the Landau subbands becomes comparable both to miniband width and the energy of thermal motion, the spread of charge carriers in energies becomes difficult. At the same time, the electron gas chemical potential starts growing much faster with magnetic field increase and the reducing role of "harmful" charge carriers is increased considerably. Exactly for this reason the Seebeck coefficient modulus in the model $\tau \propto 1/g(\varepsilon)$, as well as in the model $\tau = const$, at first starts increasing slower and with a certain magnetic field induction reaches its maximum, and then starts decreasing. With increase in miniband filling degree, the Seebeck coefficient modulus maximum becomes less pronounced. In so doing, in the effective mass approximation this maximum is more pronounced than for a real layered crystal, therefore, in the case of a transient FS, i.e. at $\zeta_0/\Delta = 2$, in a real layered crystal this maximum is *fully* smoothed over, whereas in the effective mass approximation it remains detectable. At $kT/\Delta = 1$ and $\zeta_0/\Delta = 0.5;1;1.5$, respectively, for a real layered crystal the Seebeck coefficient modulus maxima are reached at $\mu^* B/\Delta = 3.3;1.6;0.95$, respectively, and make 319;169;103μV/K, respectively. In the effective mass approximation, при $\zeta_0/\Delta = 0.5;1;1.5;2$, respectively, and the same temperature, these maxima are achieved at $\mu^* B/\Delta = 3.05;1.65;1.35;1.15$, respectively, and make 355;215;157;121μV/K, respectively, i.e. at $\zeta_0/\Delta = 0.5;1;1.5$ they are larger by 11,27 and 52%, respectively, than in a real layered crystal. In so doing, in a real layered crystal at $\zeta_0/\Delta = 0.5;1;1.5$, respectively, in the model $\tau \propto 1/g(\varepsilon)$ these maxima are lower by 10,20 and 25% than in the model $\tau = const$. As in the absence of a magnetic field, this reduction takes place because in the model $\tau \propto 1/g(\varepsilon)$ the relative contribution to the Seebeck coefficient of "harmful" charge carriers with the energies lower than chemical potential is larger than in the model $\tau = const$. In the effective mass approximation in the model $\tau = 1/g(\varepsilon)$ the maxima of the Seebeck coefficient moduli at $\zeta_0/\Delta = 0.5;1;1.5;2$ are greater by 11, 23, 27 and 31%, respectively, than in the model $\tau = const$, which is also caused by the increase in the relative contribution of "harmful" charge carriers. At $m^* = 0.01m_0; \Delta = 0.01\text{eV}$ in the model $\tau = 1/g(\varepsilon)$ the magnetic field inductions whereby the maxima are achieved make for a real layered crystal 5.69;2.76; 1.64T, respectively, and in the effective mass approximation – 5.26;2.85;2.33;1.98T, respectively. These inductions are lower than in the model $\tau = const$, since in the model $1/g(\varepsilon)$ the contribution of "harmful" charge carriers to the Seebeck coefficient with magnetic field increase grows faster than in the model $\tau = const$. Therefore, with regard to NLMR which is also the case in the model $\tau \propto 1/g(\varepsilon)$, this model all the more admits control of thermoelectric material properties by magnetic field application parallel to temperature gradient and superlattice axis.

Chapter 50

TEMPERATURE DEPENDENCE OF THE SEEBECK COEFFICIENT OF LAYERED CHARGE-ORDERED CRYSTALS IN THE ABSENCE OF A MAGNETIC FIELD

The Seebeck coefficient of layered charge-ordered crystals in the absence of a magnetic field in the approximation of constant relaxation time is determined by the following formula:

$$\alpha_{zz} = -\alpha_0 \frac{E^{(\alpha)} + F^{(\alpha)}}{G^{(\alpha)}}. \tag{50.1}$$

Dimensionless coefficients $E^{(\alpha)}, F^{(\alpha)}, G^{(\alpha)}$ are determined as follows:

$$E^{(\alpha)} = \int_{w\delta}^{\sqrt{w^2\delta^2}} y^{-1}\sqrt{\left(1+w^2\delta^2-y^2\right)\left(y^2-w^2\delta^2\right)}\left\{\ln\left[1+\exp\left(\frac{\gamma+y}{t}\right)\right]+\ln\left[1+\exp\left(\frac{\gamma-y}{t}\right)\right]\right\}dy, \tag{50.2}$$

$$F^{(\alpha)} = t^{-1}\int_{w\delta}^{\sqrt{w^2\delta^2+1}} y^{-1}\sqrt{\left(1+w^2\delta^2-y^2\right)\left(y^2-w^2\delta^2\right)}\left\{\frac{y-\gamma}{1+\exp\left(\frac{y-\gamma}{t}\right)}-\frac{y+\gamma}{1+\exp\left(-\frac{y+\gamma}{t}\right)}\right\}dy, \tag{50.3}$$

$$G^{(\alpha)} = \int_{w\delta}^{\sqrt{w^2\delta^2+1}} y^{-1}\sqrt{\left(1+w^2\delta^2-y^2\right)\left(y^2-w^2\delta^2\right)}\left\{\frac{1}{1+\exp\left(\frac{y-\gamma}{t}\right)}+\frac{1}{1+\exp\left(-\frac{y+\gamma}{t}\right)}\right\}dy. \tag{50.4}$$

The results of calculations of the Seebeck coefficient by these formulae are given in Figure 144.

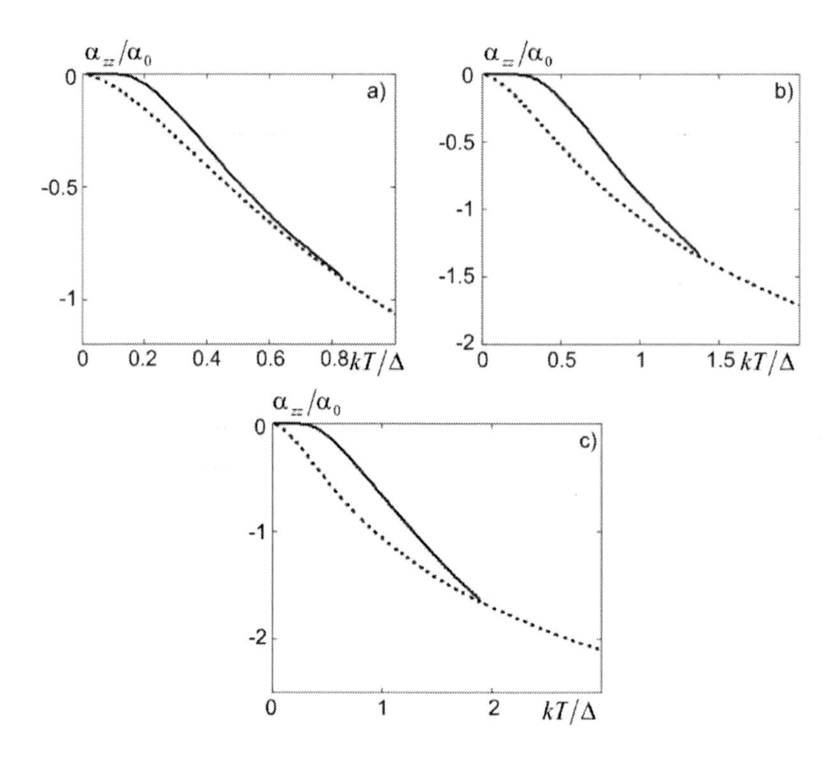

Figure 144. Temperature dependences of the Seebeck coefficient in a layered charge-ordered crystal within the framework of the model $\tau = const$ at: a) $W_0/\zeta_0 = 1.5$; b) $W_0/\zeta_0 = 2$, c) $W_0/\zeta_0 = 2.5$. Solid curves are for the ordered state, dashed curves – for the disordered state.

The figure demonstrates that for all considered values of effective interaction the Seebeck coefficient remains "purely electron", i.e. negative. As could be expected, at $T = 0$ it is equal to zero and increases with a rise in temperature. In the ordered state, the Seebeck coefficient modulus at all temperatures, except for transition temperatures, is lower than in the disordered state. The reason is that charge ordering, leading to conduction miniband narrowing and drastic increase in the density of electron states, makes difficult thermal transfer of charge carriers to the levels higher than chemical potential level. In the ordered state, for all values of effective interaction, at first, in a certain temperature range close to $T = 0$, the Seebeck coefficient modulus is practically not increased. It takes place because in the charge-ordered state at $T = 0$ the FS of crystal is open, and in the disordered state – transient. Moreover, for all values of effective interaction there exists a temperature range close to $T = 0$ wherein the order parameter is practically unchanged. Therefore, while the FS of a crystal is open, the Seebeck coefficient modulus is close to zero and virtually unchanged. However, with further temperature increase, the FS is converted from an open to a closed one, the order parameter starts decreasing drastically, thermal umklapp of charge carriers to the energy levels higher than chemical potential level is facilitated and the Seebeck coefficient modulus starts rising sharply. For each value of effective interaction the Seebeck coefficient at a phase-transition point coincides with such in the disordered phase, and the temperature dependence at a phase-transition point undergoes a break which is the more sharply pronounced and occurs at the higher temperature, the higher is the effective interaction value. Dependences of this type are typical of second-order phase transitions. At $W_0/\zeta_0 = 1.5;2;2.5$, respectively, the Seebeck coefficient moduli are 78;121;155μV/K, respectively, and are achieved at temperatures of 96;159;221K, respectively, assuming that $\Delta = 0.01\text{eV}$.

Chapter 51

Magnetic Field Dependence of the Seebeck Coefficient of the Charge-Ordered Layered Crystals under Conditions of Weak and Intermediate Degeneracy in the Approximation of Constant Relaxation Time

Under conditions of weak and intermediate degeneracy of electron gas in the presence of interlayer charge ordering in the model $\tau = const$ the coefficients $A^{(\alpha)}, B^{(\alpha)}, C^{(\alpha)}, D^{(\alpha)}$ are determined by the following formulae:

$$A^{(\alpha)} = (b/t)^2 \sum_{l=1}^{\infty} \frac{(-1)^l l \coth(bl/t)}{\sinh(bl/t)} \int_{w\delta}^{\sqrt{1+w^2\delta^2}} y^{-1} \sqrt{(1+w^2\delta^2 - y^2)(y^2 - w^2\delta^2)} \exp\left(-lt^{-1}|\gamma - y|\right) dy, \quad (51.1)$$

$$B^{(\alpha)} = \left(b/t^2\right) \sum_{l=1}^{\infty} \frac{(-1)^{l-1} l}{\sinh(bl/t)} \Bigg\{ \int_{-(y+\gamma)\le 0} y^{-1} \sqrt{(1+w^2\delta^2 - y^2)(y^2 - w^2\delta^2)} \times$$
$$\times (-y-\gamma) \exp\left[lt^{-1}(-y-\gamma)\right] dy + \int_{y-\gamma\le 0} y^{-1} \sqrt{(1+w^2\delta^2 - y^2)(y^2 - w^2\delta^2)} \times$$
$$\times (y-\gamma) \exp\left[lt^{-1}(y-\gamma)\right] dy + \int_{-y+\gamma\le 0} y^{-1} \sqrt{(1+w^2\delta^2 - y^2)(y^2 - w^2\delta^2)} \times \quad , \quad (51.2)$$
$$\times (-y+\gamma) \exp\left[lt^{-1}(-y+\gamma)\right] dy +$$
$$+ \int_{y+\gamma\le 0} y^{-1} \sqrt{(1+w^2\delta^2 - y^2)(y^2 - w^2\delta^2)} (y+\gamma) \exp\left[lt^{-1}(y+\gamma)\right] dy \Bigg\}$$

$$C^{(\alpha)} = \int_{\gamma+y\ge 0} y^{-1} \sqrt{(1+w^2\delta^2 - y^2)(y^2 - w^2\delta^2)} + \int_{\gamma-y\ge 0} y^{-1} \sqrt{(1+w^2\delta^2 - y^2)(y^2 - w^2\delta^2)}, \quad (51.3)$$

$$D^{(\alpha)} = \sum_{l=1}^{\infty} \frac{(-1)^l bl}{t \sinh(bl/t)} \left\{ \int_{-y-\gamma \le 0} y^{-1} \sqrt{(1 + w^2\delta^2 - y^2)(y^2 - w^2\delta^2)} \exp[lt^{-1}(-y-\gamma)] + \right.$$
$$+ \int_{y-\gamma \le 0} y^{-1} \sqrt{(1 + w^2\delta^2 - y^2)(y^2 - w^2\delta^2)} \exp[lt^{-1}(y-\gamma)] -$$
$$- \int_{-y+\gamma \le 0} y^{-1} \sqrt{(1 + w^2\delta^2 - y^2)(y^2 - w^2\delta^2)} \exp[lt^{-1}(-y+\gamma)] - \qquad (51.4)$$
$$\left. - \int_{y+\gamma \le 0} y^{-1} \sqrt{(1 + w^2\delta^2 - y^2)(y^2 - w^2\delta^2)} \exp[lt^{-1}(y+\gamma)] \right\}$$

The results of calculations of the field dependences of the Seebeck coefficient of a layered charge-ordered crystal by these formulae are given in Figure 145.

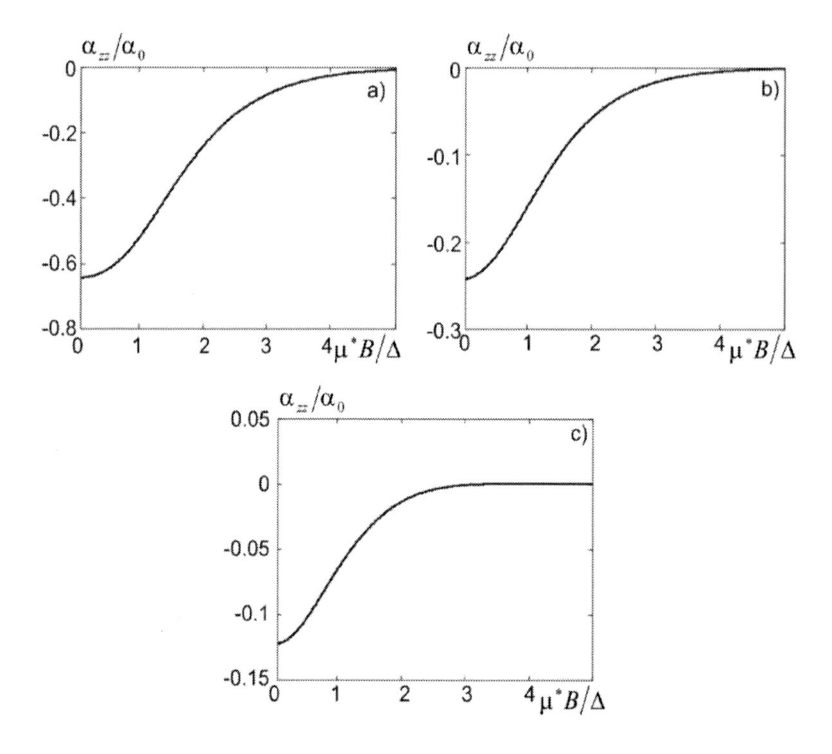

Figure 145. Field dependences of the Seebeck coefficient of a layered charge-ordered crystal at $kT/\Delta = 0.6; 0 \le \mu^* B/\Delta \le 5$ and: a) $W_0/\zeta_0 = 1.5$; b) $W_0/\zeta_0 = 2$; c) $W_0/\zeta_0 = 2.5$.

The figures show that for all considered values of effective attractive interaction the Seebeck coefficient of a layered charge-ordered crystal on the whole decreases in modulus with magnetic field increase. This occurs because with magnetic field increase the order parameter is increased, conduction minibands describing the interlayer motion of electrons are narrowed, and thermal umklapp of electrons to the energy levels higher than chemical potential level becomes difficult. Thermal umklapp is further complicated due to increasing the distance between the Landau subbands with magnetic field increase. Moreover, it is seen that with increase in the value of effective attractive interaction, the Seebeck coefficient modulus is decreased, and its drop with magnetic field increase proceeds faster. The reason is that the larger is the effective attractive interaction, the narrower are conduction minibands into which the initial miniband is split when passing into the charge-ordered state. Therefore, with regard to the order parameter growth with magnetic field increase, the Seebeck coefficient modulus must also decrease faster.

Moreover, if at $W_0/\zeta_0 = 1.5;2$ the Seebeck coefficient remains negative, i.e. "purely electron", then at $W_0/\zeta_0 = 2.5$ there is such magnetic field induction whereby the Seebeck coefficient turns to zero, and then changes its sign for the positive, i.e. becomes a "hole" one. However, due to increasing the distance between the Landau subbands and the order parameter with magnetic field increase, the Seebeck coefficient modulus in the limit $B \to \infty$ must turn to zero. Therefore, at $W_0/\zeta_0 = 2.5$ there must exist such magnetic field induction whereby the Seebeck coefficient, being positive, reaches the local maximum of the modulus. At $\Delta = 0.01$ eV this induction is 6.66T, and the respective maximum Seebeck coefficient modulus – 61nV/K. The magnetic field induction whereby the Seebeck coefficient turns to zero is 5.44T. Thus, charge-ordered crystals with the ratio $W_0/\zeta_0 > 2.5$ possess the property of the Seebeck coefficient inversion in a longitudinal quantizing magnetic field.

Chapter 52

TEMPERATURE DEPENDENCE OF THE SEEBECK COEFFICIENT OF A CHARGE-ORDERED LAYERED CRYSTAL IN THE APPROXIMATION OF ENERGY-DEPENDENT RELAXATION TIME

Under conditions of weak and intermediate degeneracy in the model $\tau \propto 1/g(\varepsilon)$ the Seebeck coefficient of a charge-ordered layered crystal is determined as follows:

$$\alpha_{zz} = -\alpha_0 \frac{E^{(\alpha)} + F^{(\alpha)}}{G^{(\alpha)}}, \tag{52.1}$$

In so doing:

$$E^{(\alpha)} = \int_0^{\pi/2} \phi(x) \left\{ \ln\left[1 + \exp\left(\frac{\gamma + \sqrt{w^2\delta^2 + \cos^2 x}}{t}\right)\right] + \ln\left[1 + \exp\left(\frac{\gamma - \sqrt{w^2\delta^2 + \cos^2 x}}{t}\right)\right]\right\} dx, \tag{52.2}$$

$$F^{(\alpha)} = \frac{1}{t}\int_0^{\pi/2} \phi(x) \left\{ \left(\sqrt{w^2\delta^2 + \cos^2 x} - \gamma\right)\left[1 + \exp\left(\frac{\sqrt{w^2\delta^2 + \cos^2 x} - \gamma}{t}\right)\right]^{-1} - \right.$$
$$\left. - \left(\sqrt{w^2\delta^2 + \cos^2 x} + \gamma\right)\left[1 + \exp\left(-\frac{\sqrt{w^2\delta^2 + \cos^2 x} + \gamma}{t}\right)\right]^{-1}\right\} dx, \tag{52.3}$$

$$G^{(\alpha)} = \int_0^{\pi/2} \phi(x) \left\{ \left[1 + \exp\left(-\frac{\gamma + \sqrt{w^2\delta^2 + \cos^2 x}}{t}\right)\right]^{-1} + \left[1 + \exp\left(\frac{\sqrt{w^2\delta^2 + \cos^2 x} - \gamma}{t}\right)\right]^{-1}\right\} dx, \tag{52.4}$$

$$\phi(x)=\frac{\sin^2 2x}{x\left(w^2\delta^2+\cos^2 x\right)}-\frac{\int_0^x \sin^2 2z\left(w^2\delta^2+\cos^2 z\right)^{-1}dz}{x^2}. \quad (52.5)$$

The results of calculation of the temperature dependences of the Seebeck coefficient of a charge-ordered layered crystal in the absence of a magnetic field in the model $\tau \propto 1/g(\varepsilon)$ by these formulae are given in Figure 146.

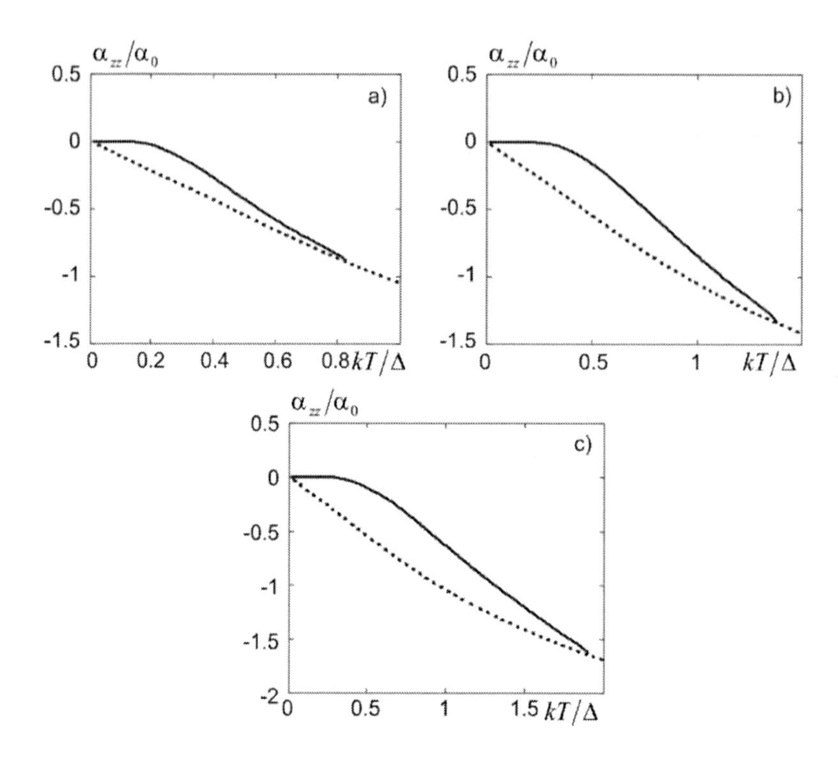

Figure 146. Temperature dependences of the Seebeck coefficient of a charge-ordered layered crystal in the model $\tau \propto 1/g(\varepsilon)$ at: a) $W_0/\zeta_{02D}=1.5$; b) $W_0/\zeta_{02D}=2$; c) $W_0/\zeta_{02D}=2.5$.

The figure demonstrates that for all considered values of effective interaction the Seebeck coefficient almost everywhere remains "purely electron", i.e. negative. As in the model $\tau = const$, with $T=0$ it is equal to zero. However, as long as in the model $\tau = const$ the contribution of low-energy states is larger, by virtue of conduction miniband splitting into two, at first, this coefficient is weakly positive, i.e. "hole". Then, on reaching certain positive maximum, it starts decreasing in modulus, turns to zero, and then, having become "purely electron", i.e. negative, it again increases in modulus. Positive maxima of the Seebeck coefficient in the model $\tau \propto 1/g(\varepsilon)$ are quite small and achieved at rather low temperatures which is attributable to increased contribution of low-energy states as compared to the model $\tau = const$. Therefore, assuming that $\Delta = 0.01\text{eV}$, it appears that inversion of the Seebeck coefficient sign occurs at temperatures of 1.67; 1.32 and 0.97K, respectively, at

$W_0/\zeta_{02D} = 1.5;2;2.5$, respectively. In the ordered state in the model $\tau \propto 1/g(\varepsilon)$, as well as in the model $\tau = const$, the Seebeck coefficient modulus at all temperatures, except for transition temperatures, is lower than in the disordered state. The main tendencies for a change in the Sebeck coefficient with temperature in the model $\tau \propto 1/g(\varepsilon)$ are determined by the same processes as in the model $\tau = const$. Therefore, the Seebeck coefficient at a phase-transition point coincides with such in the disordered phase, and the temperature dependence at a phase-transition point undergoes a break which is the more sharply pronounced and occurs at the higher temperature, the higher is the effective interaction value. At $W_0/\zeta_{02D} = 1.5;2;2.5$, respectively, the Seebeck coefficient moduli make 76;115;141 μV/K, respectively, and are achieved at temperatures of 96;159;221K, respectively, assuming that $\Delta = 0.01\text{eV}$. This slight reduction of the Seebeck coefficient moduli in the model $\tau \propto 1/g(\varepsilon)$ as compared to the model $\tau = const$ is attributable to increased contribution of low-energy states.

Chapter 53

FIELD DEPENDENCE OF THE SEEBECK COEFFICIENT OF LAYERED CHARGE-ORDERED CRYSTALS UNDER CONDITIONS OF WEAK AND INTERMEDIATE DEGENERACY WITH REGARD TO ENERGY DEPENDENCE OF RELAXATION TIME

Under conditions of weak and intermediate degeneracy in the model $\tau \propto 1/g(\varepsilon)$ the coefficients $A^{(\alpha)}, B^{(\alpha)}, C^{(\alpha)}, D^{(\alpha)}$ are determined by the formulae:

$$A^{(\alpha)} = \left(\frac{h}{t}\right)^2 \sum_{l=1}^{\infty} \frac{(-1)^l l \coth(hlt^{-1})}{\sinh(hlt^{-1})} \int_0^{\pi/2} \phi(x) \Big[\exp\left(-lt^{-1}\left|\gamma + \sqrt{w^2\delta^2 + \cos^2 x}\right|\right) + \\ + \exp\left(-lt^{-1}\left|\gamma - \sqrt{w^2\delta^2 + \cos^2 x}\right|\right)\Big] dx \quad , \quad (53.1)$$

$$B^{(\alpha)} = \frac{h}{t^2} \sum_{l=1}^{\infty} \frac{(-1)^l l}{\sinh(hlt^{-1})} \Bigg[\int_0^{\pi/2} \phi(x) \left|\gamma + \sqrt{w^2\delta^2 + \cos^2 x}\right| \exp\left(-lt^{-1}\left|\gamma + \sqrt{w^2\delta^2 + \cos^2 x}\right|\right) + \\ + \int_0^{\pi/2} \phi(x) \left|\gamma - \sqrt{w^2\delta^2 + \cos^2 x}\right| \exp\left(-lt^{-1}\left|\gamma - \sqrt{w^2\delta^2 + \cos^2 x}\right|\right) \Bigg] dx \quad , \quad (53.2)$$

$$C^{(\alpha)} = 0.5 \int_0^{\pi/2} \phi(x) \left[2 + \mathrm{sgn}\left(\gamma + \sqrt{w^2\delta^2 + \cos^2 x}\right) + \mathrm{sgn}\left(\gamma - \sqrt{w^2\delta^2 + \cos^2 x}\right)\right], \quad (53.3)$$

$$D^{(\alpha)} = \sum_{l=1}^{\infty} \frac{(-1)^l hlt^{-1}}{\sinh(hlt^{-1})} \int_0^{\pi/2} \phi(x) \Big[\operatorname{sgn}\Big(\gamma + \sqrt{w^2\delta^2 + \cos^2 x}\Big) \times$$
$$\times \exp\Big(-lt^{-1}\Big|\gamma + \sqrt{w^2\delta^2 + \cos^2 x}\Big|\Big) + \operatorname{sgn}\Big(\gamma - \sqrt{w^2\delta^2 + \cos^2 x}\Big) \times \tag{53.4}$$
$$\times \exp\Big(-lt^{-1}\Big|\gamma - \sqrt{w^2\delta^2 + \cos^2 x}\Big|\Big)\Big]$$

The results of calculation of the field dependences of the Seebeck coefficient for a layered charge-ordered crystal are given in Figure 147.

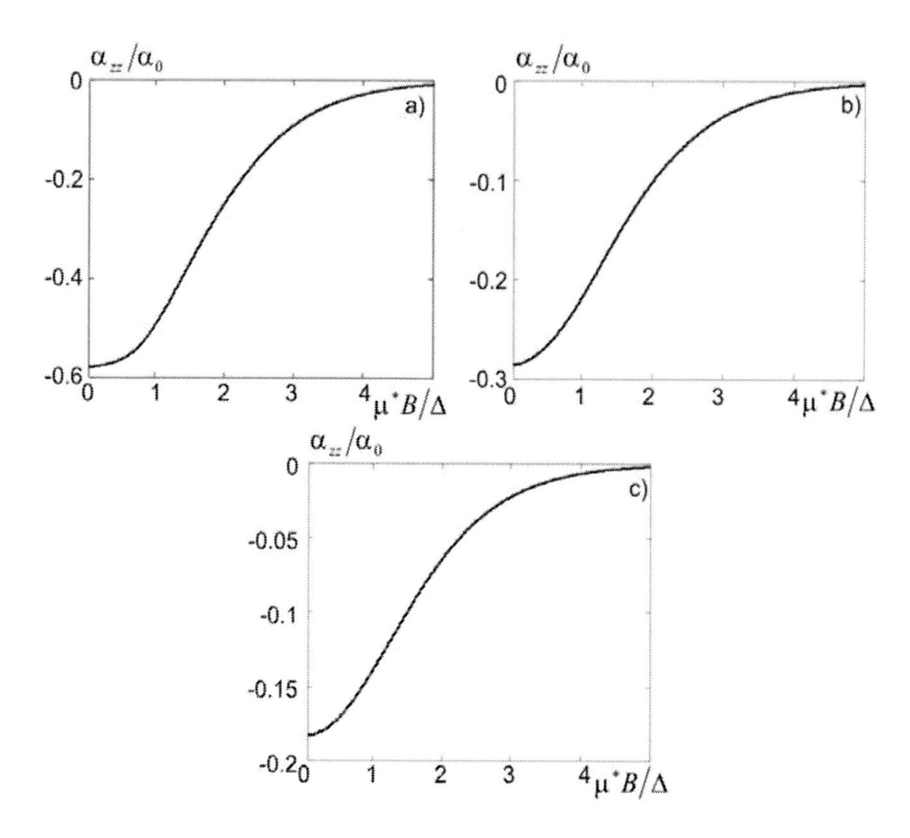

Figure 147. Field dependences of the Seebeck coefficient of a charge-ordered layered crystal in the model $\tau \propto 1/g(\varepsilon)$ at $kT/\Delta = 0.6; 0 \le \mu^* B/\Delta \le 5$ and: a) $W_0/\zeta_{02D} = 1.5$; b) $W_0/\zeta_{02D} = 2$; c) $W_0/\zeta_{02D} = 2.5$.

The figure demonstrates that in the model $\tau \propto 1/g(\varepsilon)$, as well as in the model $\tau = const$, at $kT/\Delta = 0.6; 0 \le \mu^* B/\Delta \le 5$ and for all considered values of effective attractive interaction the Seebeck coefficient remains negative, i.e. "purely electron", and its modulus monotonously decreases with magnetic field increase. Thus, this tendency is determined by electron gas thermodynamics in a magnetic field in the presence of charge ordering, rather than by subtle details of scattering mechanisms. Namely, the order parameter increase still remains the main factor determining the Seebeck coefficient modulus decrease in a magnetic field. Moreover, the Seebeck coefficient modulus is affected by the effective attractive interaction, namely, the larger it is the lower is the Seebeck coefficient modulus. Thus, in a weak magnetic field at $kT/\Delta = 0.6$ and $W_0/\zeta_{02D} = 1.5;2;2.5$, respectively, the Seebeck coefficient modulus is about 50; 25 and 16μV/K, respectively. As in the model $\tau = const$, it is due to the fact that with increase in the effective attractive interaction, conduction miniband describing the interlayer motion of electrons is narrowed, and thermal transfer of charge carriers to the levels higher than electron gas chemical potential level, becomes difficult. However, in the model $\tau \propto 1/g(\varepsilon)$ the Seebeck coefficient modulus for all magnetic field inductions is lower than in the model $\tau = const$, which is caused by increased contribution of low-energy states due to increase in their respective electron relaxation time.

Chapter 54

POWER FACTOR OF A LAYERED CRYSTAL IN THE ABSENCE OF A MAGNETIC FIELD

When calculating power factor of a layered crystal in the absence of a magnetic field we will first use the approximation of constant relaxation time and proceed from formulae (46.1) and (46.2). Then, assuming that acoustic phonon deformation potential scattering is the dominant mechanism of charge carrier scattering, we obtain:

$$P = P_0 \frac{\left[\int_0^\pi \sin^2 x \left\{ \ln\left[1+\exp\left(\frac{\gamma - 1 + \cos x}{t}\right)\right] + \frac{(1-\cos x - \gamma)}{t}\left[1+\exp\left(\frac{1-\cos x-\gamma}{t}\right)\right]^{-1} \right\} dx \right]^2}{t \int_0^\pi \sin^2 x \left[1+\exp\left(\frac{1-\cos x-\gamma}{t}\right)\right]^{-1} dx}, \quad (54.1)$$

In the effective mass approximation formula (54.1) acquires the form:

$$P = P_0 \frac{\left[\int_0^\infty x^2 \left\{ \ln\left[1+\exp\left(\frac{\gamma - 0.5x^2}{t}\right)\right] + \frac{(0.5x^2 - \gamma)}{t}\left[1+\exp\left(\frac{0.5x^2-\gamma}{t}\right)\right]^{-1} \right\} dx \right]^2}{t \int_0^\infty x^2 \left[1+\exp\left(\frac{0.5x^2-\gamma}{t}\right)\right]^{-1} dx}. \quad (54.2)$$

In these formulae, $P_0 = 4k^2a^2\rho s^2\Delta/(3\pi h\Xi^2)$. The results of calculations of the temperature dependences of power factor both for a real layered crystal and in the effective mass approximation are given in Figure 148.

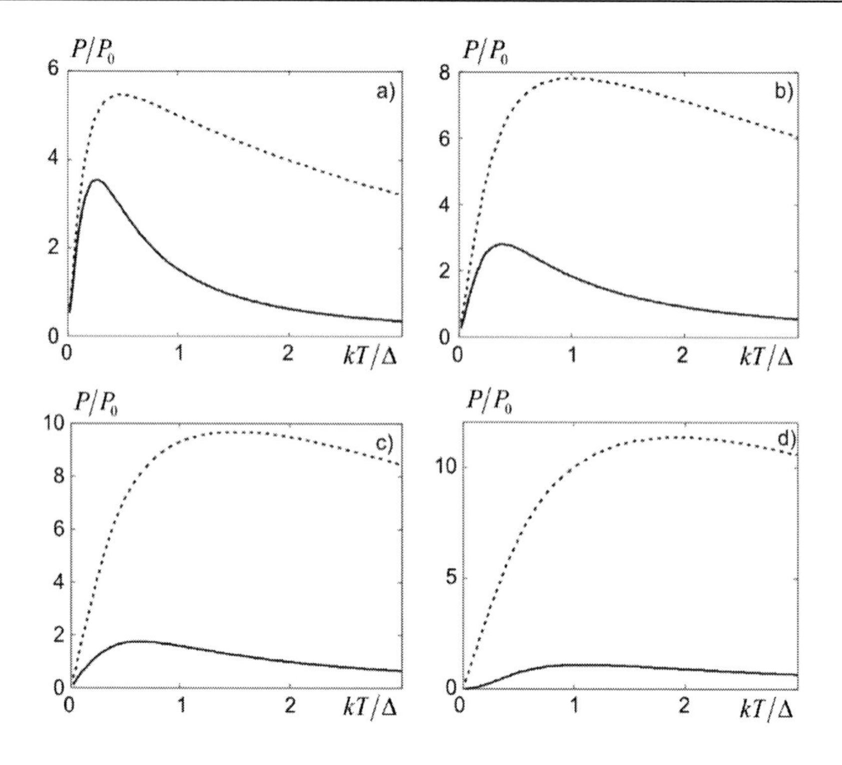

Figure 148. Temperature dependences of power factor of a layered crystal at $0 \leq kT/\Delta \leq 3$.

The figure demonstrates that in all cases the power factor at first increases from zero to certain maximum value, and then decreases. The reason is that with a rise in temperature, electric conductivity decreases if drift, and the Seebeck coefficient increases. With growing degree of miniband filling, the difference between the solid and dashed curves is increased. This occurs because the longitudinal electric conductivity is restricted by the miniband finite width more severely than the Seebeck coefficient. With previously stipulated problem parameters and at $\zeta_0/\Delta = 0.5;1;1.5;2$, respectively, power factor maxima for a real layered crystal are achieved at temperatures 30.15; 46.38; 71.89; 122.90K, respectively, and make $3.371 \cdot 10^{-5};2.661 \cdot 10^{-5};1.652 \cdot 10^{-5};1.029 \cdot 10^{-5}$ W/(m · K^2), respectively. In the effective mass approximation these maxima are achieved at temperatures 55.65; 113.62; 150.73; 173.91K, respectively, and make $5.202 \cdot 10^{-5};7.425 \cdot 10^{-5};9.048 \cdot 10^{-5};1.019 \cdot 10^{-4}$ W/(m · K^2), respectively. First, we see that at $\zeta_0/\Delta = 0.5;1;1.5;2$, respectively, in the effective mass approximation the maxima of power factor are a factor of 1.54;2.79;5.48;10.10, respectively, greater than in a real layered crystal. Second, in a real layered crystal these maxima drop with a rise in the ratio ζ_0/Δ, and in the effective mass approximation they grow. Third, in the effective mass approximation these maxima are achieved at higher temperatures than in a real layered crystal. All this occurs, first, because with a rise in the ratio ζ_0/Δ in a real layered crystal, the Seebeck coefficient drops faster and the electric conductivity increases much slower than in the effective mass approximation. Second, in a real layered crystal the electric conductivity with a rise in temperature drops faster and the Seebeck coefficient grows slower than in the effective mass approximation. At the end of the investigated temperature range, i.e. at $kT/\Delta = 3$, or, assuming that $\Delta = 0.01$ eV, at $T = 347.8$ K the values of power factor with previously stipulated problem parameters and at $\zeta_0/\Delta = 0.5;1;1.5;2$, respectively, for a real layered crystal make $3.267 \cdot 10^{-6};5.219 \cdot 10^{-6};6.114 \cdot 10^{-6};6.200 \cdot 10^{-6}$ W/(m $\cdot$ K^2), respectively, and in the effective mass approximation – $2.670 \cdot 10^{-5};4.894 \cdot 10^{-5};6.624 \cdot 10^{-5};7.959 \cdot 10^{-5}$ W/(m $\cdot$ K^2) , respectively. Therefore, at this temperature and the above ratios ζ_0/Δ the values of power factor in the effective mass approximation are a factor of 8.17; 9.38;10.83;12.8, respectively, larger than in a real layered crystal. However, at $T = 347.8$ K the values of power factor increase with a rise in the ratio ζ_0/Δ both in a real layered crystal, and in the effective mass approximation. This occurs because at sufficiently high temperatures in a real layered crystal, with increase in charge carrier concentration, the electric conductivity grows faster than at low temperatures. Therefore, we see that in strongly anisotropic layered crystals the width finiteness of a narrow conduction miniband and its related band spectrum nonparabolicity can result in considerable reduction of power factor. Hence, both thermoelectric figure of merit (Z) and dimensionless figure of merit (ZT) of material are reduced, i.e. in fact, there is a drop in thermoelectric device efficiency, particularly with increase in major carrier concentration in material. So, possible negative symptoms of layered structure effects must necessarily be taken into account in the development of thermoelectric materials and in the design of thermoelectric devices.

Chapter 55

Magnetic Field Dependence of Power Factor of a Layered Crystal in the Approximation of Constant Relaxation Time under Conditions of Weak and Intermediate Degeneracy

At first, as before, we will consider the magnetic field dependence of power factor of a layered crystal under conditions of weak and intermediate degeneracy, using approximation of constant relaxation time and assuming acoustic phonon deformation potential scattering of electrons as the dominant. In this case power factor is determined by the formula:

$$P = P_0 \frac{\left(A^{(\alpha)} + B^{(\alpha)}\right)^2}{t\left(C^{(\alpha)} + D^{(\alpha)}\right)}. \tag{55.1}$$

In this formula, the coefficients $A^{(\alpha)}, B^{(\alpha)}, C^{(\alpha)}, D^{(\alpha)}$ have the following values:

$$A^{(\alpha)} = \left(\frac{h}{t}\right)^2 \sum_{l=1}^{\infty} \frac{(-1)^l l \coth\left(hlt^{-1}\right)}{\sinh\left(hlt^{-1}\right)} \int_0^2 \sqrt{2y - y^2} \exp\left(-lt^{-1}|\gamma - y|\right) dy, \tag{55.2}$$

$$B^{(\alpha)} = \frac{h}{t^2} \sum_{l=1}^{\infty} \frac{(-1)^l l}{\sinh\left(hlt^{-1}\right)} \int_0^2 \sqrt{2y - y^2} |\gamma - y| \exp\left(-lt^{-1}|\gamma - y|\right) dy, \tag{55.3}$$

$$C^{(\alpha)} = 0.5 \int_0^2 \sqrt{2y - y^2} \left[1 + \operatorname{sgn}(\gamma - y)\right] dy, \tag{55.4}$$

$$D^{(\alpha)} = \sum_{l=1}^{\infty} \frac{(-1)^l hl}{t \sinh\left(hlt^{-1}\right)} \int_0^2 \sqrt{2y - y^2} \operatorname{sgn}(\gamma - y) \exp\left(-lt^{-1}|\gamma - y|\right). \tag{55.5}$$

In the effective mass approximation these coefficients acquire the following form:

$$A^{(\alpha)} = \left(\frac{h}{t}\right)^2 \sum_{l=1}^{\infty} \frac{(-1)^l l \coth\left(hlt^{-1}\right)}{\sinh\left(hlt^{-1}\right)} \int_0^2 \sqrt{2y}\exp\left(-lt^{-1}|\gamma - y|\right)dy, \tag{55.6}$$

$$B^{(\alpha)} = \frac{h}{t^2} \sum_{l=1}^{\infty} \frac{(-1)^l l}{\sinh\left(hlt^{-1}\right)} \int_0^2 \sqrt{2y}|\gamma - y|\exp\left(-lt^{-1}|\gamma - y|\right)dy, \tag{55.7}$$

$$C^{(\alpha)} = \frac{1}{6}(2\gamma)^{3/2}(1 + \operatorname{sgn}\gamma), \tag{55.8}$$

$$D^{(\alpha)} = \sum_{l=1}^{\infty} \frac{(-1)^l hl}{t\sinh\left(hlt^{-1}\right)} \int_0^2 \sqrt{2y - y^2}\operatorname{sgn}(\gamma - y)\exp\left(-lt^{-1}|\gamma - y|\right). \tag{55.9}$$

Moreover, $P_0 = 4k^2a^2\rho s^2\Delta/\left(3\pi h\Xi^2\right)$. The results of calculations of the field dependences of power factor by these formulae for a real layered crystal and in the effective mass approximation are given in Figure 149.

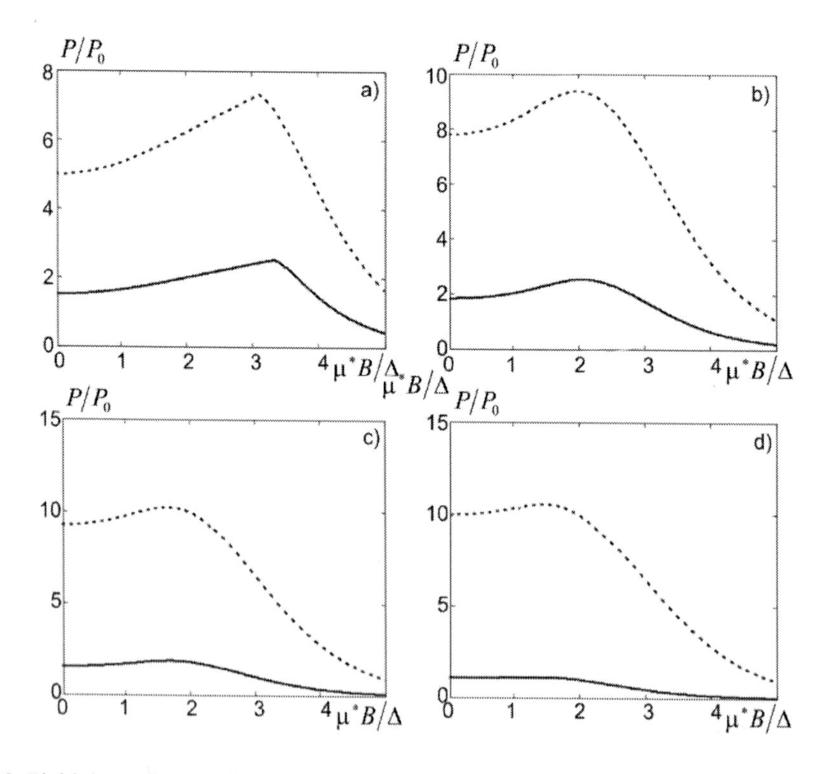

Figure 149. Field dependences of power factor under conditions of weak and intermediate degeneracy at $kT/\Delta = 1; 0 \le \mu^* B/\Delta \le 5$ and: a) $\zeta_0/\Delta = 0.5$; b) $\zeta_0/\Delta = 1$; c) $\zeta_0/\Delta = 1.5$; d) $\zeta_0/\Delta = 2$.

The figure demonstrates that for all degrees of miniband filling, the dashed curves lie above the solid ones, i.e. power factor in the effective mass approximation is essentially higher than for a real layered crystal. This occurs because both the Seebeck coefficient modulus and electric conductivity in a real layered crystal are lower than in the effective mass approximation. This, as mentioned above, is due to the effect of finite miniband width. Moreover, for each degree of miniband filling both in a real layered crystal and in effective mass approximation there exists the value of magnetic field induction depending on this filling degree, whereby the power factor is maximum. The reason is that, as was shown before, in a quantizing magnetic field under conditions of weak and intermediate degeneracy there exist the NLMR and the Seebeck coefficient modulus maximum. Therefore, for instance, with previously stipulated problem parameters and at $\zeta_0/\Delta = 0.5;1;1.5;2$, respectively, power factor maxima at $T = 116\,\mathrm{K}$ for a real layered crystal make $2.425\cdot10^{-5};2.428\cdot10^{-5};1.808\cdot10^{-5};1.088\cdot10^{-5}$ W/(m·K^2), respectively, and are achieved in magnetic fields with inductions 5.76; 3.50; 2.85 and 2.17T, respectively. Whereas in the effective mass approximation these maxima make $6.992\cdot10^{-5};8.953\cdot10^{-5};9.762\cdot10^{-5};1.002\cdot10^{-4}$ W/(m·K^2) and are achieved in magnetic fields with inductions 5.36; 3.42; 2.76 and 2.31T, respectively. Again we see that with increase in miniband filling degree, in the case of a real layered crystal power factor maxima drop, and in the effective mass approximation – increase. As in the absence of a magnetic field, this occurs because with increase in miniband filling degree, the longitudinal conductivity of a real layered crystal grows slower than in the effective mass approximation. Furthermore, the assumption is confirmed on the possibility of thermoelectric figure of merit control for at least certain materials by magnetic field application parallel to temperature gradient. Note that at $\zeta_0/\Delta = 0.5;1;1.5;2$, respectively, power factor maxima in the effective mass approximation are a factor of 2.89; 3.69; 5.40; 9.21, respectively, higher than such maxima for a real layered crystal. Thus, in the crystals where conduction band nonparabolicity is retained in a sufficiently wide range of doping levels, power factor values can be achieved on the level of best samples of cuprate thermoelectric materials [56]. However, as is evident from the results obtained, conduction band nonparabolicity, apparent at high doping levels, reduces drastically maximum achievable power factor and, hence, the thermoelectric figure of merit of materials.

Chapter 56

TEMPERATURE DEPENDENCE OF POWER FACTOR OF A LAYERED CRYSTAL UNDER CONDITIONS OF WEAK AND INTERMEDIATE DEGENERACY WITH REGARD TO ENERGY DEPENDENCE OF RELAXATION TIME

We now consider the temperature dependence of power factor of a layered crystal under conditions of weak and intermediate degeneracy, assuming that relaxation time caused by acoustic phonon scattering of electrons, is inversely proportional to temperature and density of electron states. In this case power factor of a real layered crystal is determined by the formula:

$$\alpha = P_0 \frac{\pi t^{-1}\left\{\int_0^\pi \varphi(x)\left\{\ln\left[1+\exp\left(\frac{\gamma-1+\cos x}{t}\right)\right]+t^{-1}(1-\cos x-\gamma)\left[1+\exp\left(\frac{1-\cos x-\gamma}{t}\right)\right]^{-1}\right\}dx\right\}^2}{4\int_0^\pi \varphi(x)\left[1+\exp\left(\frac{1-\cos x-\gamma}{t}\right)\right]^{-1}dx}. \quad (56.1)$$

In the effective mass approximation formula (56.1) acquires the form:

$$\alpha = P_0 \frac{\frac{2}{3}\pi t^{-1}\left\{\int_0^\infty \left\{\ln\left[1+\exp\left(\frac{\gamma-y}{t}\right)\right]+t^{-1}(y-\gamma)\left[1+\exp\left(\frac{y-\gamma}{t}\right)\right]^{-1}\right\}dy\right\}^2}{\int_0^\infty \left[1+\exp\left(\frac{y-\gamma}{t}\right)\right]^{-1}dy}. \quad (56.2)$$

In so doing, as before, $P_0 = 4k^2a^2\rho s^2\Delta/(3\pi h\Xi^2)$. The results of calculations of the temperature dependences of power factor of a layered crystal by these formulae are given in Figure 150.

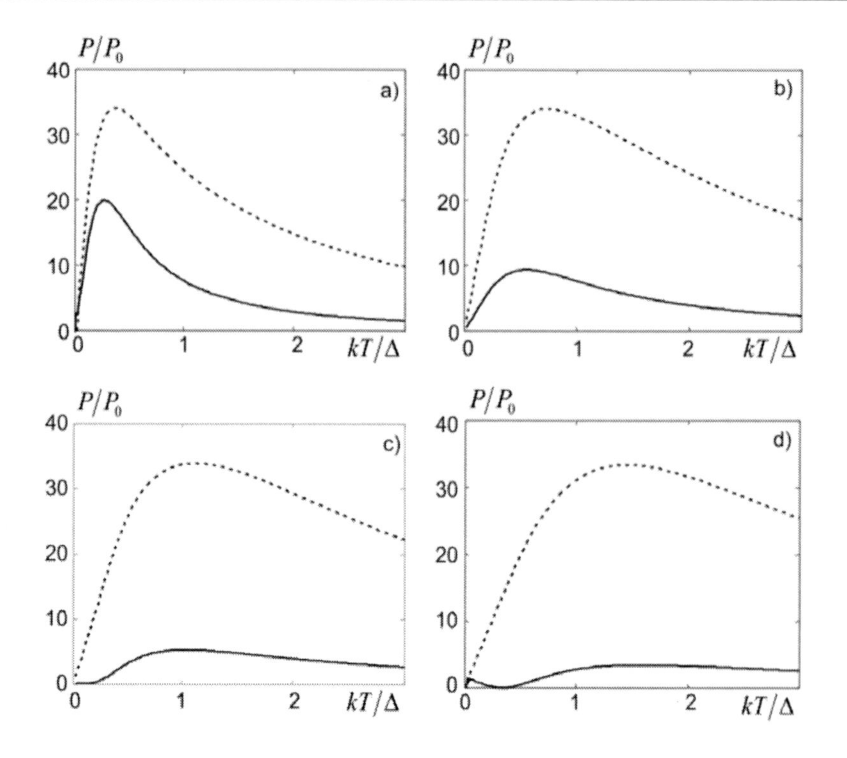

Figure 150. Temperature dependences of power factor of a layered crystal at $0 \le kT/\Delta \le 3$ in the model $\tau \propto 1/g(\varepsilon)$.

The figure shows that in the model $\tau \propto 1/g(\varepsilon)$, as well as in the model $\tau = const$, at acoustic phonon deformation potential scattering of charge carriers for each degree of miniband filling there is a temperature whereby power factor is maximum. With previously stipulated problem parameters these maxima for a real layered crystal at $\zeta_0/\Delta = 0.5;1;1.5;2$, respectively, are achieved at temperatures 30.14, 64.93; 122.90 and 178.55K, respectively, and make $4.750 \cdot 10^{-5};2.230 \cdot 10^{-5};1.250 \cdot 10^{-5};8.130 \cdot 10^{-6}\,\mathrm{W/(m \cdot K^2)}$ respectively. In the effective mass approximation for the same degrees of miniband filling these maxima are achieved at temperatures 41.74, 85.80; 132.17 and 166.96K, respectively, and are equal to $8.095 \cdot 10^{-5}\,\mathrm{W/(m \cdot K^2)}$ for all ratios ζ_0/Δ. Thus, for a real layered crystal these maxima are a factor of 1.70; 3.63; 6.48 and 9.95, respectively, lower than in the effective mass approximation. As in the model $\tau = const$, this difference is due to the fact that miniband finite width reduces both the longitudinal electric conductivity and the Seebeck coefficient modulus. As in the model $\tau = const$, in the model $\tau \propto 1/g(\varepsilon)$ power factor maxima decrease with increase of miniband filling degree, hence, of electron concentration, in a real layered crystal. In the effective mass approximation these maxima are practically constant. This occurs because in the effective mass approximation in the model $\tau \propto 1/g(\varepsilon)$ a drop in the Seebeck coefficient modulus with increase in miniband filling degree is much stronger compensated by electric conductivity growth than in the model $\tau = const$. From the figure it is also evident that at $\zeta_0/\Delta = 2$ on the temperature dependence of power factor of a real layered crystal the effect of "square-law detection" becomes apparent, when power factor vanishes at that different from absolute zero temperature whereby the Seebeck coefficient turns to zero and changes its sign from the positive to negative. And this, as already mentioned, is related to a decrease in the Seebeck coefficient modulus due to increased contribution to it of low-energy states and the presence in a real layered crystal of the density of electron states singularity at $\varepsilon = 2\Delta$. In the effective mass approximation this singularity is absent. Next, like at $\zeta_0/\Delta < 2$, power factor in the case of a real layered crystal first increases and reaches maximum, and then decreases.

At $T = 347.83$ K and $\zeta_0/\Delta = 0.5;1;1.5;2$, respectively, power factors for a real layered crystal make $3.672 \cdot 10^{-6};5.633 \cdot 10^{-6};6.331 \cdot 10^{-6};6.138 \cdot 10^{-6}\,\mathrm{W/(m \cdot K^2)}$, respectively. In the effective mass approximation these factors make $2.269 \cdot 10^{-5};4.432 \cdot 10^{-5};5.819 \cdot 10^{-5};6.790 \cdot 10^{-5}\,\mathrm{W/(m \cdot K^2)}$, respectively, and this is a factor of 6.18;7.87;9.19;11.06, respectively, higher than in a real layered crystal. Therefore, we again see the negative influence of layered structure effects on power factor, hence, on thermoelectric figure of merit of material.

Chapter 57

FIELD DEPENDENCE OF POWER FACTOR OF A LAYERED CRYSTAL UNDER CONDITIONS OF WEAK AND INTERMEDIATE DEGENERACY WITH REGARD TO THE ENERGY DEPENDENCE OF RELAXATION TIME

We now consider the field dependence of power factor of a layered crystal with regard to the energy dependence of relaxation time. In the case of a real layered crystal this factor is determined as:

$$P = \pi t^{-1} P_0 \left(A^{(\alpha)} + B^{(\alpha)}\right)^2 \left(C^{(\alpha)} + D^{(\alpha)}\right)^{-1}, \tag{57.1}$$

where the coefficients $A^{(\alpha)}, B^{(\alpha)}, C^{(\alpha)}, D^{(\alpha)}$ have the following values:

$$A^{(\alpha)} = \left(\frac{h}{t}\right)^2 \sum_{l=1}^{\infty} \frac{(-1)^l l \coth\left(hlt^{-1}\right)}{\sinh\left(hlt^{-1}\right)} \int_0^{\pi} \left(x^{-2} \sin 2x - 2x^{-1} \cos 2x\right) \exp\left(-lt^{-1}|\gamma - 1 + \cos x|\right) dx, \tag{57.2}$$

$$B^{(\alpha)} = \frac{h}{t^2} \sum_{l=1}^{\infty} \frac{(-1)^l l}{\sinh\left(hlt^{-1}\right)} \int_0^{\pi} \left(x^{-2} \sin 2x - 2x^{-1} \cos 2x\right) |\gamma - 1 + \cos x| \exp\left(-lt^{-1}|\gamma - 1 + \cos x|\right) dx, \tag{57.3}$$

$$C^{(\alpha)} = 0.5 \int_0^{\pi} \left(x^{-2} \sin 2x - 2x^{-1} \cos 2x\right) \left[1 + \operatorname{sgn}(\gamma - 1 + \cos x)\right] dx, \tag{57.4}$$

$$D^{(\alpha)} = \sum_{l=1}^{\infty} \frac{(-1)^l hlt^{-1}}{\sinh\left(hlt^{-1}\right)} \int_0^{\pi} \left(x^{-2} \sin 2x - 2x^{-1} \cos 2x\right) \operatorname{sgn}(\gamma - 1 + \cos x) \exp\left(-lt^{-1}|\gamma - 1 + \cos x|\right) dx. \tag{57.5}$$

In the effective mass approximation formulae (57.1) – (57.5) acquire the form:

$$P = (8/3)\pi t^{-1} P_0 \left(A^{(\alpha)} + B^{(\alpha)}\right)^2 \left(C^{(\alpha)} + D^{(\alpha)}\right)^{-1}, \tag{57.6}$$

$$A^{(\alpha)} = \left(\frac{h}{t}\right)^2 \sum_{l=1}^{\infty} \frac{(-1)^l l \coth\left(hlt^{-1}\right)}{\sinh\left(hlt^{-1}\right)} \int_0^{\infty} \exp\left(-lt^{-1}|\gamma - y|\right) dy, \tag{57.7}$$

$$B^{(\alpha)} = \frac{h}{t^2} \sum_{l=1}^{\infty} \frac{(-1)^l l}{\sinh\left(hlt^{-1}\right)} \int_0^{\infty} |\gamma - y| \exp\left(-lt^{-1}|\gamma - y|\right) dy, \tag{57.8}$$

$$C^{(\alpha)} = 0.25\gamma^2 [1 + \mathrm{sgn}(\gamma)], \tag{57.9}$$

$$D^{(\alpha)} = \sum_{l=1}^{\infty} \frac{(-1)^l hlt^{-1}}{\sinh\left(hlt^{-1}\right)} \int_0^{\infty} \mathrm{sgn}(\gamma - y) \exp\left(-lt^{-1}|\gamma - y|\right) dy. \tag{57.10}$$

The results of calculation of the field dependences of power factor by these formulae are given in Figure 151. Here, $P_0 = k^2 a^2 \rho s^2 \Delta / \left(3\pi h \Xi^2\right)$.

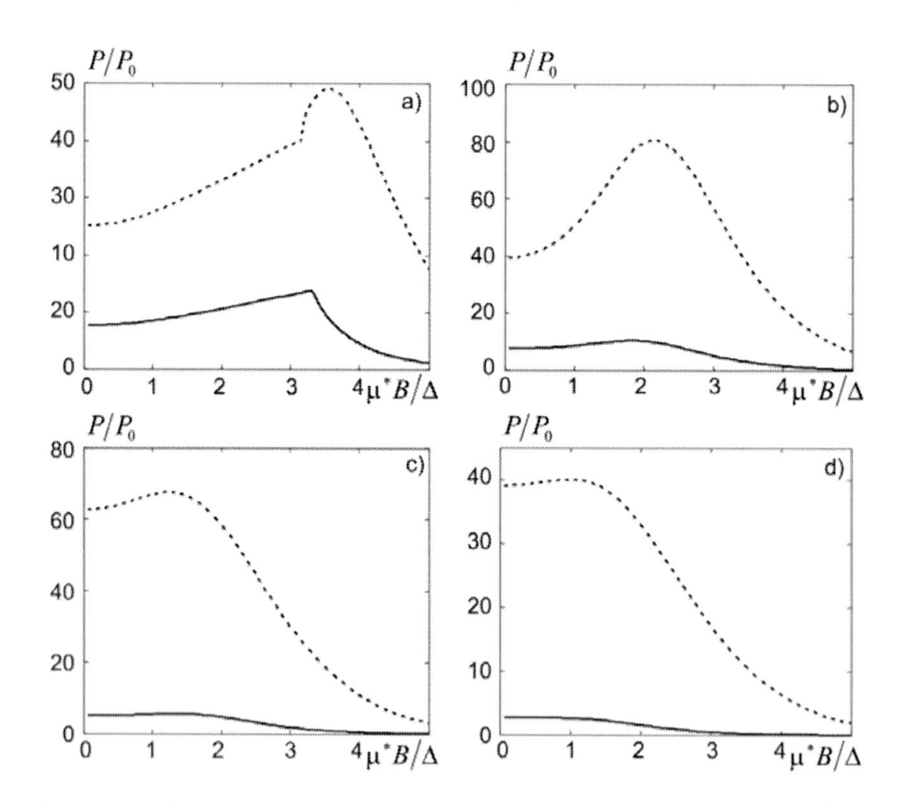

Figure 151. Field dependences of power factor of a layered crystal in the model $\tau \propto 1/g(\varepsilon)$ at $kT/\Delta = 1; 0 \le \mu^* B/\Delta \le 5$.

The figure demonstrates that in the model $\tau \propto 1/g(\varepsilon)$, as well, for all degrees of miniband filling, the dashed curves lie above the solid ones, i.e. power factor in the effective mass approximation is essentially higher than for a real layered crystal. This occurs because both the Seebeck coefficient modulus and the electric conductivity in a real layered crystal are lower than in the effective mass approximation. In the model $\tau \propto 1/g(\varepsilon)$, just as in the model $\tau = const$, it is due to the effect of miniband finite width. Moreover, in the model $\tau \propto 1/g(\varepsilon)$, as well, for each degree of miniband filling, both in a real layered crystal and in the effective mass approximation there is a value of magnetic field induction depending on this filling degree whereby power factor is maximum. The reason is that, as was shown before, in a quantizing magnetic field under conditions of weak and intermediate degeneracy there exist the NLMR and the Seebeck coefficient modulus maximum. Therefore, in the model $\tau \propto 1/g(\varepsilon)$, for instance, with previously stipulated problem parameters and at $\zeta_0/\Delta = 0.5;1;1.5$, respectively, power factor maxima at $T = 116$ K for a real layered crystal make $3.274 \cdot 10^{-5};2.474 \cdot 10^{-5};1.366 \cdot 10^{-5}$ W/(m·K^2), respectively, and are achieved in magnetic fields with inductions 5.69; 3.11; 2.33T, respectively. At $\zeta_0/\Delta = 2$ the *local* power factor maximum is not apparent, but maximum induction of a magnetic field wherein power factor does not drop yet, is 1.38T, and power factor value with this induction is $6.586 \cdot 10^{-6}$ W/(m·K^2), i.e. this value can be considered as the *absolute* maximum in the investigated range of magnetic fields. Whereas in the effective mass approximation said maxima make $1.170 \cdot 10^{-4};1.915 \cdot 10^{-4};1.608 \cdot 10^{-4};9.522 \cdot 10^{-5}$ W/(m·K^2) and are achieved in magnetic fields with inductions 6.12; 3.80; 2.33 and 1.64T, respectively. In the model $\tau \propto 1/g(\varepsilon)$ power factor maxima are considerably higher than in the model $\tau = const$, which is due to a larger contribution of low energy states to longitudinal electric conductivity in the model $\tau \propto 1/g(\varepsilon)$. However, in this model, as well, at $\zeta_0/\Delta = 0.5;1;1.5;2$, respectively, power factor maxima in the effective mass approximation are a factor of 3.6; 7.7; 11.8 and 14.5, respectively, higher than in a real layered crystal.

Chapter 58

POWER FACTOR OF A LAYERED CHARGE-ORDERED CRYSTAL IN THE ABSENCE OF A MAGNETIC FIELD UNDER CONDITIONS OF WEAK AND INTERMEDIATE DEGENERACY

Let us now consider power factor of a layered charge-ordered crystal in the absence of a magnetic field under conditions of weak and intermediate degeneracy. As before, we will assume acoustic phonon deformation potential scattering to be the dominant mechanism of charge carrier scattering. We will first consider this problem under approximation of constant relaxation time. In this case, the following expression for power factor is valid:

$$P = \pi t^{-1} P_0 \int\limits_{w\delta}^{\sqrt{w^2\delta^2+1}} dy y^{-1} \sqrt{\left(1+w^2\delta^2-y^2\right)\left(y^2-w^2\delta^2\right)} \left\{ \ln\left[1+\exp\left(\frac{\gamma+y}{t}\right)\right] + \ln\left[1+\exp\left(\frac{\gamma-y}{t}\right)\right] - \right.$$
$$\left. - \left(\frac{y+\gamma}{t}\right)\left[1+\exp\left(-\frac{y+\gamma}{t}\right)\right]^{-1} + \left(\frac{y-\gamma}{t}\right)\left[1+\exp\left(\frac{y-\gamma}{t}\right)\right]^{-1} \right\}^2 \times \qquad (58.1)$$
$$\times \left\{ \int\limits_{w\delta}^{\sqrt{w^2\delta^2+1}} dy y^{-1} \sqrt{\left(1+w^2\delta^2-y^2\right)\left(y^2-w^2\delta^2\right)} \left\{ \left[1+\exp\left(-\frac{y+\gamma}{t}\right)\right]^{-1} + \left[1+\exp\left(\frac{y-\gamma}{t}\right)\right]^{-1} \right\} \right\}^{-1}$$

In this formula, $P_0 = 4k^2a^2\rho s^2\Delta/\left(3\pi h\Xi^2\right)$. The results of calculation of the temperature dependences of power factor of a charge-ordered layered crystal by this formula are given in Figure 152.

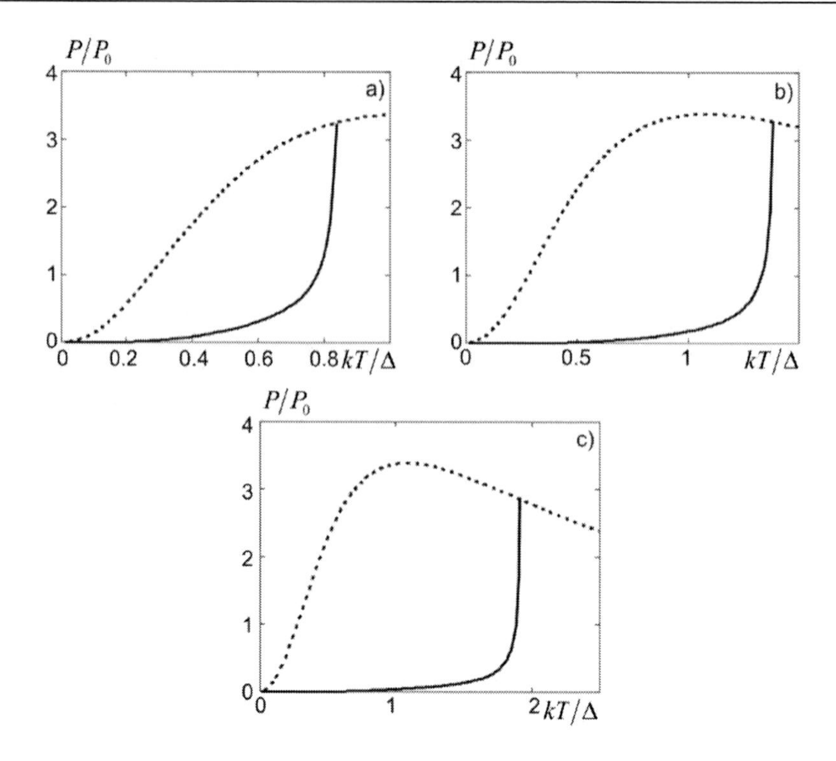

Figure 152. Temperature dependences of power factor of a layered charge-ordered crystal in the model $\tau = const$ at $\zeta_{02D}/\Delta = 1$ and: a) $W_0/\zeta_{02D} = 1.5$; b) $W_0/\zeta_{02D} = 2$; c) $W_0/\zeta_{02D} = 2.5$. The dashed curves correspond to the disordered state.

The figure demonstrates that for all values of effective attractive interaction between electrons, in the subcritical temperature region the value of power factor of a layered crystal in the ordered state is considerably lower than in the disordered one. This occurs because conduction miniband describing the interlayer motion of electrons in the charge-ordered state is narrowed. And this narrowing, as already mentioned, leads to a drop in conduction and the Seebeck coefficient modulus. Therefore, power factor is considerably reduced as well. For each value of effective attractive interaction between electrons, the value of power factor of a layered crystal is low in that temperature region wherein the order parameter δ is close to unity and varies only slightly with temperature. However, as the temperature approaches the critical value corresponding to a transition to the disordered phase, the order parameter starts decreasing sharply, and power factor, accordingly, starts increasing steeply. In so doing, the larger is the effective attractive interaction between electrons, the wider is the temperature range wherein power factor is low, and, on the other hand, the sharper is its increase as the temperature approaches the critical value. At a second-order transition point the temperature dependence of power factor undergoes a kink, and the value of power factor at a transition temperature coincides with such in the disordered state. Moreover, at the ratios W_0/ζ_{02D}, equal to 2 and 2.5, the respective power factor value at a phase-transition point is maximum for the considered temperature range, whereas at $W_0/\zeta_{02D} = 1.5$ – not. This occurs because at a phase transition temperature corresponding to the value of $W_0/\zeta_{02D} = 1.5$ (with regard to $\zeta_{02D}/\Delta = 1$) power factor in the disordered state is not maximum yet. At the same time, at the ratios W_0/ζ_{02D} equal to 2 and 2.5, power factor maximum typical of the disordered state is achieved at temperatures lower than the respective phase transition temperatures. With previously stipulated problem parameters and $W_0/\zeta_{02D} = 1.5;2;2.5$, respectively, the values of power factor at phase-transition temperatures make $3.089 \cdot 10^{-5};3.132 \cdot 10^{-5};2.730 \cdot 10^{-5}$ W/(m·K^2), respectively. Power factor maximum in the disordered state is $3.231 \cdot 10^{-5}$ W/(m·K^2). The above mentioned physical circumstances yield a weakly expressed power factor at a phase-transition point as a function of effective attractive interaction between electrons.

Chapter 59

FIELD DEPENDENCE OF POWER FACTOR OF A CHARGE-ORDERED LAYERED CRYSTAL IN THE APPROXIMATION OF CONSTANT RELAXATION TIME UNDER CONDITIONS OF WEAK AND INTERMEDIATE DEGENERACY

Under conditions of weak and intermediate degeneracy of electron gas in the model $\tau = const$ in the presence of a quantizing magnetic field, power factor of a charge-ordered layered crystal is determined as follows:

$$P = \pi t^{-1} P_0 \frac{\left(A^{(\alpha)} + B^{(\alpha)}\right)^2}{C^{(\alpha)} + D^{(\alpha)}}, \tag{59.1}$$

where the coefficients $A^{(\alpha)}, B^{(\alpha)}, C^{(\alpha)}, D^{(\alpha)}$ are determined by the formulae:

$$A^{(\alpha)} = \left(\frac{h}{t}\right)^2 \sum_{l=1}^{\infty} \frac{(-1)^l l \coth(hl/t)}{\sinh(hl/t)} \left\{ \int_{w\delta}^{\sqrt{w^2\delta^2+1}} dy y^{-1} \sqrt{\left(1 + w^2\delta^2 - y^2\right)\left(y^2 - w^2\delta^2\right)} \times \left[\exp\left(-l\frac{|\gamma + y|}{t}\right) + \exp\left(-l\frac{|\gamma - y|}{t}\right) \right] \right\}, \tag{59.2}$$

$$B^{(\alpha)} = \frac{h}{t^2} \sum_{l=1}^{\infty} \frac{(-1)^l l}{\sinh(hl/t)} \left\{ \int_{w\delta}^{\sqrt{w^2\delta^2+1}} dy y^{-1} \sqrt{\left(1 + w^2\delta^2 - y^2\right)\left(y^2 - w^2\delta^2\right)} \times \left[|\gamma + y| \exp\left(-l\frac{|\gamma + y|}{t}\right) + |\gamma - y| \exp\left(-l\frac{|\gamma - y|}{t}\right) \right] \right\}, \tag{59.3}$$

$$C^{(\alpha)} = 0.5\int_{w\delta}^{\sqrt{w^2\delta^2+1}} dyy^{-1}\sqrt{\left(1+w^2\delta^2-y^2\right)\left(y^2-w^2\delta^2\right)}\left[2+\operatorname{sgn}(\gamma+y)+\operatorname{sgn}(\gamma-y)\right], \quad (59.4)$$

$$D^{(\alpha)} = \sum_{l=1}^{\infty}\frac{(-1)^l(hl/t)}{\sinh(hl/t)}\int_{w\delta}^{\sqrt{w^2\delta^2+1}} dyy^{-1}\sqrt{\left(1+w^2\delta^2-y^2\right)\left(y^2-w^2\delta^2\right)}\times$$
$$\times\left[\operatorname{sgn}(\gamma+y)\left(-l\frac{|\gamma+y|}{t}\right)+\operatorname{sgn}(\gamma-y)\left(-l\frac{|\gamma-y|}{t}\right)\right], \quad (59.5)$$

and $P_0 = 4k^2a^2\rho s^2\Delta/\left(3\pi h\Xi^2\right)$. The results of calculation of the field dependences of power factor of a layered charge-ordered crystal by this formula are given in Figure 153.

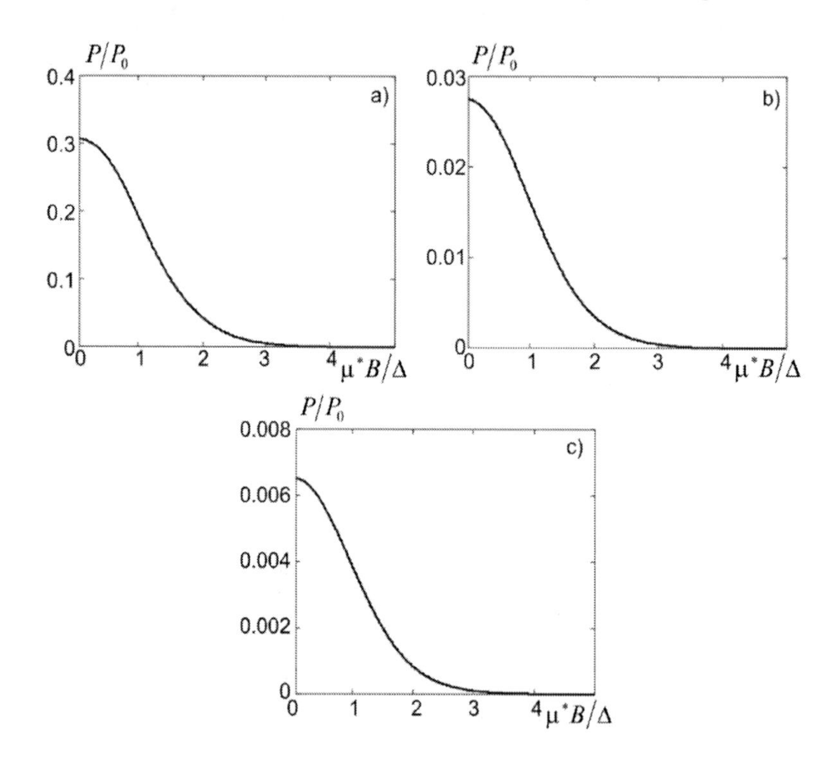

Figure 153. Field dependences of power factor at $\zeta_{02D}/\Delta = 1; kT/\Delta = 0.6; 0 \le \mu^* B/\Delta \le 5$ and: a) $W_0/\zeta_{02D} = 1.5$; b) $W_0/\zeta_{02D} = 2$; c) $W_0/\zeta_{02D} = 2.5$.

The figure demonstrates that in the ordered phase for all values of effective attractive interaction between electrons, power factor is a monotonically decreasing function of a magnetic field. The main reason is that, as already mentioned, magnetic field increase is equivalent to crystal cooling. The main factors determining the field dependences of power factor is increase in the order parameter and the ratio of the distance between the Landau levels to the energy of thermal motion with magnetic field increase. And this leads to narrowing of conduction minibands into which the initial miniband is split when passing into the charge-ordered state, hence, to power factor decrease due to simultaneous reduction of longitudinal electric conductivity and the Seebeck coefficient modulus. Moreover, it is clear that with the same temperature power factor is the lower, the higher is the value of effective interaction. With selected problem parameters and $W_0/\zeta_{02D} = 1.5;2;2.5$, respectively, the values of power factor in the absence of a magnetic field make $2.924 \cdot 10^{-6};2.571 \cdot 10^{-7};6.211 \cdot 10^{-8}$ W/(m·K^2). Therefore, control of thermoelectric figure of merit of layered charge-ordered crystals in the ordered phase with the aid of a magnetic field directed parallel to temperature gradient is impossible.

Chapter 60

TEMPERATURE DEPENDENCE OF POWER FACTOR OF LAYERED CHARGE-ORDERED CRYSTALS UNDER WEAK AND INTERMEDIATE DEGENERACY IN THE ABSENCE OF A MAGNETIC FIELD WITH REGARD TO THE ENERGY DEPENDENCE OF RELAXATION TIME

As before, we will consider acoustic phonon deformation potential scattering of charge carriers to be predominant, but relaxation time is a function of energy in conformity with the model $\tau \propto 1/g(\varepsilon)$. In this case for power factor of a charge-ordered layered crystal the following expression is valid:

$$
\begin{aligned}
P = \pi t^{-1} P_0 \int_0^{\pi/2} dx \phi(x) \Bigg\{ & \ln\left[1+\exp\left(\frac{\gamma + \sqrt{w^2\delta^2+\cos^2 x}}{t}\right)\right] + \ln\left[1+\exp\left(\frac{\gamma - \sqrt{w^2\delta^2+\cos^2 x}}{t}\right)\right] - \\
& -\left(\frac{\sqrt{w^2\delta^2+\cos^2 x}+\gamma}{t}\right)\left[1+\exp\left(-\frac{\sqrt{w^2\delta^2+\cos^2 x}+\gamma}{t}\right)\right]^{-1} + \\
& +\left(\frac{\sqrt{w^2\delta^2+\cos^2 x}-\gamma}{t}\right)\left[1+\exp\left(\frac{\sqrt{w^2\delta^2+\cos^2 x}-\gamma}{t}\right)\right]^{-1}\Bigg\}^2 \times \\
\times \Bigg\{ \int_0^{\pi/2} dx \phi(x) \Bigg\{ & \left[1+\exp\left(-\frac{\sqrt{w^2\delta^2+\cos^2 x}+\gamma}{t}\right)\right]^{-1} + \left[1+\exp\left(\frac{\sqrt{w^2\delta^2+\cos^2 x}-\gamma}{t}\right)\right]^{-1}\Bigg\}\Bigg\}^{-1}
\end{aligned}
\quad . (60.1)
$$

Here, $P_0 = k^2 a^2 \rho s^2 \Delta / (3\pi h \Xi^2)$. The results of calculation of the temperature dependences of power factor of a layered charge-ordered crystal by this formula are given in Figure 154.

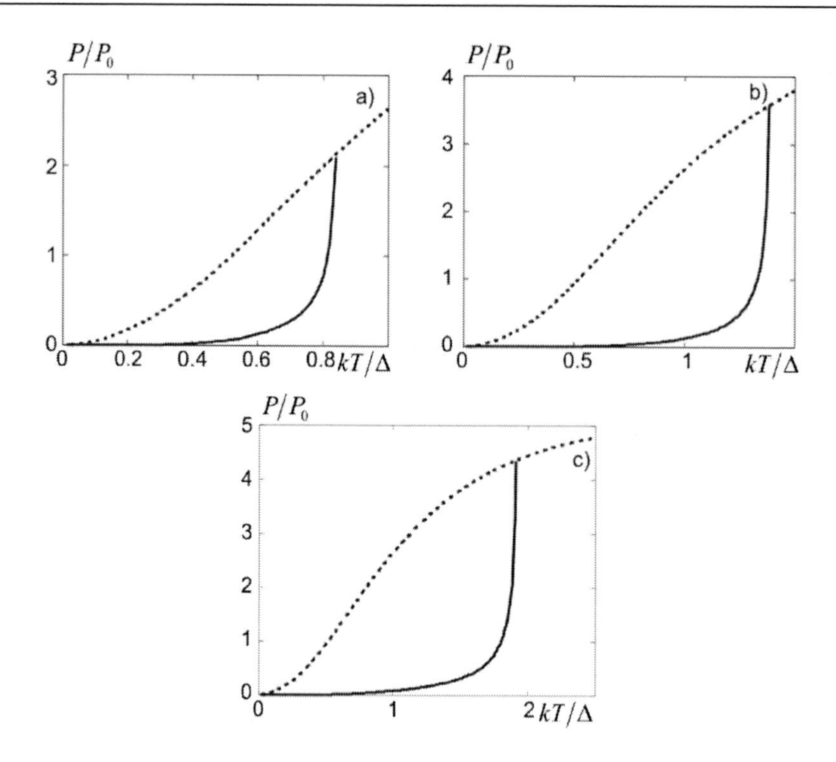

Figure 154. Temperature dependences of power factor of a layered charge-ordered crystal in the model $\tau \propto 1/g(\varepsilon)$ at $\zeta_{02D}/\Delta = 1$ and: a) $W_0/\zeta_{02D} = 1.5$; b) $W_0/\zeta_{02D} = 2$; c) $W_0/\zeta_{02D} = 2.5$. The dashed curves correspond to the disordered state.

The figure demonstrates that, as in the model $\tau = const$, for all values of effective attractive interaction between electrons in the subcritical region of temperatures the value of power factor of a layered crystal in the ordered state is considerably lower than in the disordered state. As in the model $\tau = const$, this occurs because conduction miniband describing the interlayer motion of electrons in the charge-ordered state is narrowed. And this narrowing, as already mentioned, leads to a drop of conductivity and the Seebeck coefficient modulus. Therefore, power factor also decreases considerably. For each value of effective attractive interaction between electrons, the value of power factor of a layered crystal is low in that temperature region wherein the order parameter δ is close to unity and slightly changes as a function of temperature. However, as the temperature approaches the critical value corresponding to a transition to the disordered phase, the order parameter starts decreasing sharply, and power factor, accordingly, starts growing steeply. In so doing, the larger is the effective attractive interaction between the electrons, the wider is the range of temperatures wherein power factor is low, and, on the other hand, the sharper is its increase as the temperature approaches the critical value. At second-order phase transition point the temperature dependence of power factor undergoes a kink, and the value of power factor at transition temperature coincides with that in the disordered state. But, unlike the model $\tau = const$, in the model $\tau \propto 1/g(\varepsilon)$ for all considered ratios W_0/ζ_{02D} the respective value of power factor at a point of transition is maximum. This occurs because in the model $\tau \propto 1/g(\varepsilon)$ at transition temperatures corresponding to considered values W_0/ζ_{02D} (taking into account that $\zeta_{02D}/\Delta = 1$) power factor in the disordered state is not maximum yet. With previously stipulated problem parameters and $W_0/\zeta_{02D} = 1.5;2;2.5$, respectively, the values of power factor at phase transition temperatures are $4.980 \cdot 10^{-6};9.050 \cdot 10^{-6};1.036 \cdot 10^{-5}$ W/(m·K^2), respectively. These values are a factor of 6.20;3.46;2.63, respectively, lower than in the model $\tau = const$. The reason for this reduction is that increased contribution of low-energy states in the model $\tau \propto 1/g(\varepsilon)$ as compared to the model $\tau = const$ decreases sharply the Seebeck coefficient modulus. Besides, as long as in the model $\tau \propto 1/g(\varepsilon)$ as compared to the model $\tau = const$ power factor maximum in the disordered state is shifted towards higher temperatures, power factor at second-order phase transition point monotonically increases with increase in effective interaction.

Chapter 61

FIELD DEPENDENCE OF POWER FACTOR OF A CHARGE-ORDERED LAYERED CRYSTAL WITH REGARD TO THE ENERGY DEPENDENCE OF RELAXATION TIME UNDER CONDITIONS OF WEAK AND INTERMEDIATE DEGENERACY

Under conditions of weak and intermediate degeneracy of electron gas in the model $\tau \propto 1/g(\varepsilon)$ in the presence of a quantizing magnetic field the value of power factor of a charge-ordered layered crystal is determined as:

$$P = \pi t^{-1} P_0 \frac{\left(A^{(\alpha)} + B^{(\alpha)}\right)^2}{C^{(\alpha)} + D^{(\alpha)}}, \tag{61.1}$$

where the coefficients $A^{(\alpha)}, B^{(\alpha)}, C^{(\alpha)}, D^{(\alpha)}$ are determined by the formulae:

$$A^{(\alpha)} = \left(\frac{h}{t}\right)^2 \sum_{l=1}^{\infty} \frac{(-1)^l l \coth(hl/t)}{\sinh(hl/t)} \int_0^{\pi/2} dx\, \phi(x) \left[\exp\left(-l \frac{\left|\gamma + \sqrt{w^2\delta^2 + \cos^2 x}\right|}{t} \right) + \right. \\ \left. + \exp\left(-l \frac{\left|\gamma - \sqrt{w^2\delta^2 + \cos^2 x}\right|}{t} \right) \right], \tag{61.2}$$

$$B^{(\alpha)} = \frac{h}{t^2} \sum_{l=1}^{\infty} \frac{(-1)^l l}{\sinh(hl/t)} \int_0^{\pi/2} dx\, \phi(x) \left[\left|\gamma + \sqrt{w^2\delta^2 + \cos^2 x}\right| \exp\left(-l \frac{\left|\gamma + \sqrt{w^2\delta^2 + \cos^2 x}\right|}{t} \right) + \right. \\ \left. + \left|\gamma + \sqrt{w^2\delta^2 + \cos^2 x}\right| \exp\left(-l \frac{\left|\gamma - \sqrt{w^2\delta^2 + \cos^2 x}\right|}{t} \right) \right], \tag{61.3}$$

$$C^{(\alpha)} = 0.5\int_0^{\pi/2} dx\phi(x)\left[2+\operatorname{sgn}\left(\gamma+\sqrt{w^2\delta^2+\cos^2 x}\right)+\operatorname{sgn}\left(\gamma-\sqrt{w^2\delta^2+\cos^2 x}\right)\right], \quad (61.4)$$

$$D^{(\alpha)} = \sum_{l=1}^{\infty}\frac{(-1)^l (hl/t)}{\sinh(hl/t)}\int_0^{\pi/2} dx\phi(x)\left[\operatorname{sgn}\left(\gamma+\sqrt{w^2\delta^2+\cos^2 x}\right)\times\right.$$

$$\times\exp\left(-l\frac{\left|\gamma+\sqrt{w^2\delta^2+\cos^2 x}\right|}{t}\right)+\operatorname{sgn}\left(\gamma-\sqrt{w^2\delta^2+\cos^2 x}\right)\times, \quad (61.5)$$

$$\left.\times\exp\left(-l\frac{\left|\gamma-\sqrt{w^2\delta^2+\cos^2 x}\right|}{t}\right)\right]$$

and $P_0 = k^2a^2\rho s^2\Delta/\left(3\pi h\Xi^2\right)$. The results of calculation of the field dependences of power factor of a layered charge-ordered crystal by this formula are given in Figure 155.

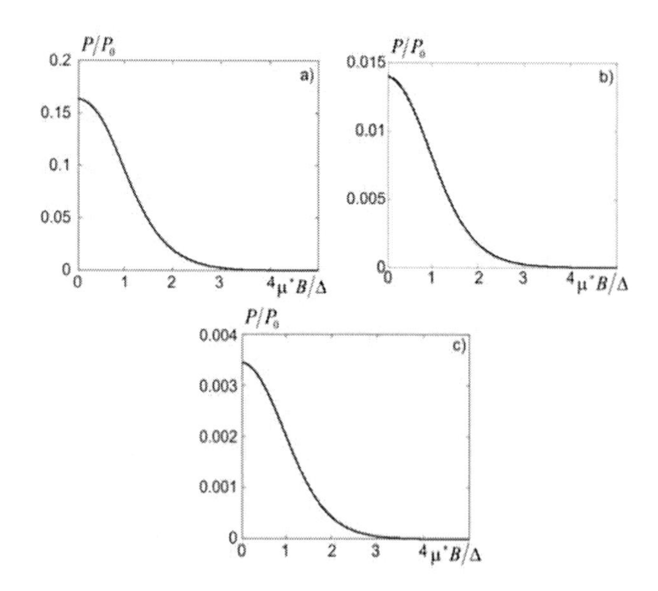

Figure 155. Field dependences of power factor in the model $\tau \propto 1/g(\varepsilon)$ at $\zeta_{02D}/\Delta = 1; kT/\Delta = 0.6; 0 \le \mu^* B/\Delta \le 5$ and: a) $W_0/\zeta_{02D} = 1.5$; b) $W_0/\zeta_{02D} = 2$; c) $W_0/\zeta_{02D} = 2.5$.

The figure demonstrates that in the model $\tau \propto 1/g(\varepsilon)$, as in the model $\tau = const$, for all values of effective attractive interaction between electrons, power factor is a monotonically decreasing function of a magnetic field. The main reason is that, as already mentioned, magnetic field increase is equivalent to crystal cooling. The basic factors determining the field dependences of power factor is increase in the order parameter and the ratio of the distance between the Landau levels to the energy of thermal motion with magnetic field increase. And this results in narrowing of conduction minibands into which the initial miniband is split when passing into the charge-ordered state, hence, in power factor decrease due to simultaneous reduction of longitudinal electric conductivity and the Seebeck coefficient modulus. Moreover, it is clear that with the same subcritical temperature, power factor is the lower the higher is the value of effective interaction. With selected problem parameters and $W_0/\zeta_{02D} = 1.5; 2; 2.5$, respectively, the values of power factor in the absence of a magnetic field are $3.881 \cdot 10^{-7}; 3.333 \cdot 10^{-8}; 8.214 \cdot 10^{-9}$ W/(m·K^2). And these values are a factor of 7.53;7.77;7.56, respectively, lower than in the model $\tau = const$. This is caused by a sharp decrease in the Seebeck coefficient modulus in the model $\tau \propto 1/g(\varepsilon)$ as compared to the model $\tau = const$ due to increased contribution of low-energy states. Therefore, in the model $\tau \propto 1/g(\varepsilon)$, as in the model $\tau = const$, control of thermoelectric figure of merit of layered charge-ordered crystals in the ordered phase with the aid of a magnetic field directed parallel to temperature gradient is impossible.

Chapter 62

CLASSIFICATION AND GENERAL CHARACTERIZATION OF SIZE EFFECTS USED FOR FIGURE OF MERIT IMPROVEMENT OF THERMOELECTRIC MATERIALS

Nowadays, for fabrication of thermoelectric products, increasing use is made of materials prepared by hot pressing, extrusion or spark plasma sintering (SPS) methods, rather than single-crystal thermoelectric materials (TEM). They are produced of various-size powders that are prepared from suitable single crystals by grinding in the ball mill or (less frequently) by deposition from solutions. At first, the main benefit of such materials was considered to be their purely technological advantage that, unlike single crystals, they can be much easier configured into a shape dictated by device purpose and design [64]. However, later it was observed that thermoelectric figure of merit of such materials is little or almost not reduced as compared to a single crystal, and, in a number of cases, it even grows considerably [65]. This brought about a surge of theoretical and experimental works dedicated to research of the influence of size effects on the figure of merit of thermoelectric materials. Concurrently, studies were pursued on the behavior of thermoelectric materials under extreme conditions of large temperature gradients and warming electrical fields, including those caused by thermoEMF, [66], because such studies are deemed to be topical for solving the problems of microminiaturization of thermoelectric products. Though, the possibility as such of creating temperature gradients leading to heat up of charge carriers by *thermoelectric* fields in a relevant for thermoelectricity temperature region is somewhat dubious for the following reason. Suppose that temperature difference ΔT has been created on a small particle of TEM. If the *thermoelectric* field is warming, then, without regard to energy losses due to scattering which can be low in a particle which is small compared to "cooling length" of charge carriers, the electron in thermoelectric field will acquire the energy of the order $\alpha e \Delta T$, where α – the Seebeck coefficient. But if the field is warming, this energy must be comparable to the energy of electron thermal motion kT caused, for instance, by average particle temperature. This yields the following condition for temperature difference:

$$\frac{\Delta T}{T} > \frac{k}{\alpha e}. \tag{62.1}$$

Therefore, for instance, for thermoelectric material Bi_2Te_3, for which $\alpha = 200$ μV/K, there must be $\Delta T/T > 0.431$. At $T = 300$ K and characteristic particle size of the order, for instance, 40nm, we obtain temperature gradient of the order $3.23 \cdot 10^9$K/m. Such gradients are hardly possible in real thermoelectric instruments and devices, even though superminiature.

Size effects that can influence the figure of merit of thermoelectric materials can be divided into two large classes – classical and quantum. The former take place when the size of material particles and/or contacts between them is comparable to the mean free path of charge carriers or phonons, but far exceeds de Broglie wavelength of charge carriers. The latter are realized when the size of material particles and/or contacts between them is comparable to or less than de Broglie wavelength of charge carriers. In this case, alongside with the drift mechanism or instead of it, the tunneling conduction mechanism is operating which in a sense is similar to wave diffraction effect.

The physical mechanism of figure of merit increase when passing from a single crystal to material prepared by hot pressing, extrusion or SPS methods can be that thermal conductivity in this case is restricted severely and electric conductivity – weakly. It is considered that the Seebeck coefficient here remains unvaried or changes downwards considerably less than the electric to thermal conductivity ratio is increased. The basic mechanism of certain reduction of the Seebeck coefficient when passing from a single crystal to material prepared by hot pressing, extrusion or SPS methods, most probably, is formation of structural defects in the process of material grinding, hence, of their related surface electron (hole) states. However, the Seebeck coefficient can be partially or completely restored as a result of annealing [65]. It is the combination of these factors that results in retention or increase of the thermoelectric figure of merit of material.

In terms of size effects, both classical and quantum, the structures of thermoelectric materials can be divided into three-dimensional (single crystals), two-dimensional (films, layers), single-dimensional (wires) and zero-dimensional structures ("dots" or "powders"). They are classified with regard to in how many directions (or along how many coordinate axes) the characteristic size of structural shape-forming element exceeds considerably some of characteristic mean free path lengths of charge carriers, phonons or De Broglie wavelength of charge carriers. In real thermoelectric materials, films (layers), wires or "dots" can form random or ordered (oriented) structures. In the case of ordered (oriented) structures we speak of the respective superlattices.

Chapter 63

TEM LATTICE THERMAL CONDUCTIVITY REDUCTION THROUGH OPTIMIZATION OF ITS SHAPE-FORMING ELEMENT

The most widely used TEM nowadays are Bi-Te based alloys (Bi_2Te_3 compounds). The respective single crystals are prepared by different methods, namely zone recrystallization, Czochralski pulling and oriented crystallization process. Extruded and SPS materials are prepared with the use of powders. These methods yield thermoelectric figure of merit Z in the range of $(2.8 \div 3.1)\cdot 10^{-3}$ K^{-1}. Such Z values are achieved with heat flux and electric current orientations in the directions normal to Bi_2Te_3 trigonal axis. Whereas in the direction parallel to trigonal axis the Z values are essentially lower. This situation is due to the fact that Bi_2Te_3 is uniaxial anisotropic crystal whose conductivity values σ_{11} in the direction normal to trigonal axis are $\sigma_{11} = (800 \div 1000)$ S/cm and are considerably higher than σ_{33} – conductivity along trigonal axis. The values of thermal conductivity χ_{ij} are also anisotropic and make χ_{11}=1.45W/m·K and χ_{33} = 0.58W/m·K. At the same time, the Seebeck coefficients α_{11} and α_{33} are little different and make 210 ÷ 220 μV/K. Thus, $Z_{11} = (2.4 \div 2.5)\cdot 10^{-3} K^{-1}$. For this reason, practical use is found by materials oriented normal to trigonal axis.

Thermoelectric instruments and devices are also manufactured with the use of Bi_2Te_3 based materials prepared by extrusion method whose thermoelectric figure of merit is about $3\cdot 10^{-3}$ K^{-1}, that is, rather close to that of single crystal materials. It should be noted that the macroscopic structure of extruded materials is a combination of arbitrarily oriented powder particles of size (40 ÷ 80) μm whole properties are close to those of oriented crystalline thermoelectric materials. For extruded thermoelectric materials $\sigma_{ef} = \sqrt{\sigma_{11}\sigma_{33}}$ and $\chi_{ef} = \sqrt{\chi_{11}\chi_{33}}$. Therefore, thermoelectric figure of merit of extruded material must be lower than that of a single crystal. Taking into account that electric conductivity anisotropy of Bi_2Te_3 depending on conductivity type is 2.7 for *p*-type and 4 ÷ 6 for *n*-type, and thermal conductivity anisotropy is 2 ÷ 3, the figure of merit can be reduced by a factor of $\sqrt{2}$ – $\sqrt{3}$, that is, by 30 – 40%. However, in the best case for *p*-type material it can even grow by about 5%. In practice, no figure of merit reduction is observed. Hence, there must be a mechanism leading to thermoelectric figure of merit increase in going from single crystal to extruded material structure due to a change in the character of phonon and charge carrier scattering.

Research on this mechanism would provide for helpful information as to the ways of radical improvement of thermoelectric figure of merit of said materials.

The physical concept of thermoelectric figure of merit improvement in extruded material is that thermal conductivity in going from a bulk to porous or fine-dispersed structure is reduced considerably and the electric conductivity – essentially weaker.

The authors of [67] who were among the first who paid attention to this fact, made evaluation calculations of the electric and thermal conductivity of model structure of thermoelectric material rods divided by vacuum gaps. From the evaluation formulae it follows that if the characteristic dimensions of the rods are small, then the electric and thermal conductivity of the structure is proportional to these dimensions. However, with the large rod dimensions said characteristics tend to parameters of the bulk material. Moreover, in the estimation of thermoelectric figure of merit in this work it was considered that the lattice conductivity of such structure is zero. The electric and thermal conductivity of the structure caused by free charge carriers is essentially dependent on the coefficient of electrons passage through the vacuum gap which does not affect, however, the thermoelectric figure of merit. Said approach did not yield quantitative estimates of rod dimensions and gaps between them that are optimal in terms of thermoelectric figure of merit. In [66] it is shown that in going from single-crystal to fine-dispersed germanium with the average grain radius $2.0 \div 2.5$ μm (of which samples with porosity 70% were made), the ratio of electric conductivity to thermal conductivity increased by a factor of 100 as compared to single crystal, and thermoelectric figure of merit – only by a factor of 4 – 6, which, in the authors' opinion is attributable to incomplete restoration of negative thermoEMF after annealing. In [68], formulae were obtained for the determination of the electric and thermal conductivity of a dispersed medium comprising spherical particles of thermoelectric material, but electron and phonon scattering on the boundaries of spheres and contacts between them was not considered.

Paper [69] is a theoretical study of the thermal conductivity of the bulk nanostructured bismuth telluride samples which, nevertheless, does not take into account that phonon scattering on the boundaries of individual nanoparticles occurs at all phonon frequencies, rather than at "selected" ones.

In patent [70] for the efficient phonon drag it is proposed to use small-area contacts between relatively large parts of thermoelectric material. In so doing, said contacts must have dimensions of the order of several nanometers.

The possibilities were also considered of creating such thermoelectric materials that would be "phonon glasses", remaining in this case "electron crystals" due to the fact that lattice thermal conductivity with a large concentration of structural defects is reduced more than the electric conductivity owing to the peculiarity of electron density of states [71]. The researchers' attention is also focused on the whiskers of organic conductors of the type TTF-TCNQ and the like [72]. It is considered that a high degeneracy level of free charge carriers gas is attainable in these crystals, owing to which lattice thermal conductivity cannot affect considerably the thermoelectric figure of merit of material, and the latter can be regarded as the integral characteristic of free charge carriers subsystem in material [73,74], that is, the lower limit of thermal conductivity in these crystals has already been achieved, and the only opportunity of thermoelectric figure of merit improvement is the Lorentz number increase.

In the manufacture of thermoelectric modules of conventional material powders by hot pressing or extrusion methods, a question arises as to the optimal in terms of thermoelectric figure of merit size of powder grains and contacts between them. According to [75], particles of source powder can be adequately considered spherical. In the course of pressing they can acquire the shape of hemispheres with a circular contact between them. The shape-forming element of such structure can be approximated by two equal-radius hemispheres contacting in a circle. Research on the generalized conductivities of such shape-forming element should be a preliminary to a research on the above mentioned characteristics of structure as a whole. It is this that motivates the relevance of the problem solved in this work.

Our purpose in this work is to calculate changes in the lattice thermal conductivity of shape-forming element of extruded thermoelectric material structure due to phonon scattering on the boundaries of contact between two osculating hemispheres, and estimate the radius of contact necessary for 30-40% reduction of lattice thermal conductivity of shape-forming element.

63.1. CONSIDERATION OF THE PROBLEM OF PHONON SCATTERING ON THE BOUNDARIES OF SHAPE-FORMING ELEMENT IN THE APPROXIMATION OF CONSTANT RELAXATION TIME

As will be shown below, for consideration of this problem it is reasonable to involve a model of unit sphere placed in a heat flux. This will enable a more transparent physical interpretation of quantitative estimates. With a constant phonon relaxation time, the following expression for the resulting phonon mean free path in a limited sample is valid [76]:

$$l_{tp} = \frac{l_p L}{l_p + L}. \tag{63.1}$$

In this formula, l_p is phonon mean free path in material caused by all scattering mechanisms, except for the boundaries of contact spot or sample as a whole, L is effective phonon mean free path due to sample boundaries. As long as the effective mean free paths in a sample due to the boundaries are not equal for all the phonons, the thermal conductivity of material at boundary scattering is:

$$\chi_l^{ef} = \frac{1}{3}\rho s c_V l_p \left\langle \frac{L}{L + l_p} \right\rangle. \tag{63.2}$$

In this formula, ρ is material density, v is sound velocity in it, c_V is specific heat of material with a constant volume. The angular brackets mean averaging of respective expression over possible effective lengths L of phonon mean free path in a sample, including the shortest ones, because theirs is the major contribution to general possibility of phonon scattering [77]. In the case of a circular contact which is small as compared to hemisphere diameters, it can be considered that phonon drag takes place only in its vicinity. Moreover, all points of contact boundary are equivalent by virtue of its symmetry. Hence, formula (63.2) implies the following ratio between thermal conductivity of shape-forming element and that of the bulk sample:

$$\chi_l^{ef} \Big/ \chi_l = \pi^{-1} \int_0^1 \int_0^{2\pi} x \frac{k_l \sqrt{1 + x^2 - 2x\cos\phi}}{1 + k_l \sqrt{1 + x^2 - 2x\cos\phi}} d\phi dx . \tag{63.3}$$

In this formula, $k = r / l_p$, l_p is phonon mean free path. As it must be, at $k = 0$ formula (63.3) gives zero, and at $k \to \infty$ – thermal conductivity of a bulk sample. The results of these calculations are shown in Figure 156.

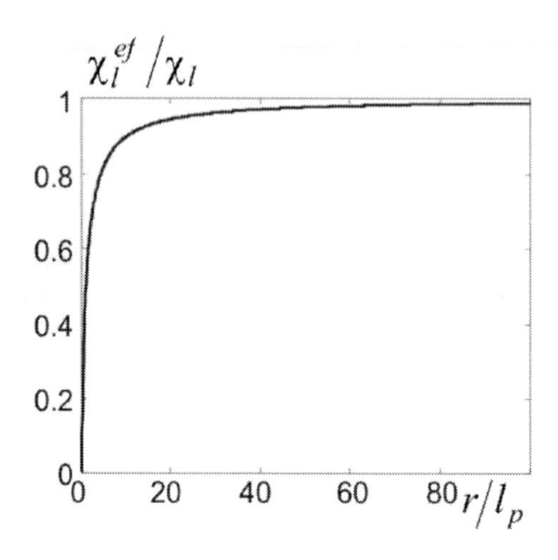

Figure 156. Dependence of thermal conductivity of a system of two hemispheres contacting in a circle on contact radius.

From the results of calculations it follows that for thermal conductivity reduction, for instance, by 30 – 40%, the contact radius must not exceed $(1.3 \div 2.5)l_p$. Taking into account that according to [64] the phonon mean free path corresponding to a greater thermal conductivity value is 4.16 nm, we obtain that contact radius must not exceed (5 ÷ 10) nm. The mean free path corresponding to the lower thermal conductivity value is, however, 1.4 nm. Therefore, for the same reduction of the lower thermal conductivity value the contact radius must not exceed (1.8 ÷ 3.3) nm. By analogy, this problem can be solved for a unit sphere. The respective formula is given by:

$$\chi_l^{ef} \Big/ \chi_l = 1.5 \int_0^1 \int_0^{2\pi} x^2 \frac{k_l \sqrt{1 + x^2 - 2xy}}{1 + k_l \sqrt{1 + x^2 - 2xy}} dy dx. \tag{63.4}$$

Double integral in this formula is caused by averaging the expression for thermal conductivity over the effective phonon mean free paths inside the sphere. In this formula, $k = R / l_p$, where R is sphere radius. The corresponding plot is presented in Figure 157.

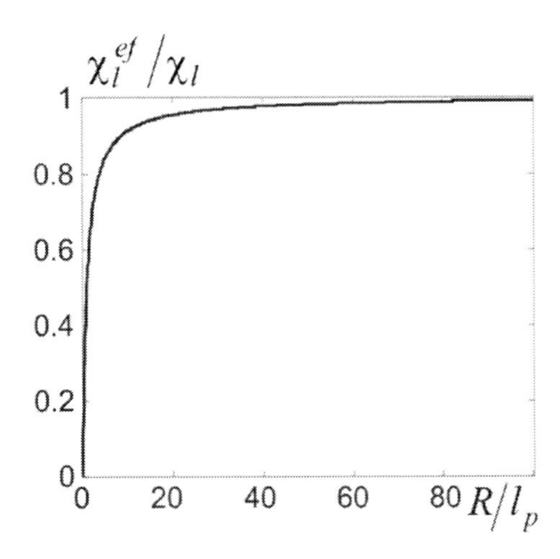

Figure 157. Sphere radius dependence of a relative decrease in lattice thermal conductivity due to phonon drag size effect.

It is seen that for the above discussed thermal conductivity reduction the sphere radius should not exceed $(1.2 \div 2.2)\, l_p$. For a greater thermal conductivity value it makes $(5.1 \div 9.3)$ nm, and for the lower – $(1.6 \div 2.9)$ nm.

63.2. Problem Consideration with Regard to Frequency Dependence of Phonon Relaxation Time

All previous calculations have been valid in the approximation of constant phonon relaxation time. Next, we consider the problem with regard to frequency dependence of phonon relaxation time.

If we normalize phonon relaxation time to the time of normal processes, then, taking into account [77], components of thermal conductivity tensor of the bulk sample of such layered material as bismuth telluride can be written as:

$$\chi_{l\|,\perp} = \frac{3\hbar M s_{\|,\perp}^4 k_B}{32\gamma^2 a_{\|,\perp}^3 (kT_D)^2 \theta^3 \pi} \int_0^1 \frac{x^4 \exp(x/\theta)}{[\exp(x/\theta)-1]^2} \left(\frac{1}{Q_{l\|,\perp}(x)} + \frac{2}{Q_{t\|,\perp}(x)} \right) dx \,. \qquad (63.5)$$

In these formulae, χ_l is lattice thermal conductivity, M is average atom mass in bismuth telluride, v is sound velocity in it, k_B is the Boltzmann constant, γ is the Gruneisen parameter, T_D is the Debye temperature of material, $\theta = T/T_D$, $Q_l(x)$ and $Q_t(x)$ are frequency polynomials introduced by the author that are to a power not higher than the fourth and caused by scattering mechanisms for the longitudinal and transverse phonons, indexes $\perp$ and $\|$ refer to thermal conductivity and sound velocity normal and parallel to the layers, the rest of designations are generally accepted.

At room temperatures and higher the thermal conductivity of thermoelectric material is mainly determined by Umklapp processes (U-processes). Therefore, polynomials $Q_l(x)$ and $Q_t(x)$ are determined as:

$$Q_{l\|,\perp}(x)=Q_{t\|,\perp}(x)=\mu_{\|,\perp}x\,. \tag{63.6}$$

Coefficient μ in the analytical form was calculated by Leibfried and Shleman [77] for a cubic lattice. However, according to experimental data [77], the value μ is not universal. Therefore, we will "retrieve" $\mu_{\|}$ and $\mu_{\perp}$ coefficients from the real values of components of thermal conductivity tensor of bismuth telluride [64], on condition of their coincidence with the theoretical values (63.5). At $\chi_{l\perp}=0.58$ W/m·K, $\chi_{l\|}=1.45$ W/m·K, M=158.8 a.m.u, $a_{\perp}=3\cdot10^{-9}$ m, $v_{\perp}=1867$ m/s, $a_{\|}=7\cdot10^{-10}$ m, $v_{\|}=2952$ m/s, $TD=155$ K and $T=300$ K we obtain $\mu_{\|}=0.131$, $\mu_{\perp}=6.657\cdot10^{-4}$.

Based on these coefficients, it is easy to calculate a relative reduction of thermal conductivity due to scattering on the boundaries of circular contact and sphere. By analogy with formula (63.3) in the case of a circular contact:

$$\chi_{l\|,\perp}^{ef}\Big/\chi_{l\|,\perp}=\pi^{-1}\int_0^1\int_0^1\int_0^{2\pi}\frac{zx^4\exp(x/\theta)}{[\exp(x/\theta)-1]^2}\left(\frac{k_{\|,\perp}^{*}\sqrt{z^2-2z\cos\phi+1}}{1+k_{\|,\perp}^{*}Q_{l\|,\perp}(x)\sqrt{z^2-2z\cos\phi+1}}+\frac{2k_{\|,\perp}^{*}\sqrt{z^2-2z\cos\phi+1}}{1+k_{\|,\perp}^{*}Q_{t\|,\perp}(x)\sqrt{z^2-2z\cos\phi+1}}\right)d\phi dz dx\left\{\int_0^1\frac{x^4\exp(x/\theta)}{[\exp(x/\theta)-1]^2}\left(\frac{1}{Q_{l\|,\perp}(x)}+\frac{2}{Q_{t\|,\perp}(x)}\right)dx\right\}^{-1}. \tag{63.7}$$

Here $k_{\|,\perp}^{*}=\dfrac{r_{\|,\perp}\gamma^2}{a_{\|,\perp}}\left(\dfrac{kT_D a_{\|,\perp}}{\hbar s_{\|,\perp}}\right)^4\left(\dfrac{kT_D}{Ms_{\|,\perp}^2}\right)$.

The afore-mentioned figures of thermal conductivity reduction for its greater value are obtained at k^{*}=17.37÷33.02. With previously stipulated problem parameters we get $r_{\|}=(3.5\div6.7)\cdot10^{-9}$ m. The same figures of thermal conductivity reduction for its lower value are obtained at k^{*}=3419÷6498. Therefore, $r_{\perp}=(0.6\div1.2)\cdot10^{-9}$ m. Such radii of contacts between particles of diameter 60÷80μm are hardly feasible.

In the case of a sphere, by analogy with formula (63.7) we have:

$$\chi_{l\|,\perp}^{ef}\Big/\chi_{l\|,\perp}=1.5\int_0^1\int_0^1\int_{-1}^1\frac{z^2x^4\exp(x/\theta)}{[\exp(x/\theta)-1]^2}\left(\frac{k_{\|,\perp}^{*}\sqrt{z^2-2zy+1}}{1+k_{\|,\perp}^{*}Q_{l\|,\perp}(x)\sqrt{z^2-2zy+1}}+\frac{2k_{\|,\perp}^{*}\sqrt{z^2-2zy+1}}{1+k_{\|,\perp}^{*}Q_{t\|,\perp}(x)\sqrt{z^2-2zy+1}}\right)dydzdx\left\{\int_0^1\frac{x^4\exp(x/\theta)}{[\exp(x/\theta)-1]^2}\left(\frac{1}{Q_{l\|,\perp}(x)}+\frac{2}{Q_{t\|,\perp}(x)}\right)dx\right\}^{-1}. \tag{63.8}$$

where $k_{\|,\perp}^{*}=\dfrac{R_{\|,\perp}\gamma^2}{a_{\|,\perp}}\left(\dfrac{kT_D a_{\|,\perp}}{\hbar v_{\|,\perp}}\right)^4\left(\dfrac{kT_D}{Mv_{\|,\perp}^2}\right)$.

Hence, in case of a sphere, to obtain the above reduction of greater thermal conductivity value, there must be k^{*}=15.57÷29.07, whence $R_{\|}=(3.2\div5.9)\cdot10^{-9}$ m. To obtain the same reduction of lower thermal conductivity value, there must be k^*=3420÷6500, whence $R_\perp$ = (0.6÷1.2)·10^{-9} m. Such particle dimensions are realizable only in nanostructured thermoelectric materials. Thus, the approach that takes into account only U-processes, cannot explain yet a small change in thermoelectric figure of merit in going from a single crystal to extruded thermoelectric material.

Therefore, it is worthwhile to consider phonon scattering on the boundaries of a circular contact and sphere with regard to not only U-processes, but normal processes as well. For this purpose, frequency polynomials $Q_l(x)$ and $Q_t(x)$ may be written as follows:

$$Q_{l\|,\perp}(x) = x^4 + \mu_{\|,\perp} x \, , \tag{63.9}$$

$$Q_{t\|,\perp}(x) = \left(\mu_{\|,\perp} + 3.125\theta^3\right) x \, . \tag{63.10}$$

Hence we get $\mu_\| = 4.142 \cdot 10^{-5}$, $\mu_\perp = 5.917 \cdot 10^{-12}$. In this case, to obtain the above reduction of greater thermal conductivity value with phonon scattering on the boundaries of a circular contact, there must be k^*=(1.52÷4.37)·105. Thus contact radius $r_\|$ = 31÷89μm. To obtain the same reduction of lower thermal conductivity value, there must be k^*=(1.839÷5.454)·1014, whence $r_\perp$ = 32÷97m. Quite similarly, in the case of phonon scattering on the boundaries of a sphere, to obtain the above reduction of greater thermal conductivity value, there must be k^*=(1.37÷3.90)·105, whence $R_\|$ = 28÷80μm. For the same reduction of lower thermal conductivity value, there must be k^*= (1.66÷4.88)·1014, whence $R_\perp$ = 29÷87m.

From the absurd, on the face of it, results for $r_\perp$ and $R_\perp$ parameters it follows that thermal conductivity anisotropy of macroscopic (e.g. meter long) samples cut from Bi_2Te_3 single crystal must be essentially dependent on their size, which is not the case. So, such an approach needs to be modified. Its main disadvantage introducing an excessive error lies in a forced replacement of real crystal lattice of material by a simple cubic lattice with one atom in the unit cell. However, in this case it is clear that neither $a_\|$ nor $a_\perp$ can serve as cube edges, since $M/a_\|^3$ and $M/a_\perp^3$ quantities yield evidently understated material density values.

In conclusion, we consider an approach based on the substitution of a real Bi_2Te_3 crystal lattice by a simple cubic lattice of the same density. According to this approach, the value of dimensionless parameter k^* for the case of a circular contact should be redefined as:

$$k^*_{\|,\perp} = \frac{r_{\|,\perp}\gamma^2}{\rho}\left(\frac{kT_D}{\hbar v_{\|,\perp}}\right)^4\left(\frac{kT_D}{v^2_{\|,\perp}}\right), \tag{63.11}$$

and for the case of a sphere as:

$$k^*_{\parallel,\perp} = \frac{R_{\parallel,\perp}\gamma^2}{\rho}\left(\frac{kT_D}{\hbar v_{\parallel,\perp}}\right)^4\left(\frac{kT_D}{v^2_{\parallel,\perp}}\right) \tag{63.12}$$

As to formula (63.5), it must be re-written as:

$$\chi_{l\parallel,\perp} = \frac{3\hbar\rho s^4_{\parallel,\perp}}{32\gamma^2 kT_D^2\theta^3\pi}\int_0^1 \frac{x^4\exp(x/\theta)}{[\exp(x/\theta)-1]^2}\left(\frac{1}{Q_{l\parallel,\perp}(x)}+\frac{2}{Q_{t\parallel,\perp}(x)}\right)dx\,, \tag{63.13}$$

Just as before, with regard to formula (63.13), we obtain $\mu_{\parallel} = 0.022$, $\mu_{\perp} = 2.177\cdot 10^{-3}$. According to this approach, for the reduction of either thermal conductivity values by 30-40% in the case of phonon scattering on the circular contact boundaries, $k_{\parallel}^*$ must be 69.6÷167.7 and $k_{\perp}$ – 1008÷2691. Therefore, contact radius must be 0.4÷1.1μm. In the case of phonon scattering on the sphere boundaries, $k_{\parallel}^*$ must be 62.5÷149.1 and $k_{\perp}$ – 908÷2400. Therefore, the sphere radius is 0.35÷1μm. Note that in this case the coefficients $k_{\parallel}^*$ and $k_{\perp}$ are taken at the Debye temperature of material. For arbitrary temperature these coefficients should be multiplied by θ, so that, for instance, at T = 300 K we get contact radius 0.21÷0.57μm, and sphere radius – 0.18÷0.52μm.

Contacts of said dimensions can occur between particles of diameter 40÷80μm at extrusion, which can account for the absence of a considerable decrease in thermoelectric figure of merit in going from a single crystal to extruded material.

Chapter 64

EFFECT OF CHARGE CARRIER SCATTERING ON THE BOUNDARIES ON THE ELECTRIC CONDUCTIVITY OF TEM CONTACTING PARTICLES

Let us now determine the effective electric conductivity of a system of two half-spheres of radius R of TEM contacting in a circle of radius r as a ratio of current through the system to potential difference between the large circles of half-spheres [78]. A physical model of this problem is shown in Figure 158.

At first we consider the problem purely phenomenologically. In the formulation of the problem it is assumed that the surfaces of half-spheres are electrically isolated, their bases (large circle planes) are maintained at given potentials φ_1 and φ_2, and quantum tunnelling of charge carriers to a gap between the spherical surfaces is ignored.

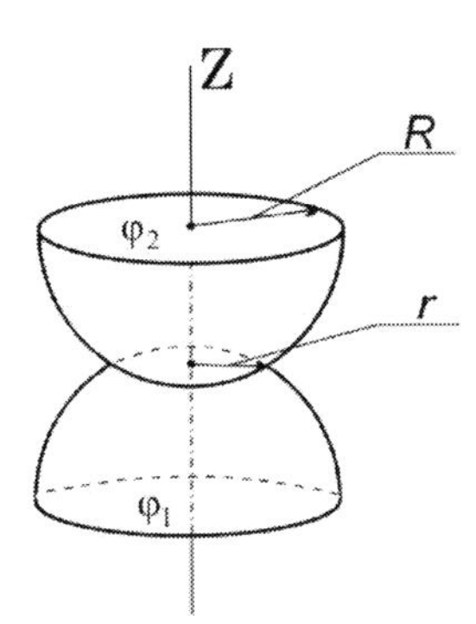

Figure 158. Physical model of the problem.

For the analytical calculation of potential distribution in such a system we will make direct use of Ohm's law. Let us direct the Z axis of coordinate system along the common axis of half-spheres. Then in the area of half-sphere with a larger potential from Ohm's law follows the equation:

$$-\sigma_0 \pi \left(R^2 - z^2\right)\frac{d\varphi}{dz} = I , \qquad (64.1)$$

where σ_0 is known electric conductivity of half-sphere material, φ is potential, I is current through the system to be determined from the boundary conditions. The solution of Eq.(1) is of the form:

$$\varphi = \varphi_1 - \frac{I}{2\sigma_0 \pi R} \ln \frac{R+z}{R-z}. \tag{64.2}$$

Hence we find the contact potential:

$$\varphi_s = \varphi_1 - \frac{I}{2\sigma_0 \pi R} \ln \frac{R+\sqrt{R^2-r^2}}{R-\sqrt{R^2-r^2}}. \tag{64.3}$$

If, however, current coordinate varies within a half-sphere with a smaller potential $\sqrt{R^2-r^2} \le z \le 2\sqrt{R^2-r^2}$, from Ohm's law follows the equation:

$$-\sigma_0 \pi \left(R^2 - \left(2\sqrt{R^2-r^2} - z \right)^2 \right) \frac{d\varphi}{dz} = I. \tag{64.4}$$

Solution of Eq.(64.4) is given by:

$$\varphi = \varphi_1 - \frac{I}{\sigma_0 \pi R} \ln \frac{R+\sqrt{R^2-r^2}}{R-\sqrt{R^2-r^2}} - \frac{I}{2\sigma_0 \pi R} \ln \frac{R-2\sqrt{R^2-r^2}+z}{R+2\sqrt{R^2-r^2}-z}. \tag{64.5}$$

Satisfying condition $\varphi|_{z=2\sqrt{R^2-r^2}} = \varphi_2$, we get the following expression for current through the system:

$$I = \frac{\pi R \sigma_0 (\varphi_1 - \varphi_2)}{\ln\left(\frac{R+\sqrt{R^2-r^2}}{R-\sqrt{R^2-r^2}} \right)}. \tag{64.6}$$

Therefore, the effective electric conductivity of a system of two half-spheres in S is equal to:

$$\sigma_{ef} = \frac{\pi R \sigma_0}{\ln\left(\frac{R+\sqrt{R^2-r^2}}{R-\sqrt{R^2-r^2}} \right)}. \tag{64.7}$$

At $r/R \ll 1$ this formula goes over into:

$$\sigma_{ef} = \frac{\pi R \sigma_0}{\ln\left(4R^2/r^2\right)}. \tag{64.8}$$

The final expression for potential distribution in a half-sphere with a larger potential is as follows:

$$\varphi = \varphi_1 - 0.5(\varphi_1 - \varphi_2)\frac{\ln\dfrac{R+z}{R-z}}{\ln\dfrac{R+\sqrt{R^2-r^2}}{R-\sqrt{R^2-r^2}}}. \tag{64.9}$$

Whereas in a half-sphere with a smaller potential this expression is given by:

$$\varphi = \varphi_2 + 0.5(\varphi_1 - \varphi_2)\frac{\ln\dfrac{R+2\sqrt{R^2-r^2}-z}{R-2\sqrt{R^2-r^2}+z}}{\ln\dfrac{R+\sqrt{R^2-r^2}}{R-\sqrt{R^2-r^2}}}. \tag{64.10}$$

Examples of potential fields in a system of two half-spheres are shown in Figures 159 and 160. The Z axis is directed as in Figure 158, i.e. from the base with a smaller potential to that with a larger potential. For simulation the values φ_1 = 10V, φ_2 = 0V, R = 3 and 4mm, r = 500 and 25μm were taken

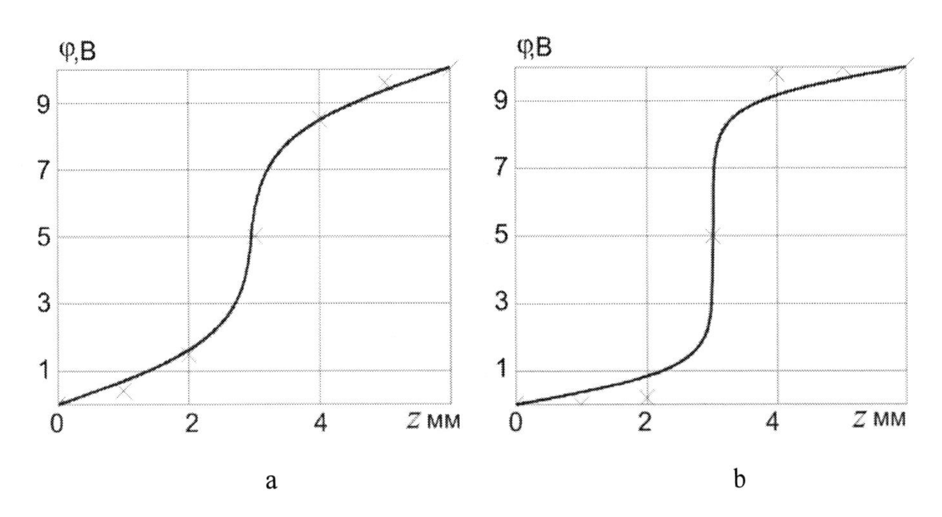

Figure 159. Results of a correlation between the numerical and analytical solutions at R = 3 mm and r = 500μm (a), 25 μm (b).

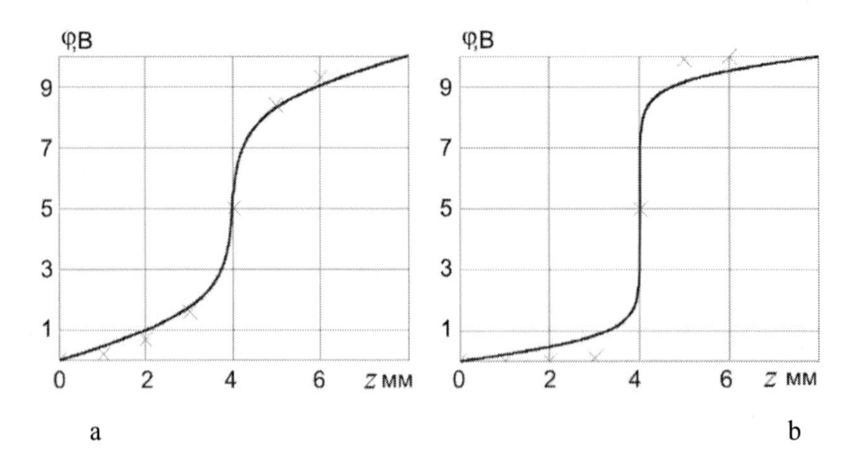

Figure 160. Results of a correlation between the numerical and analytical solutions at R = 4 mm and r = 500 μm (a) and 25 μm (b).

For comparison, crosses on the same plots are used to show the results of a numerical solution of Laplace's equation for a system of two half-spheres with the aid of "Comsol Multiphysics" program. The distinctions are mostly due to the error of difference approximation of differential operators with a numerical solution of second-order differential equations in partial derivatives by method of networks.

Dependence of current through the system (and its effective electric conductivity) on the ratio $b^* = r / R$ with a fixed radius of half-spheres is shown in Figure 161.

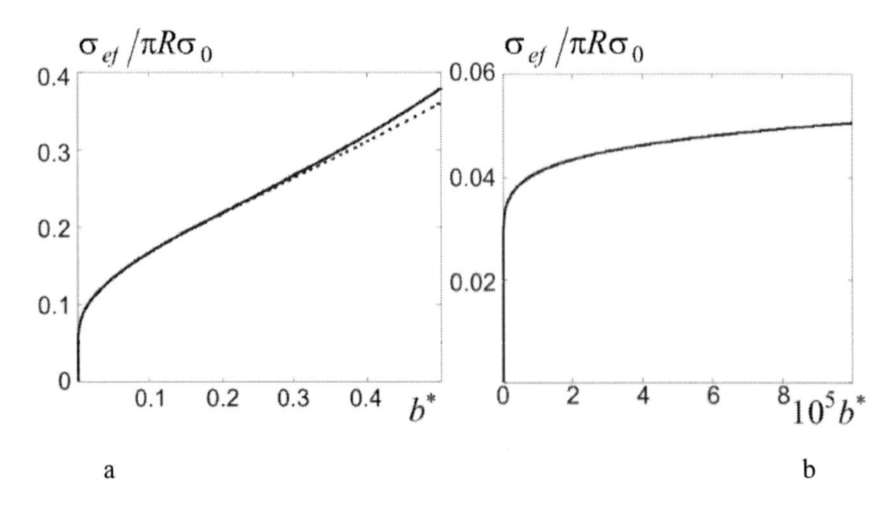

Figure 161. Dependence of effective electric conductivity of the system (and current through it) on the relative contact radius for moderate (a) and particularly small (b) radii. The dashed curve in Figure a is constructed by a simplified formula (64.8).

Thus, it follows from the results of phenomenological consideration of the problem that by virtue of similarity between thermal conductivity and electric conductivity effects, no gain in thermoelectric figure of merit can be obtained due to purely "geometric" factor, hence, it is necessary to consider the microscopic mechanism of system electric conductivity retention with a reduction of its thermal conductivity. In this book we will consider purely drift mechanism, leaving aside quantum tunneling.

In consideration of the foregoing, let us consider a "microscopic" problem of electron scattering on the boundaries of contact between TEM particles in the framework of the model of two half-spheres contacting in the circular spot of radius r described at the beginning of this paragraph. In the bulk material the mean free path l_{cc} of electron or hole depends on energy by the power law $l_{cc}(\varepsilon)=A\varepsilon^q$, where A is a certain coefficient of proportionality, q is power exponent. The values of these quantities are defined by specific scattering mechanism, where q, according to general concepts of quantum mechanics varies from 0 at low to 4 at high energies. Therefore, for the electric conductivity of the bulk sample in the isotropic approximation in the case of a nondegenerate gas of current carriers the following expression is valid:

$$\sigma_0 = D(kT)^{q+1}\int_0^\infty \exp(-x)x^{q+1}dx = D(kT)^{q+1}\Gamma(q+2). \tag{64.11}$$

In this formula, D=$ABC\exp(\zeta/kT)$, B is coefficient of proportionality between the density of states of current carriers and the square root of their energy, C is coefficient of proportionality between the rate of current carriers and the square root of their energy, ζ is chemical potential, T is temperature, $\Gamma(x)$ is gamma-function.

For carrier scattering on the contact spot boundaries the following expression for the resulting mean free path of current carriers is valid:

$$l_{cct}(\varepsilon) = \frac{l_{cc}(\varepsilon)L}{L + l_{cc}(\varepsilon)} \tag{64.12}$$

In this formula, $l_{cc}(\varepsilon)$ is mean free path of current carrier (electron or hole) in material determined by all scattering mechanisms, except for the contact spot boundaries, L is effective mean free path of current carrier determined by the sample boundaries. Now we introduce the mean free path of current carrier, for instance, electron, by the formula:

$$l_e = \frac{\int_0^\infty l_{cc}(\varepsilon)f_0(\varepsilon)g(\varepsilon)d\varepsilon}{\int_0^\infty f_0(\varepsilon)g(\varepsilon)d\varepsilon}. \tag{64.13}$$

In this formula, $f_0(\varepsilon)$ is the Maxwell—Boltzmann distribution function, $g(\varepsilon)$ is the electron density of states. From (12) follows the following relation for A:

$$A = l_e \frac{\Gamma(1.5)}{(kT)^q\Gamma(q+1.5)}. \tag{64.14}$$

In the case of a circular contact which is small as compared to half-sphere diameters, it can be considered that a drag of current carriers, for instance, electrons, takes place only in the area of this contact. Moreover, all points of contact boundary are equivalent due to its symmetry. Therefore, a general formula for electric conductivity [64] with regard to (64.12) and (64.14) yields the following expression for the ratio between the electric conductivity of a system of half-spheres and the electric conductivity of the bulk sample:

$$\frac{\sigma_{bs}}{\sigma_0} = \frac{1}{\pi\Gamma(q+2)} \int_0^{\infty}\int_0^{1}\int_0^{2\pi} \frac{k^* \sqrt{y^2+1-2y\cos\phi}\, yx^{q+1}\exp(-x)}{x^q + k^*\sqrt{y^2+1-2y\cos\phi}} d\phi dy dx \,. \tag{64.15}$$

In this formula, σ_{bs} is the electric conductivity of the system, $k^*=[\Gamma(q+1.5)/\Gamma(1.5)](r/l_e)$. As it must be, at $k^*=0$ formula (64.15) gives zero, and at $k^*\to\infty$ – the electric conductivity of the bulk sample. The results of this calculation are depicted in Figure 162.

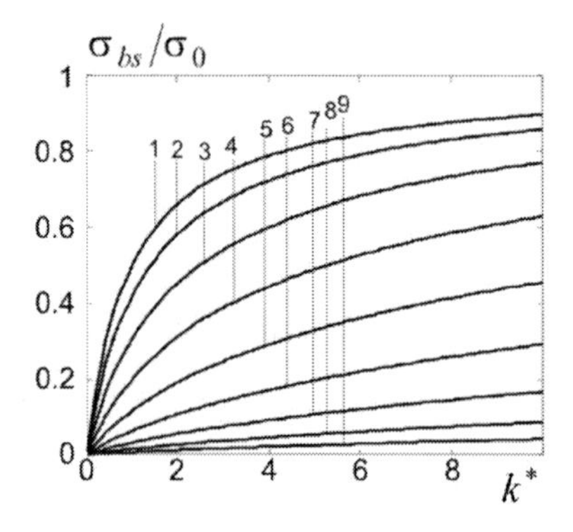

Figure 162. Dependence of the electric conductivity of a system of two half-spheres contacting in a circular spot on its radius. Curves 1 – 9 are constructed for the values of q from 0 to 4 with a step of 0.5 .

Dependence of r/l_e ratio on q following from considerations of retention of at least 90% of the electric conductivity of the bulk sample is depicted in Figure 163.

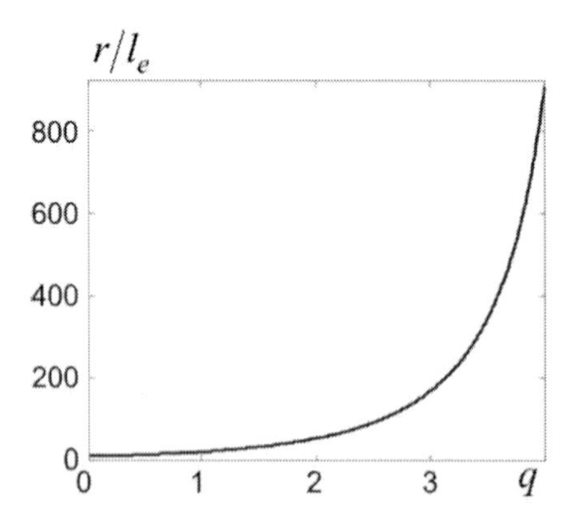

Figure 163. Dependence of r/l_e ratio on q following from considerations of retention of at least 90% of the electric conductivity of the bulk sample .

From this figure it is evident that with increase in q after the value equal to 2, r/l_e ratio increases rather sharply. However, in semiconductors the most frequent q values are equal to 0, which corresponds to approximation of constant mean free path, or 0.5, which is equivalent of constant relaxation time approximation. In the temperature range of 300K and above which is relevant for thermoelectricity it can be considered that q=0. In this case the mean free path of charge carriers, i.e. electrons (holes) l_{cc} at temperature T is expressed through their mobility b_{cc} and density-of-state effective mass m_g^* as follows:

$$l_{cc} = \sqrt{\frac{\pi}{2}} \frac{b_{cc}}{e} \sqrt{m_g^* kT} . \tag{64.16}$$

Therefore, the estimate of mean free paths at this temperature, based on the mobilities and density-of-state effective masses of electrons and holes в Bi_2Te_3 [64], yields l_e=38.7nm, l_h=20.4nm. Minimum contact radius at q=0, necessary for the retention of 90% of electric conductivity is 10.4 of the mean free path of electron or hole. And this corresponds (with a larger path) to 0.4μm.

Chapter 65

EFFECT OF TEM ANISOTROPY ON THE ELECTRIC CONDUCTIVITY OF ITS CONTACTING PARTICLES

As mentioned above, the specific feature of Bi_2Te_3 is well-expressed electric conductivity and thermal conductivity anisotropy. Taking into account that this crystal possesses *R3m* group symmetry and cleavage planes along which it easily splits, its thermal conductivity and electric conductivity tensors have two independent components each. In particular, in the absence of a magnetic field, electric conductivity tensor has component σ_{11} in cleavage planes and component σ_{33} in a direction normal to them. The ratio σ_{11}/σ_{33} is 2.7 for *p*-type material and $4 \div 6$ for *n*-type material. Bi_2Te_3 is intermediate in the electric conductivity value between high-resistance semiconductors traditionally used in radio electronics and computer technique, such as germanium and silicon, and semimetals, such as bismuth. The band spectrum of this crystal is anisotropic and described by different models. However, the most approved and frequently used is a six-ellipsoid Drabble-Wolfe model [10].

Due to conductivity anisotropy, thermoelectric modules of solid single crystals Bi_2Te_3 are made so that temperature gradient and electric current are parallel to cleavage planes, where conductivity value is higher than in the direction perpendicular to them. Alongside with single crystals, extruded materials that may consist of particles with oriented or random cleavage planes are used for the manufacture of thermoelectric modules. With a random arrangement of cleavage planes, the electric conductivity of material in conformity with the Odelevsky formula will make $\sigma = \sqrt{\sigma_{11}\sigma_{33}}$, i.e. will be lower than the largest value. Further electric conductivity reduction can be due to current carrier scattering on the boundaries of small contacts between particles. These factors should have resulted in thermoelectric figure of merit reduction. However, in real practice this reduction is not observed. Hence, a mechanism must exist which assures electric conductivity retention and lattice thermal conductivity reduction at charge carrier and phonon scattering on the boundaries of contacts between the particles. Without a detailed account of electric conductivity anisotropy, but with regard to lattice thermal conductivity anisotropy this mechanism was considered in paragraphs 63 and 64. In the present paragraph this mechanism is considered with regard to real anisotropy of charge carrier band spectrum and the electric conductivity of Bi_2Te_3.

Consider this problem of anisotropic scattering of electrons (holes) on the boundaries of contact between TEM particles within the framework of earlier described model of two half-spheres of radius R ($r << R$) contacting in a circle of radius r. This model can approximate the shape-forming element of extruded thermoelectric material structure [75]. For this purpose, using the results given in [64], we first write general formulae for single crystal electric conductivity components σ_{11} and σ_{33}. To accomplish this, we predetermine relaxation time tensor components in the approximation of constant mean free paths in the directions of principal axes of ellipsoids by means of full energy of current carriers. These components are:

$$\tau_{1,2,3} = \frac{l_{1,2,3}\sqrt{m^*}}{\sqrt{2\varepsilon}}. \tag{65.1}$$

In this formula, l_1, l_2, l_3 –charge carrier mean free paths in the respective directions, m^* – density-of-state effective mass, ε – full energy of charge carriers. Such an approach corresponds to "nearly isotropic scattering" whose anisotropy is taken into account by means of different lengths l_1, l_2, l_3. Substitution of density-of-state effective mass into formula (65.1) unambiguously follows from the model assumption of relaxation time tensor components dependence on full energy of charge carriers.

With this relaxation time tensor, according to [64], components of electric conductivity tensor for nondegenerate charge carrier gas are equal to:

$$\sigma_{11} = \frac{4e^2 n_0\sqrt{m_1 m_2 m_3}}{m_2\sqrt{\pi m^*}(kT)^{1/2}}\left(l_2 + \frac{m_2}{m_1}l_1\cos^2\vartheta + \frac{m_2}{m_3}l_3\sin^2\vartheta\right). \tag{65.2}$$

$$\sigma_{33} = \frac{8e^2 n_0\sqrt{m_1 m_2 m_3}}{m_2\sqrt{\pi m^*}(kT)^{1/2}}\left(\frac{m_2}{m_1}l_1\sin^2\vartheta + \frac{m_2}{m_3}l_3\cos^2\vartheta\right). \tag{65.3}$$

In these formulae, m_1, m_2, m_3 - effective masses of charge carriers along the principal axes of ellipsoid, ϑ – the smallest rotation angle of ellipsoid up to coincidence of its long axis with crystal trigonal axis, k – the Boltzmann constant, T – absolute temperature, n_0 – charge carrier concentration, the rest of designations are generally accepted or explained above.

In the temperature region relevant for thermoelectric applications scattering mainly occurs on the deformation potential of acoustic phonons. In this region, $l_1, l_2, l_3 \propto T^{-1}$, so finally we get the known dependence $\sigma \propto T^{-3/2}$, which in reality is somewhat distorted by the temperature dependence of respective effective masses.

Thus, formulae (65.2) and (65.3) fully determine electric conductivity tensor of Bi_2Te_3 single crystal in the absence of a magnetic field. Account in these formulae of scattering on contact boundaries presents no special problems. However, of crystal parameters in these formulae only all the effective masses and angle ϑ are known with certainty, for they are parameters of band structure which is reliably studied through measurement of de Haas-van-Alphen and de Haas-Shubnikov effects. As to mean free paths l_1, l_2, l_3, they depend on acoustic phonon deformation potential tensor components. The band structure determines only the bulk component of this tensor [36], whereas in crystal with well-expressed cleavage planes the shear and bend components are also essential. So, it is worthwhile to write formulae (65.2) and (65.3) in such a form in which unknown parameters could be determined, for instance, from the data on electron and hole mobility.

Passing from the electric conductivity tensor to charge carrier mobility tensor, we write its components as follows:

$$b_{11,33} = \frac{eL_{11,33}\sqrt{2}}{\sqrt{\pi m^* kT}} . \quad (65.4)$$

In these formulae, in conformity with (65.2) and (65.3), charge carrier mean free paths determined by known mobility values are equal to:

$$L_{11} = \frac{4\sqrt{m_1 m_3}}{\sqrt{2m^* m_2}}\left(l_2 + \frac{m_2}{m_1}l_1\cos^2\vartheta + \frac{m_2}{m_3}l_3\sin^2\vartheta\right). \quad (65.5)$$

$$L_{33} = \frac{8\sqrt{m_1 m_3}}{\sqrt{2m^* m_2}}\left(\frac{m_2}{m_1}l_1\sin^2\vartheta + \frac{m_2}{m_3}l_3\cos^2\vartheta\right). \quad (65.6)$$

Let us now discuss charge carrier scattering on contact boundaries [79]. Using the summation rule of inverse mean free paths, we find the ratio between mobilities $\tilde{b}_{11}$ and $\tilde{b}_{33}$ determined with regard to scattering on contact boundaries and mobilities determined by formula (65.4):

$$\tilde{b}_{11,33}\Big/ b_{11,33} = \frac{1}{\pi}\int_0^1\int_0^{2\pi}\frac{k_{11,33}\sqrt{z^2+1+2z\cos\phi}}{1+k_{11,33}\sqrt{z^2+1+2z\cos\phi}}zd\phi dz . \quad (65.7)$$

In these formulae, $k_{11} = r/L_{11}$, $k_{33} = r/L_{33}$. Double integrals in them are due to averaging the expression for mobility over phonon mean free paths inside a circle where half-spheres are contacting. From formula (65.7) it follows that to maintain the mobilities in the form-shaping structural element at a level of 90 % of their values in a single crystal, coefficients k_{11} and k_{33} must be at least 10.4. At $T = 300$ K for electrons substituting to (6) $b_{11} = 1200$ cm^2/C·s, $b_{11}/b_{33} = 5$, $m^* = 0.45\ m_0$ [64], we obtain $L_{11} = 38.7$ nm, $L_{33} = 7.7$ nm, whence $r = 400$ nm. Similarly for holes, substituting $b_{11} = 510$ cm^2/V·s, $b_{11}/b_{33} = 2.7$, $m^* = 0.69\ m_0$ we obtain $L_{11} = 20.4$ nm, $L_{33} = 7.6$ nm, whence $r = 212$ nm. So, finally $r = 400$ nm. Contacts of such dimensions can appear between particles of diameter 40 ÷ 80 μm.

Chapter 66

The Mechanism of Thermoelectric Figure of Merit Increase in the Bulk Nanostructured TEM

Consider now the matter of possible figure of merit increase in the bulk nanostructured TEM due to the use of phonon and charge carrier scattering on the boundaries of nanoparticles of which the bulk nanostructured TEM consists. Note first of all that in the approximation of constant mean free path the Seebeck coefficient of the bulk nanostructured TEM must be the same as in a single crystal. This occurs because the relaxation time of charge carriers, hence their effective mean free path length, corrected by scattering on the boundaries, enters the expressions for thermal diffusion flow and electric current similarly. The ratio of electric conductivity of the bulk nanostructured TEM consisting of identical spherical nanoparticles of radius r to electric conductivity of a single crystal in the approximation of constant mean free path is equal to:

$$\sigma_n/\sigma_m = \int_0^1\int_{-1}^1 \frac{(r/l_e)\sqrt{y^2+2zy+1}\,y^2dzdy}{(r/l_e)\sqrt{y^2+2zy+1}+1}, \tag{66.1}$$

With regard to expression (63.4) for the respective thermal conductivity ratio, we obtain the following expression for thermoelectric figure of merit of the bulk nanostructured TEM in the approximation of constant mean free paths of charge carriers and phonons:

$$Z_n\Big/Z_m = \left[\int_0^1\int_{-1}^1 \frac{(r/l_e)\sqrt{y^2+2zy+1}\,y^2dzdy}{(r/l_e)\sqrt{y^2+2zy+1}+1}\right]\left[\int_0^1\int_{-1}^1 \frac{(r/l_p)\sqrt{y^2+2zy+1}\,y^2dzdy}{(r/l_p)\sqrt{y^2+2zy+1}+1}\right]^{-1}. \tag{66.2}$$

However, taking into account the dependence of phonon mean free path on their frequency yields the following expression for the respective figure of merit ratio:

$$Z_n \Big/ Z_m = 1.5\left[\int_0^1\int_{-1}^1 \frac{(r/l_e)\sqrt{y^2+2zy+1}\,y^2 dzdy}{(r/l_e)\sqrt{y^2+2zy+1}+1}\right]\left[\chi_l^{(n)} \big/ \chi_{\|lm}\right]^{-1}. \quad (66.3)$$

Here

$$\chi_l^{(n)} \Big/ \chi_{l\|m} = 1.5\int_0^1\int_0^1\int_{-1}^1 \frac{z^2 x^4 \exp(x/\theta)}{[\exp(x/\theta)-1]^2}\left(\frac{(r/L^*)\sqrt{z^2-2zy+1}}{1+(r/L^*)Q_{l\|}(x)\sqrt{z^2-2zy+1}}+\right.$$
$$\left.\frac{2(r/L^*)\sqrt{z^2-2zy+1}}{1+(r/L^*)Q_{t\|}(x)\sqrt{z^2-2zy+1}}\right)dydzdx\left\{\int_0^1 \frac{x^4\exp(x/\theta)}{[\exp(x/\theta)-1]^2}\left(\frac{1}{Q_{l\|}(x)}+\frac{2}{Q_{t\|}(x)}\right)dx\right\}^{-1}. \quad (66.4)$$

Assuming that $L^* = \rho\hbar^4 s_\|^6 / \gamma^2 (k_B T_D)^5$ is weakly dependent on temperature and is the same as at the Debye temperature of TEM, the dependence of dimensionless thermoelectric figure of merit of the bulk nanostructured TEM on nanoparticle radius is of the form shown in Figure 164

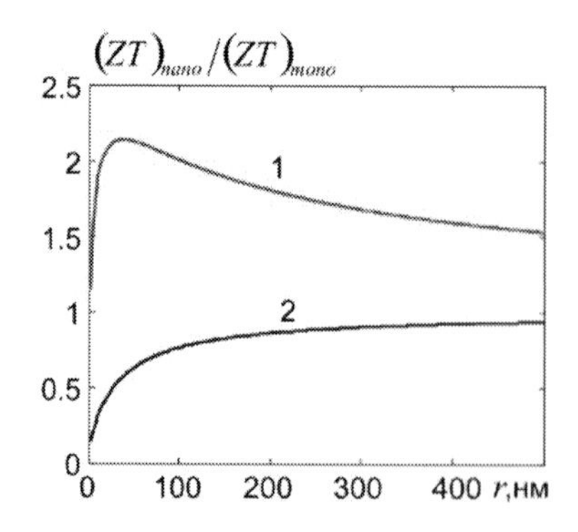

Figure 164. Dependence of the relative thermoelectric figure of merit of the bulk nanostructured TEM on nanoparticle radius: 1) – with regard to frequency dependence of phonon relaxation time; 2) – in the approximation of constant electron and phonon mean free paths.

When constructing curve 2, in this figure, based on the reference data [64] it was assumed that l_{ph}=4.16nm, l_e=38.6nm at a temperature of 300K. In connection with this figure a question may arise as to the validity of transferring the properties of individual nanoparticle to the properties of material as a whole. Therefore, we note that if pores in material structure are vacuum, tunneling of charge carriers is absent and the pores are not communicating, then material porosity, both in the framework of percolation theory, and within the approach set forth, for instance, in [77], enters the expressions for thermal conductivity and electric conductivity through the same multiplier, hence, it does not produce a direct effect on the thermoelectric figure of merit. Thus, abstracting from size distribution of nanoparticles, the thermoelectric figure of merit of material as a whole is the same as the thermoelectric figure of merit of an individual particle.

From the figure it is seen that with regard to frequency dependence of phonon relaxation time the relative thermoelectric figure of merit has a maximum 2.14 which is achieved in the range of nanoparticle radii 35÷40 nm. However, it is possible only with oriented pressing. Whereas with a random orientation of cleavage planes the relative thermoelectric figure of merit of the bulk nanostructured material based on Bi_2Te_3 will be a factor of $\sqrt{3}$ lower, i.e. it will remain on the level approximately 23% higher than the thermoelectric figure of merit of a single crystal. Even with nanoparticle radius of the order of 5nm with correction for random orientation of cleavage planes, the thermoelectric figure of merit of the bulk nanostructured material should remain on the level of at least 97% of the thermoelectric figure of merit of a single crystal. These results are contradict to the results of work [80], according to which the thermoelectric figure of merit of the bulk nanostructured material of nanoparticles with the radius 5÷20nm based on Bi_2Te_3 at 300K is as low as about 82% of the thermoelectric figure of merit of a single crystal. Thus, when passing from a single crystal to the bulk nanostructured material, power factor is scarcely ever retained, which permits calling in question the presence of the energy filtration of charge carriers that should have resulted in Seebeck coefficient increase. If, however, such filtration exists, then, apparently, it does not always contribute to power factor retention, since the electric conductivity is reduced more than Seebeck coefficient square is increased. Whereas in the approximation of constant electron and phonon mean free paths, with increase in nanoparticle radius, the thermoelectric figure of merit monotonously increases from a low value to 1. Therefore, in such approximation the thermoelectric figure of merit values of the bulk nanostructured materials exceeding unity are mainly attributable to tunnelling effects.

Chapter 67

METHOD FOR KINETIC COEFFICIENTS AVERAGING OVER THE SIZE OF PARTICLES AND ITS IMPACT ON THE PREDICTED FIGURE OF MERIT OF THE BULK NANOSTRUCTURED TEM

Methods for determination of the effective kinetic coefficients of thermoelectric material produced by hot pressing, extrusion or spark plasma sintering techniques via the kinetic coefficients of its component particles or, in other words, its shape-forming elements, have been considered in a variety of theoretical papers [67, 81-83]. These methods can be divided into two big classes, namely the methods based on solving phenomenological equations of electric and thermal conductivity for shape-forming elements and the methods based on constructing from shape-forming elements of the so-called "effective thermoelectric medium". The former in the strict sense of the word are applicable only when the characteristic sizes of shape-forming elements of material structure are much in excess of the characteristic mean free paths of charge carriers and phonons, and, hence, their application to nanostructured thermoelectric materials based on Bi_2Te_3 provokes certain objections from the author of this book. These objections are as follows. First, the mean free path, for instance, of electrons in Bi_2Te_3 at 300 K is of the order of 36 nm [5], hence, phenomenological equations for electric potential distribution are hardly applicable to particles of comparable or, the more so, smaller dimensions. Second, the lattice conductivity of Bi_2Te_3 at 300 K and higher is mainly caused by phonon Umklapp processes [64,77]. If such processes are available, the relaxation time of phonons with their mutual scattering is a function of frequency ω, and, hence, introduction of phonon mean free path in the usual sense of this concept is hardly possible, though some formal parameter having dimensions of length and depending on scattering mechanisms, namely $l_p(\omega) = \nu\tau(\omega)$, where ν – average sound velocity in material, $\tau(\omega)$ – phonon relaxation time, can be introduced. The disadvantage of this method is also the fact that the analytical solution of phenomenological equations for arbitrary-shaped particles, the more so with account of thermoelectric effects, is difficult or impossible, and one has to resort to somewhat artificial simulation of this shape. On the other hand, the effective medium method also requires certain modification with regard to thermoelectric phenomena. Some aspects of this problem have been considered in [83].

Taking into account the foregoing, our purpose in this paper is to consider and compare a number of methods for determination of the effective kinetic coefficients of nanostructured thermoelectric material with regard to size distribution of its particles. The methods in hand do not call for solution of phenomenological equations of thermal conductivity and thermal conductivity for an individual nanoparticle.

We first determine the electric conductivity of an individual nanoparticle as a function of its size. To simplify the calculations, the nanoparticle will be considered to be spherical. Moreover, we will take into account that the electric conductivity of Bi_2Te_3 at temperatures 300 K and higher is determined by acoustic phonon deformation potential scattering of charge carriers. And for this case the approximation of constant mean free path is valid. Therefore, the ratio between the electric conductivity $\sigma_n(r)$ of a nanoparticle of radius r and the electric conductivity σ_0 of a single crystal will be determined by formula (66.1).

A change in thermoelectric figure of merit in this case is completely governed by a change in the electric conductivity - thermal conductivity ratio.

We now turn to determination of lattice thermal conductivity of a nanoparticle with regard to both phonon Umklapp processes and normal processes. The ratio between the thermal conductivity $\chi_n(r)$ of a nanoparticle of radius r and the thermal conductivity χ_{lm} of a single crystal, provided its cleavage planes are oriented parallel to temperature gradient in conformity with the results of [76] will be determined by formula (66.4).

In the determination of the effective kinetic coefficients of nanostructured thermoelectric material as a whole, for simplicity of calculations we will ignore the influence of material pores on these coefficients, i.e. we will not consider the effects of tunnelling or charge carrier emission, as well as convective and radiation mechanisms of energy transfer and the effects related to intercommunication of pores. Then, with account of particle size distribution function $w(r)$, these coefficients can be determined by the four different methods.

The first method consists in using the Odelevsky formula [82] with regard to the volume share of various-size particles. In this case the effective electric conductivity σ_{ef} and thermal conductivity κ_{ef} of material as a whole are determined as solutions of the following equations:

$$\int_0^\infty \frac{\sigma_{ef} - \sigma_n(r)}{2\sigma_{ef} + \sigma_n(r)} r^3 w(r) dr = 0, \tag{67.1}$$

$$\int_0^\infty \frac{\chi_{ef} - \chi_n(r)}{2\chi_{ef} + \chi_n(r)} r^3 w(r) dr = 0. \tag{67.2}$$

The second method consists in using the Odelevsky formula with regard to the fraction of various-size particles. In this case equations (67.1) and (67.2) acquire the form:

$$\int_0^\infty \frac{\sigma_{ef} - \sigma_n(r)}{2\sigma_{ef} + \sigma_n(r)} w(r) dr = 0, \tag{67.3}$$

$$\int_0^\infty \frac{\chi_{ef}-\chi_n(r)}{2\chi_{ef}+\chi_n(r)}w(r)dr=0\,. \tag{67.4}$$

The third method consists in averaging the electric conductivity and lattice thermal conductivity over the volume share of various-size particles. In this case the effective kinetic coefficients of nanostructured material are determined directly:

$$\begin{pmatrix}\sigma_{ef}\\ \chi_{ef}\end{pmatrix}=\int_0^\infty \begin{pmatrix}\sigma_n(r)\\ \chi_n(r)\end{pmatrix} r^3 w(r)dr\,. \tag{67.5}$$

The fourth method consists in averaging the electric conductivity and lattice thermal conductivity over the fraction of various-size particles. In this case the ratio (67.5) acquires the form:

$$\begin{pmatrix}\sigma_{ef}\\ \chi_{ef}\end{pmatrix}=\int_0^\infty \begin{pmatrix}\sigma_n(r)\\ \chi_n(r)\end{pmatrix} w(r)dr. \tag{67.6}$$

In Eq. (67.1) - (67.4) normalization factors are omitted because they have no effect on the results of calculation of the effective kinetic coefficients, and in formulae (67.5) and (67.6) – because they have no effect on the dimensionless thermoelectric figure of merit of nanostructured material with respect to a single crystal whose calculation in the framework of the outlined methods is the purpose of this paper.

For concrete calculations we take particle size distribution function $w(r)$ as follows:

$$w(r)=\frac{r}{r_0^2}\exp\left(-r^2/2r_0^2\right). \tag{67.7}$$

This particle size distribution function is called the Rayleigh distribution and has been assumed because it is the simplest single-parameter distribution. Parameter r_0 is the most probable particle radius. This function, as it must be, satisfies normalization condition $\int_0^\infty w(r)\,\mathrm{d}r=1$.

With regard to (67.7), relations (67.1) – (67.6) acquire the form as follows:

$$\int_0^\infty \frac{\sigma_{ef}-\sigma_n\left(r_0\sqrt{2t}\right)}{2\sigma_{ef}+\sigma_n\left(r_0\sqrt{2t}\right)}\sqrt{t^3}\exp(-t)dt=0\,, \tag{67.8}$$

$$\int_0^\infty \frac{\kappa_{ef}-\kappa_n\left(r_0\sqrt{2t}\right)}{2\kappa_{ef}+\kappa_n\left(r_0\sqrt{2t}\right)}\sqrt{t^3}\exp(-t)dt=0\,, \tag{67.9}$$

$$\int_0^\infty \frac{\sigma_{ef} - \sigma_n\left(r_0\sqrt{2t}\right)}{2\sigma_{ef} + \sigma_n\left(r_0\sqrt{2t}\right)} \exp(-t)dt = 0, \quad (67.10)$$

$$\int_0^\infty \frac{\kappa_{ef} - \kappa_n\left(r_0\sqrt{2t}\right)}{2\kappa_{ef} + \kappa_n\left(r_0\sqrt{2t}\right)} \exp(-t)dt = 0, \quad (67.12)$$

$$\begin{pmatrix} \sigma_{ef} \\ \kappa_{ef} \end{pmatrix} = \int_0^\infty \begin{pmatrix} \sigma_n\left(r_0\sqrt{2t}\right) \\ \kappa_n\left(r_0\sqrt{2t}\right) \end{pmatrix} \sqrt{t^3} \exp(-t)dt, \quad (67.13)$$

$$\begin{pmatrix} \sigma_{ef} \\ \kappa_{ef} \end{pmatrix} = \int_0^\infty \begin{pmatrix} \sigma_n\left(r_0\sqrt{2t}\right) \\ \kappa_n\left(r_0\sqrt{2t}\right) \end{pmatrix} \exp(-t)dt. \quad (67.14)$$

The results of calculation of the dimensionless thermoelectric figure of merit of the bulk nanostructured material based on Bi_2Te_3 with respect to a single crystal with different methods of determination of the effective kinetic coefficients are shown in Figure 165

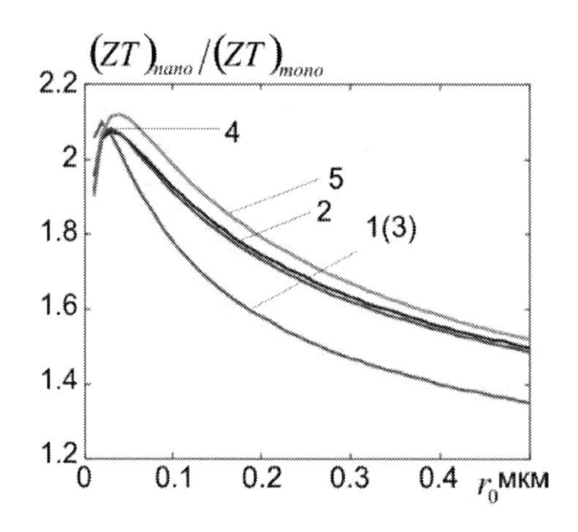

Figure 165. Dependence of the dimensionless thermoelectric figure of merit of the bulk nanostructured thermoelectric material based on Bi_2Te_3 on the most probable particle radius with different methods of determination of the effective kinetic coefficients. Curve numbers 1 – 4 correspond to conventional serial numbers of averaging methods described in the text. Curve 5 corresponds to the case of zero particle size dispersion, where there is no need in averaging.

From the figure it is seen that curves 1 and 3 coincide to a high degree of accuracy. This would imply that in determining the effective kinetic coefficients of thermoelectric material via the kinetic coefficients of shape-forming elements, instead of the Odelevsky formulae on condition of α = const one can employ the usual averaging over the volume share of particles with regard to size distribution function. In so doing, the maximum dimensionless thermoelectric figure of merit of the bulk nanostructured thermoelectric material, on condition that particle cleavage planes are oriented parallel to electric current and thermal flux, is a factor of 2.06 ÷ 2.1 larger than the dimensionless thermoelectric figure of merit of a single crystal and is achieved with the most probable particle radius 0.02 μm. Curves 2 and 4 are also close to each other which means that in determining the effective kinetic coefficients of thermoelectric material by the fraction of particles with regard to size distribution function, instead of the Odelevsky formulae with a reasonable degree of accuracy one can also use the usual averaging of the kinetic coefficients by the fraction of various-size particles. With such method of averaging, the maximum dimensionless thermoelectric figure of merit of the bulk nanostructured thermoelectric material is a factor of 2.076 ÷ 2.082 greater than the dimensionless thermoelectric figure of merit of a single crystal and is achieved with the most probable nanoparticle radius 0.02 ÷ 0.03 μm. Curve 5 has been constructed on the assumption of equal size of all nanoparticles. In this case the maximum dimensionless thermoelectric figure of merit of the bulk nanostructured thermoelectric material is a factor of 2.12 greater than the thermoelectric figure of merit of a single crystal and is achieved with nanoparticle radius 0.03 μm. Hence it follows that the method of averaging in determining the effective kinetic coefficients of the bulk nanostructured thermoelectric material scarcely affects the predicted value of its dimensionless thermoelectric figure of merit, but markedly affects the estimation of the optimal value of the most probable particle radius. If equivalent-results methods for averaging over the volume share of particles with regard to size distribution function are assumed to be most correct, then the optimal value of the most probable particle radius of the bulk nanostructured thermoelectric material is 0.02 μm.

Chapter 68

ON THE POSSIBILITY OF USING TEM POWDERS OF VARIABLE GRANULOMETRIC COMPOSITION FOR THE FABRICATION OF THERMOELECTRIC MODULES

It is known that with a view to increase the thermoelectric figure of merit and the efficiency of thermoelectric modules based on single crystals, they are doped with various impurities, for instance, in such a way that each portion of module leg has maximum possible figure of merit at given portion temperature. However, using hot pressed, extruded or SPS materials based on powders raises another possibility of control over the figure of merit and efficiency of thermoelectric modules [83]. It lies in the fact that by virtue of the temperature dependence of charge carrier and phonon mean free paths, the nanoparticle radius, optimal in terms of thermoelectric figure of merit and efficiency, must be temperature dependent. Therefore, powders of variable granulometric composition can be used for the fabrication of thermoelectric modules. In so doing, on each module portion the size of TEM-based powder should be such as to assure maximum figure of merit at the operating portion temperature. To calculate the temperature dependence of nanoparticle optimal size and it respective thermoelectric figure of merit of the bulk nanostructured TEM, we will take into account the temperature dependences of parameters l_e and L^* in conformity with the formulae:

$$l_e(T) = 300 l_e(300)/T \, , \tag{68.1}$$

$$L^*(T) = L^*(T_D)/\theta \, . \tag{68.2}$$

The respective dependences for *n*-type Bi_2Te_3, calculated with regard to formulae (68.1) and (68.2) on the assumption that particle size distribution function is the Rayleigh distribution are shown in Figure 166.

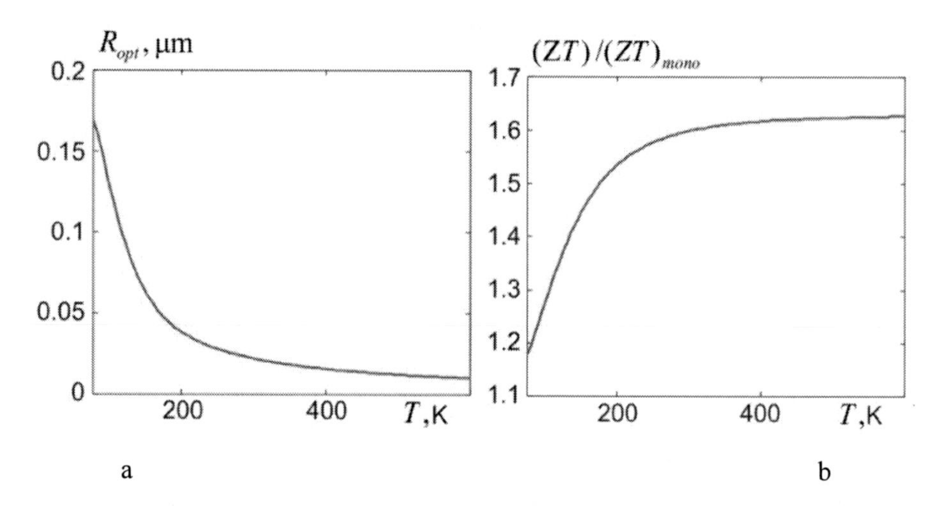

Figure 166. Temperature dependences: a) of optimal most probable radius of a spherical particle of Bi_2Te_3 powder; b) of thermoelectric figure of merit of material with respect to a single crystal.

With temperature variation from 75 to 600K the optimal most probable radius of a spherical powder particle drops from 0.168 to 0.010μm, and maximum thermoelectric figure of merit with respect to a single crystal increases from 1.19 to 1.63.

Chapter 69

ON THE IMPORTANCE OF A KNOWLEDGE OF A REAL FREQUENCY DEPENDENCE OF THE DENSITY OF PHONON STATES FOR PREDICTION OF THE THERMOELECTRIC FIGURE OF MERIT OF TEM

Earlier we have discussed phonon scattering on the boundaries of a shape-forming element of TEM structure in the framework of the Debye model for the density of phonon states. However, it is fairly well known that this model is applicable only in the case of a linear law of phonon dispersion and a spherical shape of constant frequency surface. In this case the density of phonon states is a quadratic function of phonon frequency. In all other cases the density of phonon states depends on frequency in a more complicated fashion.

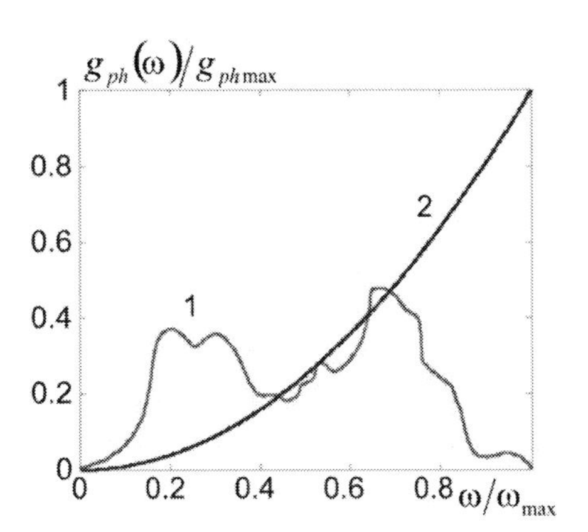

Figure 167. Frequency dependence of the density of phonon states for Bi_2Te_3: 1 – according to [85]; 2 – according to the Debye model. The density of states in both cases is normalized for its maximum in conformity with the Debye model.

Consider now the effect of the shape of the density of phonon states on the lattice thermal conductivity of a shape-forming structure element, hence, on the thermoelectric figure of merit of TEM, in the framework of two approaches. Within the first approach we will take into account the effect of the shape of the density of phonon states only on the lattice heat capacity of material, leaving aside its effect on phonon scattering on phonons and on the boundaries of a shape-forming element. Within the second approach we will consider the effect of the shape of the density of phonon states both on the lattice heat capacity of material and on phonon scattering on phonons and on the boundaries of a shape-forming element. We will consider this problem for the frequency dependence of the experimental density of phonon states for Bi_2Te_3 shown in Figure 167 and presented in [85].

If we take into account the effect of a real frequency dependence of the density of phonon states only on the lattice heat capacity, then the thermal conductivity of Bi_2Te_3 single crystal is defined by the formula:

$$\chi_{l\|,\perp}^{(mono)} = \frac{3\hbar\rho s_{\|,\perp}^4 k}{32\gamma^2 (kT_D)^2 \theta^3 \pi} \int_0^1 \frac{f(x) x^2 \exp(x/\theta)}{[\exp(x/\theta)-1]^2} \left(\frac{1}{Q_{l\|,\perp}(x)} + \frac{2}{Q_{t\|,\perp}(x)} \right) dx . \quad (69.1)$$

In this formula, $f(x)$ is functional dependence reflected by curve 1 in Figure 167, $x = \omega/\omega_{\max}$. In so doing, the Debye temperature T_D must be determined from the ratio $kT_D = \hbar\omega_{\max}$. Therefore, considering a shape-forming element as two half-spheres contacting in a circle of a small radius, we get the following relation for the lattice thermal conductivity of TEM with respect to a single crystal:

$$\chi_{l\|,\perp} \Big/ \chi_{l\|,\perp}^{(mono)} = \pi^{-1} \int_0^1 \int_0^1 \int_0^{2\pi} \frac{z x^2 f(x) \exp(x/\theta)}{[\exp(x/\theta)-1]^2} \left(\frac{k_{\|,\perp}^* \sqrt{z^2 - 2z\cos\phi + 1}}{1 + k_{\|,\perp}^* Q_{l\|,\perp}(x) \sqrt{z^2 - 2z\cos\phi + 1}} + \right.$$
$$\left. \frac{2k_{\|,\perp}^* \sqrt{z^2 - 2z\cos\phi + 1}}{1 + k_{\|,\perp}^* Q_{t\|,\perp}(x) \sqrt{z^2 - 2z\cos\phi + 1}} \right) d\phi dz dx \left\{ \int_0^1 \frac{x^4 \exp(x/\theta)}{[\exp(x/\theta)-1]^2} \left(\frac{1}{Q_{l\|,\perp}(x)} + \frac{2}{Q_{t\|,\perp}(x)} \right) dx \right\}^{-1} . \quad (69.2)$$

In this formula, $k_{\|,\perp}^* = \frac{r_{\|,\perp}\gamma^2\theta}{\rho} \left(\frac{kT_D}{\hbar s_{\|,\perp}} \right)^4 \left(\frac{kT_D}{s_{\|,\perp}^2} \right)$. Here, in conformity with [85], guided by the maximum value of phonon frequency, we consider that $T_D = 226.77\text{K}$, the other parameters of Bi_2Te_3 crystal are left the same as before. The adjustable parameter based on the requirement of agreement between the theoretical value of lattice thermal conductivity at 300K and the experimental one turns out to be equal to $\mu_{\|} = 0.375$. Dependences of lattice thermal conductivity and dimensionless thermoelectric figure of merit of TEM with respect to a single crystal on the radius of a contact between half-spheres at $T = 300\,\text{K}$ obtained in the framework of different approaches are depicted in Figure 168.

From this figure it is seen that optimal in terms of dimensionless thermoelectric figure of merit radius of a contact between half-spheres exists in the framework of the Debye model for the density of phonon states with account of the frequency dependence of phonon relaxation time. Whereas in the framework of the other two approaches, with increase in contact radius, the dimensionless thermoelectric figure of merit of TEM is monotonously increased, tending to the dimensionless thermoelectric figure of merit of a single crystal.

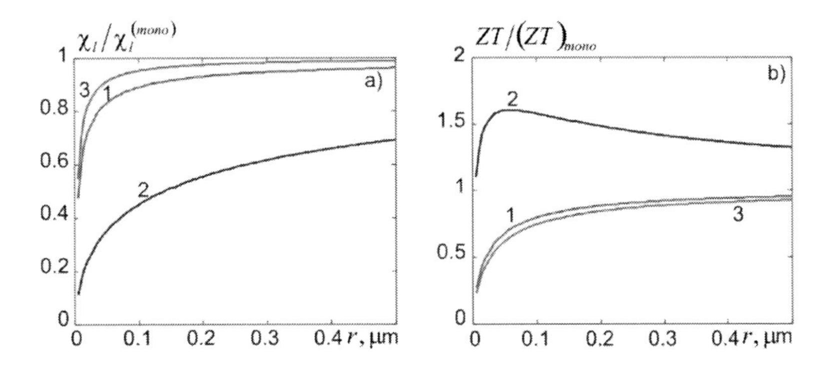

Figure 168. Dependences on the radius of a contact between half-spheres: a) of lattice thermal conductivity of TEM with respect to a single crystal; b) of the dimensionless thermoelectric figure of merit of TEM with respect to a single crystal. In the figures a), b) curves 1 correspond to account of the effect of the density of phonon states on lattice heat capacity, curves 2 – to the Debye model with account of the frequency dependence of phonon relaxation time, curves 3 – to approximation of constant phonon mean free path.

Consider now this problem with regard to the effect of the density of states shown in Fig 166 (curve 1) on phonon scattering on phonons and on the boundaries of a shape-forming element. In so doing, the constant frequency surface, as in the Debye model, will be considered spherical. Then, for thermal conductivity of a single crystal with regard to formulae for relaxation times related to Umklapp processes and normal processes, the following relation will be valid:

$$\chi_{l\|,\perp} = \frac{3h^2 \rho s_{f\|,\perp}^4 \omega_{\max}}{16\gamma^2 k^2 T^3} \int_0^1 \frac{f(x) x^2 \exp(x/\theta)}{[\exp(x/\theta)-1]^2} \left(\frac{1}{Q_{l\|,\perp}(x)} + \frac{2}{Q_{t\|,\perp}(x)} \right) dx . \tag{69.3}$$

In this formula, $s_{f\|,\perp}$ – phase sound velocity in layers plane or perpendicular to it, and functions $Q_{l\|,\perp}(x)$ and $Q_{t\|,\perp}(x)$ are defined as:

$$Q_{l\|,\perp}(x) = f(x) K^2(x) + \mu_{\|,\perp} \frac{x^2}{K(x)} , \tag{69.4}$$

$$Q_{t\|,\perp}(x) = 3.125\theta^3 \frac{f^4(x)}{K^7(x)} + \mu_{\|,\perp} \frac{x^2}{K(x)} . \tag{69.5}$$

Here,

$$K(x) = \sqrt[3]{3 \int_0^x f(x) dx} . \tag{69.6}$$

So, similar to formula (69.2) with regard to phonon scattering on the boundaries of a contact between half-spheres we get the following relation for thermal conductivity of TEM with respect to a single crystal:

$$\chi_{l\|,\perp}^{ef} \Big/ \chi_{l\|,\perp} = \pi^{-1}\int_0^1\int_0^1\int_0^{2\pi}\frac{zx^2 f(x)\exp(x/\theta)}{[\exp(x/\theta)-1]^2}\left(\frac{k_{\|,\perp}^*\sqrt{z^2-2z\cos\phi+1}}{K^2(x)[f(x)]^{-1}+k_{\|,\perp}^* Q_{l\|,\perp}(x)\sqrt{z^2-2z\cos\phi+1}}+\right. \tag{69.7}$$

$$\left.\frac{2k_{\|,\perp}^*\sqrt{z^2-2z\cos\phi+1}}{K^2(x)[f(x)]^{-1}+k_{\|,\perp}^* Q_{t\|,\perp}(x)\sqrt{z^2-2z\cos\phi+1}}\right)d\phi dz dx\left\{\int_0^1\frac{x^4\exp(x/\theta)}{[\exp(x/\theta)-1]^2}\left(\frac{1}{Q_{l\|,\perp}(x)}+\frac{2}{Q_{t\|,\perp}(x)}\right)dx\right\}^{-1}$$

In this case $k^* = \dfrac{16r\gamma^2 kT\omega_{\max}^4}{3\pi\rho s_{f\|,\perp}^6}$.

If all parameters of Bi_2Te_3 crystal, except for $\omega_{\max}$, are left the same as before, and $\omega_{\max}$ is taken the same as in [60], then for the adjustable parameter we get the value $\mu_\| = 1.097\cdot 10^3$. If, however, with regard to the order of constant lattice in layers plane, for phase velocity we assume the estimate $s_{f\|} = 2a_\|\omega_{\max}$, then $s_{f\|} = 6608$ m/s. In so doing, for the adjustable parameter we get the value $\mu_\| = 2.807\cdot 10^4$. Dependences of the lattice thermal conductivity of TEM and its dimensionless thermoelectric figure of merit with respect to a single crystal on the contact radius for these cases are depicted in Figures 169 and 170.

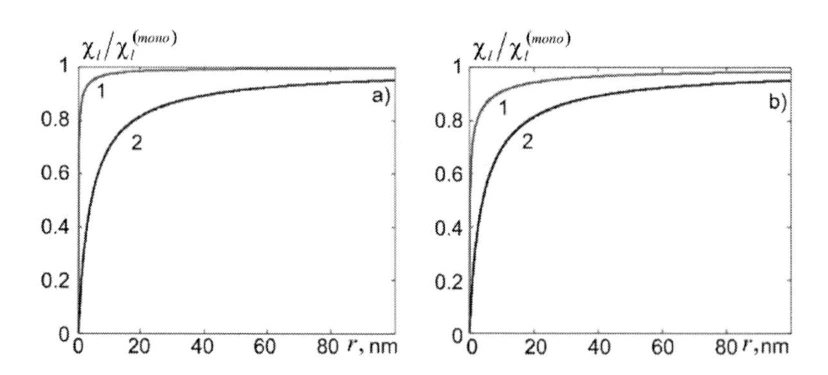

Figure 169. Dependence of lattice thermal conductivity of TEM with respect to a single crystal on the radius of a contact between half-spheres and with regard to the effect of a real density of states on phonon scattering on phonons and on the boundaries of a contact between half-spheres (curves 1) and in the approximation of constant phonon mean free path (curves 2) at: a) uncorrected sound velocity in Bi_2Te_3; b) corrected sound velocity in Bi_2Te_3.

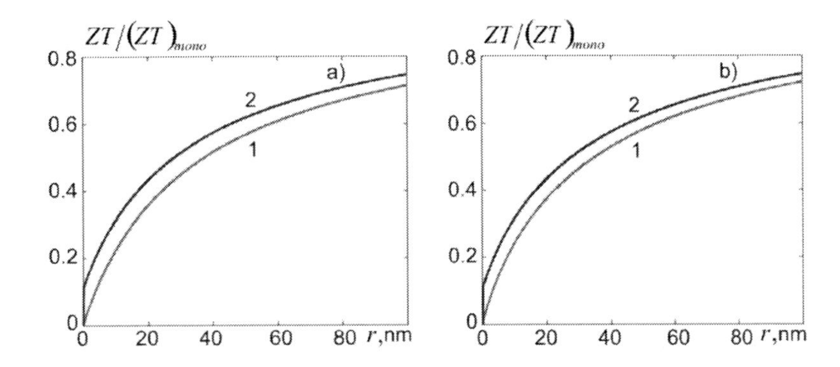

Figure 170. Dependence of dimensionless thermoelectric figure of merit of TEM with respect to a single crystal on the radius of a contact between half-spheres with regard to the effect of a real density of states on phonon scattering on phonons and on the boundaries of a contact between half-spheres (curves 1) and in the approximation of constant phonon mean free path (curves 2) at: a) uncorrected sound velocity in Bi_2Te_3; b) corrected sound velocity in Bi_2Te_3 .

The figures demonstrate that, first, with regard to the effect of a real density of phonon states on phonon scattering on phonons and on the boundaries of a contact between half-spheres, the thermal conductivity of TEM with respect to a single crystal with increase in contact radius increases faster than in the approximation of constant mean free path, whereas in the framework of the Debye model for the density of phonon states – on the contrary. Moreover, with regard to the effect of a real density of phonon states on phonon scattering on phonons and on the boundaries of a contact between half-spheres, there exists no radius of this contact which is optimal in terms of dimensionless thermoelectric figure of merit. At the same time, correction of sound velocity in Bi_2Te_3 has no great impact on the value of dimensionless thermoelectric figure of merit of TEM. Thus, for instance, with contact radius of 100nm, for sound velocities equal to 2952 and 6608m/s, respectively, the dimensionless thermoelectric figure of merit of TEM with respect to a single crystal at 300K is 0.714 and 0.722, respectively. Thus, for a final solution of a problem on possible improvement of thermoelectric figure of merit of TEM based on powders due to classical dimensional effects it is necessary to consider several, different from parabolic, variants of the frequency dependence of the density of phonon states in Bi_2Te_3. This, however, is beyond the scope of the present book.

CONCLUSION

1. The layered structure effects consisting in the nonparabolicity of electron band spectrum are manifested in the temperature and field dependences of the diamagnetic susceptibility of electron gas, electric conductivity, the Seebeck coefficient and power factor of a layered crystal, even if its FS is closed. This manifestation is the more apparent the higher is the degree of filling a narrow miniband characterizing the interlayer motion of electrons.

Basically, crystal layered structure becomes apparent, as with the same degree of filling a narrow miniband the density of electron states in a layered crystal is larger than in the effective mass approximation. The layered structure and the ensuing band spectrum nonparabolicity give rise to a situation when at low temperatures and in a weak quantizing magnetic field full diamagnetic susceptibility of electron gas in a bulk real layered crystal is larger than in the effective mass approximation. However, at rather high temperatures, when electron gas degeneracy is eliminated, the type of interlayer electron motion produces no effect on the diamagnetic susceptibility of electron gas in weak magnetic fields, and its temperature dependence is described by the Curie law. In rather thin films based on layered crystals the diamagnetic susceptibility of electron gas is higher than in the bulk crystals and has a local maximum at low temperatures. This maximum is due to discreteness of energy levels in a thin film placed in a magnetic field, as well as due to a greater value of the density of electron states in a thin film as compared to a bulk crystal. The greatest dissimilarity of the diamagnetic susceptibility of a layered crystal from the effective mass approximation occurs with a completely filled narrow miniband, i.e. in the case of a transient FS. This occurs because the density of electron states in the case of a real bulk crystal has two singularities, whereas in the effective mass approximation – only one.

2. In the charge-ordered state the temperature dependence of electron gas diamagnetic susceptibility in a weak magnetic field is typical of second-order phase transitions, namely it undergoes a kink at a point of phase transition. In so doing, the type of temperature dependence of electron gas diamagnetic susceptibility in a charge-ordered layered crystal below the point of second-order phase transition is defined by the ratio of the energy of effective attractive interaction between electrons leading to charge ordering, to the Fermi energy of an ideal two-dimensional Fermi gas with a square dispersion law at the absolute zero of temperature. Namely, if the above ratio is more than unity but less than two, the diamagnetic susceptibility of electron gas below the transition point increases with a rise in temperature. If the above ratio is equal to two, the diamagnetic susceptibility of electron gas

in a weak magnetic field below the transition point is equal to half its maximum value and does not depend on temperature. If, however, the above ratio is more than two, the diamagnetic susceptibility below transition point drops with a rise in temperature. After charge ordering destruction, i.e. above the transition point, the diamagnetic susceptibility of electron gas always drops with a rise in temperature.

3. In the field dependence of dHvA diamagnetic susceptibility in the quasi-classical region of magnetic fields with closed FS the layered structure effects are primarily manifested in phase delay of dHvA oscillations and the reduction of their relative contribution as compared to the effective mass approximation. Phase delay of the oscillations is due to the fact that with the same electron concentration the Fermi energy in a layered crystal is smaller than in the effective mass approximation. A decline in the relative contribution of dHvA oscillations occurs because in the case of a layered crystal the FS sections by planes normal to a magnetic field, with increase in the longitudinal quasi-pulse modulus, decrease slower than in the effective mass approximation. In so doing, even in the quasi-classical magnetic fields the shape of dHvA oscillations in a thin film is more complicated than in a bulk crystal, which is due to size quantization of energy levels. Therefore, layered structure effects in thin films produce a stronger effect on the shape of dHvA oscillations than in the bulk crystals.

In greater than quasi-classical magnetic fields the shape of dHvA oscillations is more complicated, because in that event, first, the field dependence of the Fermi energy becomes apparent, second, electrons from the entire FS and not only from its extreme sections by planes normal to a magnetic field, start to contribute to diamagnetic susceptibility oscillations. However, in the ultraquantum limit the difference between a real layered crystal and the effective mass approximation is leveled out by virtue of electron condensation on the bottom of the lowest filled Landau subband.

4. Charge ordering in quasi-classical range of magnetic fields is manifested in the presence of an additional set of dHvA oscillation frequencies as compared to a disordered layered crystal. This occurs because a transition to the charge-ordered state is topological, i.e. such whereby a closed or a transient FS is transformed into an open one. For the same reason the shape of dHvA oscillations in stronger magnetic fields in a charge-ordered layered crystal is different from the shape of these oscillations in the absence of ordering. This difference is also due to the fact that by virtue of magnetic field dependence of the order parameter and chemical potential in a strong quantizing magnetic field there occurs an inverse topological transition from a strongly open FS to a closed one.

5. The Langevin diamagnetic susceptibility becomes apparent in the area of weak and intermediate degeneracy of electron gas and is caused by induction of a diamagnetic moment in a crystal due to cyclotron rotation of electrons in a magnetic field. Its sign and value are defined by competing contributions of charge carriers with the energy smaller than electron gas chemical potential and with the energy larger than electron gas chemical potential. With a dominant contribution of charge carriers and the energy smaller than electron gas chemical potential, the Langevin diamagnetic susceptibility is negative. However, with a dominant contribution of charge carriers and the energy larger than electron gas chemical potential, the Langevin diamagnetic susceptibility is positive. Therefore, inversion of the Langevin magnetic susceptibility sign is governed not only by crystal band structure, charge carrier concentration and temperature, but also by the magnetic field value. Moreover, in thin films by virtue of energy spectrum discreteness, there is the oscillation dependence of the Langevin diamagnetic susceptibility on the magnetic field induction. Oscillations of the Langevin

diamagnetic susceptibility in the case of a real layered crystal are manifested the stronger the higher is charge carrier concentration.

6. In a charge-ordered layered crystal the inversion of the Langevin diamagnetic susceptibility takes place depending on the effective attractive interaction. Namely, if the value of this interaction is less than doubled Fermi energy of an ideal two-dimensional Fermi gas with a square dispersion law at the absolute zero temperature, then the Langevin diamagnetic susceptibility is negative, if it is equal to doubled Fermi energy – the Langevin diamagnetic susceptibility is zero, and if it is higher – the Langevin diamagnetic susceptibility is positive.

7. In all cases in a strong quantizing magnetic field the Langevin diamagnetic susceptibility decreases in modulus, tending to zero.

8. In the absence of degeneracy, magnetization of a layered crystal, as well as of a thin film, does not depend on the type of electron motion in a direction normal to layers and is determined by the Langevin formula.

9. According to SdH effect, in the quasi-classical range of magnetic fields the layered structure effects are manifested in phase delay of SdH oscillations and in reduced relative contribution of oscillations to full electric conductivity. The reasons for this are the same as in the case of dHvA oscillations. The shape of SdH oscillations and their relative contribution to full electric conductivity depend on the way of simulation of relaxation time dependence on the longitudinal quasi-pulse. In the range of stronger magnetic fields there is an optimal range of magnetic field inductions where layered structure effects are most pronounced. This manifestation largely consists in the fact that in the case of a real layered crystal there is local maximum of longitudinal electric conductivity which is absent in the effective mass approximation. With a closed FS the layered structure effects in the field dependence are most pronounced in the model of relaxation time proportional to full electron velocity on the FS which is applicable at acoustic phonon deformation potential scattering of electrons. These effects are least pronounced in the model of constant mean free path which is applicable at electron scattering on random arranged charged impurities potential. Here, the layered structure effects are manifested the stronger the larger is electron concentration.

10. In the presence of interlayer charge ordering due to a topological transition from a closed or transient FS into an open one, SdH oscillations in a quasi-classical range of magnetic fields become biperiodic. In stronger magnetic fields the dependence of longitudinal electric conductivity acquires the form typical of a single-dimensional crystal with a finite width of conduction band. With further magnetic field increase there is an inverse topological transition from an open FS into a closed one and the last maximum of longitudinal electric conductivity is attained.

11. In the ultraquantum limit the longitudinal electric conductivity tends to zero by the asymptotic law of the form $\sigma_{zz} \propto T^{-p}B^{-q}$ where specific positive values p and q are mainly governed by scattering mechanism.

12. In the field dependence of the longitudinal Seebeck coefficient in the quasi-classical region the layered structure effects are largely manifested in phase delay of oscillations and their amplitude reduction as compared to the effective mass approximation. In the case of predominant acoustic phonon deformation potential scattering, in the model of relaxation time proportional to full electron velocity the oscillating function of a magnetic field is the Seebeck coefficient itself. In the rest of considered relaxation time models the oscillating

function of a magnetic field is the first derivative of the Seebeck coefficient with respect to a magnetic field. 13. In the case of a closed FS the Seebeck coefficient oscillates without sign modification, and in the case of a transient FS – with the change of sign.

14. In stronger magnetic fields the Seebeck coefficient at first grows, reaching maximum, and then decays, tending to zero by the law $\alpha_{zz} \propto T^{-1}B^{-2}$ for all considered scattering mechanisms. Scattering mechanisms, layered structure effects and the degree of miniband filling affect scarcely the Seebeck coefficient maximum of a layered crystal, and this maximum is 108÷121μV/K. The layered structure effects are manifested in a slight increase of the Seebeck coefficient maximum in a real layered crystal as compared to the effective mass approximation and a shift of this maximum towards weaker magnetic fields. In so doing, in each case there is the optimal range of magnetic field inductions wherein layered structure effects are most pronounced. The layered structure effects are most pronounced in the model of relaxation time proportional to longitudinal velocity and least pronounced in the model of constant mean free path.

15. In the presence of charge ordering owing to a topological transition from an open FS into a closed one, the Seebeck coefficient oscillations occur with a change in sign and become biperiodic. In stronger magnetic fields there occurs repeated reversal of the Seebeck coefficient polarity synchronized with chemical potential changes in a magnetic field. After the point of an inverse topological transition from an open FS into a closed one the last maximum of the Seebeck coefficient modulus is achieved, and then the Seebeck coefficient decays, tending to zero by the law $\alpha_{zz} \propto T^{-1}B^{-2}$.

16. In the field dependence of power factor of a layered crystal in the quasi-classical range of magnetic fields under conditions of strong degeneracy the layered structure effects are manifested in phase delay of power factor oscillations and its value reduction as compared to the effective mass approximation. A reduction of power factor value of a real layered crystal is mainly due to a reduction of its electric conductivity due to a restricted charge carrier motion in the direction normal to layers. In stronger magnetic fields where power factor attains its maximum, the layered structure effects become apparent in the reduction of this maximum value and "rounding" its respective peak as compared to the effective mass approximation. After the maximum, power factor of a layered crystal starts to decrease tending to zero by the law $P \propto T^{-(2+p)}B^{-(4+q)}$, the values p and q being the same as for conduction.

17. In the presence of interlayer charge ordering, in the field dependence of power factor of a layered crystal the effect of "square-law detection" becomes apparent which lies in the availability of many local maxima of power factor and the points of its passing through zero. It is due to a bipolar character of the Seebeck coefficient oscillations in the quasi-classical range of magnetic fields and its polarity reversal in stronger magnetic fields. After the last maximum which is attained after the point of an inverse topological transition and is the largest, power factor of charge-ordered layered crystal decays, tending to zero by the law $P \propto T^{-(2+p)}B^{-(4+q)}$.

18. The longitudinal galvanomagnetic effects are possible either when electron gas is degenerate or when charge carrier relaxation time is a function of a magnetic field.

19. In the temperature dependence of longitudinal electric conductivity, in the absence of a magnetic field at acoustic phonon deformation potential scattering of electrons, the layered

structure effects are manifested in the reduction of electric conductivity value of a real layered crystal as compared to the effective mass approximation and its slower decrease with a rise in temperature.

20. In the model of constant relaxation time under conditions of weak and intermediate degeneracy the longitudinal magnetoresistance of real layered crystals is negative and considerably higher than in the effective mass approximation.

21. In the model of relaxation time inversely proportional to the density of electron states in the absence of a magnetic field, under conditions of weak and intermediate degeneracy the longitudinal magnetoresistance of a real layered crystal with a closed FS is always negative, whereas in the effective mass approximation in weak magnetic fields it can be positive.

22. The temperature-dependent Kapitsa effect under conditions of weak and intermediate degeneracy is attributable to a linear magnetic field dependence of acoustic phonon deformation potential scattering of electrons. This dependence takes place under suppressed transitions between the Landau subbands.

23. The temperature dependence of longitudinal electric conductivity of charge-ordered layered crystals in the absence of a magnetic field under conditions of predominant acoustic phonon deformation potential scattering of electrons is characterized by the presence of a local minimum below the point of second-order phase transition and by a local minimum at this point.

24. In charge-ordered layered crystals an inversion of magnetoresistance takes place depending on the value of effective attractive interaction between electrons. Namely, until the ratio of the value of effective attractive interaction between electrons to the Fermi energy of an ideal two-dimensional Fermi gas with a square dispersion law at the absolute zero does not exceed two, the longitudinal magnetoresistance of a layered charge-ordered crystal is positive. If, however, this ratio is more than two, the magnetoresistance of a layered charge-ordered crystal becomes negative. In so doing, the value of magnetoresistance decreases with increase in the effective attractive interaction.

25. In the absence of a magnetic field the Seebeck coefficient modulus of a layered crystal is always lower than in the effective mass approximation, which is due to restriction of the temperature spread of charge carriers along the narrow conduction miniband.

26. In the presence of interlayer charge ordering in the subcritical temperature region the Seebeck coefficient modulus increases with a rise in temperature at first slower and then faster than in the absence of ordering. At a point of second-order phase transition the temperature dependence of the Seebeck coefficient undergoes a kink. With increase in the value of the effective attractive interaction between electrons, the Seebeck coefficient modulus at a point of second-order phase transition is increased.

27. Under conditions of weak and intermediate degeneracy there exists such a value of magnetic field induction whereby the Seebeck coefficient modulus attains its maximum, which is due to increase in chemical potential and the distance between the Landau subbands with magnetic field increase. With regard to negative longitudinal magnetoresistance this creates conditions for control of the power factor and figure of merit of at least some of thermoelectric materials. In so doing, in the effective mass approximation the Seebeck coefficient modulus maximum is more pronounced than in a real layered crystal.

28. In a strong quantizing magnetic field, with its induction increase, the longitudinal Seebeck coefficient drops in modulus, tending to zero. However, charge-ordered crystals wherein the ratio of the energy of effective attractive interaction between electrons to the

Fermi energy of an ideal two-dimensional Fermi gas with a square dispersion law is more than two possess the property of the Seebeck coefficient sign inversion in a magnetic field.

29. The temperature and field dependences of the Seebeck coefficient of a charge-ordered layered crystal in the model of relaxation time inversely proportional to the density of electron states in the absence of a magnetic field are fundamentally the same as in the model of constant relaxation time. However, in the model of relaxation time inversely proportional to the density of electron states the Seebeck coefficient modulus for all magnetic field inductions is lower than in the model of constant relaxation time, which is related to increased contribution of low-energy states due to increase in their respective electron relaxation time.

30. The nonparabolicity of a layered crystal conductivity band results in drastic reduction of the power factor of a real layered crystal in the absence of a magnetic field. It is due to reduction of both the Seebeck coefficient modulus and the longitudinal electric conductivity as compared to their values in the effective mass approximation.

31. Under conditions of weak and intermediate degeneracy for each value of electron concentration there is such a value of magnetic field induction whereby crystal power factor attains its maximum. However, in a real layered crystal this maximum is much lower and much weaker expressed than in the effective mass approximation. Moreover, in the case of a transient FS in a real layered crystal this maximum is absent. Thus, the nonparabolicity of conduction band can reduce considerably the figure of merit of thermoelectric material and narrow down the possibilities of its properties control.

32. In the absence of a magnetic field the power factor of a charge-ordered layered crystal in the subcritical region increases with a rise in temperature at first slower, then faster than in the absence of charge ordering. At a second-order phase transition point the temperature dependence of power factor undergoes a kink. With increase in effective interaction, the power factor value at a transition point is increased.

33. Under conditions of weak and intermediate degeneracy of electron gas, control of thermoelectric figure of merit of layered charge-ordered crystals by application of a quantizing magnetic field parallel to temperature gradient is impossible.

34. The mechanism of retention or increase of the figure of merit of thermoelectric material when passing from a single crystal to pressed, extruded or SPS-material based on powder lies in the fact that in going from a single crystal to powder the thermal conductivity of material is restricted severely, and the electric conductivity – considerably weaker.

35. The mechanism of lattice thermal conductivity reduction when passing from a single crystal to powder lies in phonons drag due to their scattering on the boundaries of shape-forming structure elements and (or) on small contacts between these elements. In a relevant for thermoelectricity temperature region this scattering is essentially modified both by Umklapp processes and normal phonon-phonon scattering processes.

36. In the approximation of constant mean free paths of charge carriers and phonons in the presence of only classical size defects the figure of merit of thermoelectric material in passing from a single crystal to powder should drop.

37. With account of frequency dependence of phonon relaxation time in the framework of the Debye model for the density of phonon states for each temperature there is an optimal characteristic size of a shape-forming element or a contact between such elements whereby the figure of merit of thermoelectric material is maximal. With a rise in temperature, this size decreases, and its respective thermoelectric figure of merit increases. This enables using

powders of variable granulometric composition for the manufacture of thermoelectric modules.

38. On condition of the Seebeck coefficient independence of the size of shape-forming elements of thermoelectric material structure, transition from the properties of a shape-forming element to the properties of thermoelectric material as a whole can be realized by averaging the kinetic coefficients over the volume share of shape-forming elements with regard to size distribution function.

39. The conclusion at large on the existence of optimal structure of powder-based thermoelectric material is essentially dependent on the employed model of the density of phonon states. In the framework of some models of the density of phonon states, different from the Debye parabolic model, the size of shape-forming elements of thermoelectric material structure that would be optimal in terms of thermoelectric figure of merit does not exist, and in passing from a single crystal to powder the thermoelectric figure of merit drops.

References

[1] Fivaz R.F. *J. Phys. Chem. Solids.* 1967, 28, 839.

[2] Gorskyi P.V. *Ukr. J. Phys.* 2010, 55, 1296-1304.

[3] Askerov B.M., Figarova S.R., Mahmudov M.M. *Physica E.* 2006, 33, 303-307.

[4] Bender A.S., Young D.A. *Phys. Stat. Sol. B* 1974, 47K, 95-97.

[5] Bender A.S., Young D.A. *J. Phys. C.* 1972, 5, 2163-2178.

[6] Pavlovich V.V., Epshtein E.M. *Fiz. Tverd. Tela* (Soviet Union) 1977, 19, 3456-3458.

[7] Koshkin V.M., Katrrunov K.A. *Pis'ma v JETP* (Soviet Union) 1979, 29, 205-209.

[8] Bass F.G., Bulgakov A.A., Tetervov A.P. High Frequency Properties of Semiconductors with Superlattices; *Nauka:* Moscow, 1989.

[9] Kartsovnik M.V., Laukhin V.N., Nijankovsky V.I., Ignat'ev A.A. *Pis'ma v JETP* (Soviet Union) 1988, 47, 302.

[10] Drabble J.R., Wolfe R. *Proc. Phys. Soc.* 1956, 69B, 1101.

[11] Testardi L.R., Stiles P.J., Burstein E. *Sol. St. Comm.,* 1963, 1, 28.

[12] Maslyuk V.T., Bercha D.M. *Fiz. Nizk. Temp.* (Soviet Union) 1977, 3, 1025-1035.

[13] Zayachkovsky M.P., Bercha D.M., Zayachkovskaya N.F. *Ukr. J. Phys.* 1978, 23, 1119-1124.

[14] Peschansky V.G. *JETP,* 2002, 2002, 1204.

[15] Lifshits I.M., Kosevich A.M. Izv. *Akadem. Nauk SSSR. Ser. Fiz.* (Soviet Union) 1955, 19, 395-403.

[16] Lifshits I.M., Kosevich A.M. *Doklady Akadem. Nauk SSSR.* (Soviet Union) 1954, 96, 963.

[17] Lifshits I.M., Kosevich A.M. *JETP* (Soviet Union). 1955, 29, 730.

[18] Tavger B.A., Demikhovsky V. Ya. *Uspekhi Fiz. Nauk* (Soviet Union) 1968, Issue 1, 61-85

[19] Nedorezov S.S. *JETP* (Soviet Union) 1969, 56, 388-399.

[20] Ellenson W.D., Semmigsen D., Guerard D. *Mat. Sci. Eng.,* 1977, 31, 137-140.

[21] Di Salvo F.J., Wilson J.A., Bugley B. G. *Phys. Rev. B.* 1975, 12, 2220-2235.

[22] Pashitsky E.A., Spiegel A.S. *Fiz. Nizk. Temp.* (Soviet Union) 1978, 4, 976-983.

[23] Harper J.M.E., Geballe T. H. *Phys. Let. A,* 1974, 54, 27-28.

[24] Zeller C., Foley G.M.T., Falardeau E.R. *Mat. Sci. Eng.,* 1977, 31, 255-259.

[25] Vogel F.L., Foley G.M.T., Zeller C. *Mat. Sci. Eng.,* 1977, 31, 261-265.

[26] Gorskyi P.V. *Ukr. J. Phys*. 2005, 50, 1252-1260.

[27] Gorskyi P. V. *Fiz. Tekh. Poluprov* (in Russian), 2011, 45, 928-935.

[28] Shoenberg D. Magnetic oscillations in metals; *Cambridge University Press:* Cambridge, 2009
[29] Gorskyi P.V. *IJRRAS* (Islamabad), 2012, 13, 686-694.
[30] Gorskyi P.V., Nitsovich B.M. Investigation of magnetic susceptibility of the layered crystals (in Russian); Institute of Physics of Ukrainian National Academy of Sciences: Kiev, 1983.
[31] Gorskyi P.V., Nitsovich V.M. *Ukr. J. Phys.* 1986, 26, 1528-1533 (in Russian).
[32] Gorskyi P.V. *Fiz. Tekh. Poluprov* (in Russian), 2011, 45, 928-935.
[33] Gorskyi P.V., Nitsovich V.M. *Fiz. Tekh. Poluprov* (in Russian), 1983, 17, 936-938.
[34] Kapitsa P.L. Strong Magnetic Fields (in Russian); *Nauka:* Moscow, 1988.
[35] Abrikosov A.A. Fundamentals of theory of metals; *Elsevier:* Amsterdam, 1988.
[36] Gantmakher V.F., Levinson I.B. Current carriers scattering in metals and semiconductors (in Russian); *Nauka:* Moscow, 1984.
[37] Laikhman B., Menashe D. *Phys. Rev. B,* 2011, 52, 8974-8979.
[38] Gorskyi P.V. *Metallofiz. Noveishie Tekhnol* (in Russian), 2010, 32, 1545-1554.
[39] Landau L. D., Lifshits E. M. Quantum mechanics (in Russian); *Nauka:* Moscow, 1988.
[40] Kartsovnik M.V., Kononovich P.A., Laukhin V.N., Schegolev I.F. *Pis'ma v JETP* (Soviet Union) 1988, 48, 498.
[41] Gorskyi P.V. *Fiz. Nizk. Temp.* (in Russian), 1986, 12, 584-590.
[42] Gorskyi P.V. *Fiz. Tekh. Poluprov* (in Russian), 2005, 39, 343-348.
[43] Gorskyi P.V. Izvestija vysshikh uchebnikh zavedenij. *Fizika* (Tomsk)(in Russian), 2006, No6, 34-43.
[44] Gorskyi P.V. *Ukr. J. Phys.* 2002, 47, 385-389.
[45] Gorskyi P.V. *Metallofiz. Noveishie Tekhnol* (in Russian), 2002, 24, 1611-1623.
[46] Gorskyi P.V. Izvestija vysshikh uchebnikh zavedenij. *Fizika* (Tomsk)(in Russian), 2003, No5, 77-83.
[47] Gorskyi P.V. *Fiz. Tekh. Poluprov* (in Russian), 2003, 37, 166-168.
[48] Gorskyi P.V. *Fiz. Tekh. Poluprov* (in Russian), 2006, 40, 343-348.
[49] Gorskyi P.V. Izvestija vysshikh uchebnikh zavedenij. *Fizika* (Tomsk)(in Russian), 2006, No5, 34-43.
[50] Gadzhialiev M. M., Pirmagomedov Z. Sh. *Fiz. Tekh. Poluprov* (in Russian), 2009, 43, 1032-1033.
[51] Aliev F.F. *Fiz. Tekh. Poluprov* (in Russian), 2003, 37, 1082-1084.
[52] Kulbachinsky V.A., Kytin V.G., Blank V.D., Buga S.G., Popov M.Ju. *Fiz. Tekh. Poluprov* (in Russian), 2001, 45, 1241-1245.
[53] Lukyanova L.N., Kutasov V.A., Popov V.V., Konstantinov P.P. *Fiz. Tverd. Tela*, 2004, 46, 1366-1371.
[54] Shelykh A.I., Smirnov B.I., Orlova T.S., Smirnov I.A., de Arellano-Lopez A.R., Martinez-Fernandes J., Varrela-Feria F.M. *Fiz. Tverd. Tela* 2006, 48, 214-217.
[55] Gabor N.M., Song J.C.W., Ma Q., Nair N.L., Taychatanapat T., Watanabe K., Taniguchi T., Levitov L.S., Jarilo-Herrero P. *Science* 2011, 334, 6056.
[56] Gorskyi P.V. *Ukr. J. Phys.* 2013, 58, 370-378.
[57] Gorskyi P.V. *Journal of Thermoelectricity,* 2012, No4, 14-24.
[58] Borisenko S.I. *Fiz. Tekh. Poluprov* (in Russian), 2004, 38, 207-212.
[59] Gorskyi P.V. *Fiz. Tekh. Poluprov* (in Russian), 2004, 38, 864-866.
[60] Kittel Ch. Introduction to Solid State Physics (in Russian); *Nauka:* Moscow, 1978.

[61] Zlobin A.M., Zyrjanov P.S. *JETP,* 1970, 58, 952.

[62] Gorskyi P.V. Izvestija vysshikh uchebnikh zavedenij. *Fizika* (Tomsk)(in Russian), 2004, No2, 60-63.

[63] Gorskyi P.V. *Fiz. Nizk. Temp.* (in Russian), 2002, 28, 1072-1077.

[64] Goltsman B.M., Kudinov V.A., Smirnov I.A. Semiconductor Thermoelectric Materials Based on Bi_2Te_3 (in Russian); *Nauka:* Moscow, 1972.

[65] Stilbance L.S., Terekhov A.D., Sher E.M. Some Issues of the Transport Phenomena in Heterogeneous systems. In: Thermoelectric materials and films; Materials of All-Union Symposium on Deformation and Size effects in Thermoelectric Materials and Films, Films Technology and Application: Leningrad, 1976; p.199.

[66] Anatychuk L.I., Bulat L.P. Semiconductors under Extreme Temperature Conditions. (in Russian); *Nauka:* St. Petersburg, 2001.

[67] Lidorenko N.S., Andriyako V.A., Dudkin L.D., Nagayev E.L., Narva O.M. Doklady Akademii Nauk SSSR (in Russian), 1969, 186, 1295.

[68] Terekhov A.D., Sher E.M. The Structure of Disperse Medium and Effective Values of Thermal and Electric Conductivities. In: Thermoelectric materials and films; Materials of All-Union Symposium on Deformation and Size effects in Thermoelectric Materials and Films, Films Technology and Application: Leningrad, 1976; p.211.

[69] Bulat L.P., Drabkin I.A., Karatayev V.V., Osvensky V.B., Pshenai-Severin D.A. *Fiz. Tverd. Tela* (in Russian), 2010, 52, 1712-1716.

[70] Green M. Patent of USA No3524771. Patented Aug.19, 1970, Int. Cl. H0117/00, H01v1/28.

[71] Dmitriyev A.V., Zvyagin I.P. *Uspekhi Fiz.Nauk* (in Russian), 2010, 180, 821.

[72] Casian A., Dusciak V., Coropceanu I. *Phys. Rev. B,* 2002, 66,165404 (1-5).

[73] Casian A.I., Balmush I.I., Dusciak V.G. *Journ. of Thermoelectricity,* 2011, No3, 19.

[74] Dusciak V. *Journ. of Thermoeleectricity,* 2004, No4, 5.

[75] Misnar A. Thermal Conductivity of Solids, Liquids, Gases and Their Compositions (in Russian); *Mir:* Moscow, 1968.

[76] Gorsky P.V., Mikhalchenko V.P. *Journ. of Thermoelectricity,* 2013, No1, 18-25.

[77] Klemens P.G. Lattice Thermal Conductivity. In: Solid State Physics. Advances in Research and Applications; New York Academic Press: New York, NY, 1958; Vol.7, pp 1-98.

[78] Gorsky P.V., Mikhalchenko V.P. *Journ. of Thermoelectricity*, 2013, No2, 18-25.

[79] Gorsky P.V., Mikhalchenko V.P. *Journ. of Thermoelectricity,* 2013, No3, 5-10.

[80] Fan S., Zhao J., Guo J., Yan Q., Ma J., Hang H.H. *Journ. of Electron. Mat.,* 2011, 40, 1018-1023.

[81] Bulat L.P., Pshenai-Severin D.A. *Fiz. Tverd. Tela* (in Russian), 2010, 52, 452-458.

[82] Bulat L.P., Osvensky V.B., Parkhomenko Ju.N., Pshenai-Severin D.A. *Fiz. Tverd. Tela* (in Russian), 2012, 54, 20-26.

[83] Snarsky A.A., Sarychev A.K., Bezsudnov I.V., Lagar'kov A.N. *Fiz. Tekh. Poluprov* (in Russian), 2012, 46, 677-683.

[84] Gorskyi P.V. On the Possibility of Using Powders of Variable Grain Size Composition for Preparation of Thermoelectric Materials. In: Physics and Technology of Thin Films and Nanostructures; XIV International Conference. *Materials: Ivano-Frankivsk* (Ukraine), 2013; p.169.

[85] Rauh H., Geick R., Köhler H., Nücker N., Lehner N. *Sol. St. Phys.,* 1981, 14, 2705-2712.

INDEX

E

F

G

H

I

K

L